LONDON MATHEMATICAL SOCIETY LECTURE NOTE SERIES

Managing Editor: Professor J.W.S. Cassels, Department of Pure Mathematics and Mathematical Statistics, University of Cambridge, 16 Mill Lane, Cambridge CB2 1SB, England

The books in the series listed below are available from booksellers, or, in case of difficulty, from Cambridge University Press.

34 Representation theory of Lie groups, M.F. ATIYAH *et al*
36 Homological group theory, C.T.C. WALL (ed)
39 Affine sets and affine groups, D.G. NORTHCOTT
46 p-adic analysis: a short course on recent work, N. KOBLITZ
49 Finite geometries and designs, P. CAMERON, J.W.P. HIRSCHFELD & D.R. HUGHES (eds)
50 Commutator calculus and groups of homotopy classes, H.J. BAUES
57 Techniques of geometric topology, R.A. FENN
59 Applicable differential geometry, M. CRAMPIN & F.A.E. PIRANI
66 Several complex variables and complex manifolds II, M.J. FIELD
69 Representation theory, I.M. GELFAND *et al*
74 Symmetric designs: an algebraic approach, E.S. LANDER
76 Spectral theory of linear differential operators and comparison algebras, H.O. CORDES
77 Isolated singular points on complete intersections, E.J.N. LOOIJENGA
79 Probability, statistics and analysis, J.F.C. KINGMAN & G.E.H. REUTER (eds)
80 Introduction to the representation theory of compact and locally compact groups, A. ROBERT
81 Skew fields, P.K. DRAXL
82 Surveys in combinatorics, E.K. LLOYD (ed)
83 Homogeneous structures on Riemannian manifolds, F. TRICERRI & L. VANHECKE
86 Topological topics, I.M. JAMES (ed)
87 Surveys in set theory, A.R.D. MATHIAS (ed)
88 FPF ring theory, C. FAITH & S. PAGE
89 An F-space sampler, N.J. KALTON, N.T. PECK & J.W. ROBERTS
90 Polytopes and symmetry, S.A. ROBERTSON
91 Classgroups of group rings, M.J. TAYLOR
92 Representation of rings over skew fields, A.H. SCHOFIELD
93 Aspects of topology, I.M. JAMES & E.H. KRONHEIMER (eds)
94 Representations of general linear groups, G.D. JAMES
95 Low-dimensional topology 1982, R.A. FENN (ed)
96 Diophantine equations over function fields, R.C. MASON
97 Varieties of constructive mathematics, D.S. BRIDGES & F. RICHMAN
98 Localization in Noetherian rings, A.V. JATEGAONKAR
99 Methods of differential geometry in algebraic topology, M. KAROUBI & C. LERUSTE
100 Stopping time techniques for analysts and probabilists, L. EGGHE
101 Groups and geometry, ROGER C. LYNDON
103 Surveys in combinatorics 1985, I. ANDERSON (ed)
104 Elliptic structures on 3-manifolds, C.B. THOMAS
105 A local spectral theory for closed operators, I. ERDELYI & WANG SHENGWANG
106 Syzygies, E.G. EVANS & P. GRIFFITH
107 Compactification of Siegel moduli schemes, C-L. CHAI
108 Some topics in graph theory, H.P. YAP
109 Diophantine analysis, J. LOXTON & A. VAN DER POORTEN (eds)
110 An introduction to surreal numbers, H. GONSHOR
111 Analytical and geometric aspects of hyperbolic space, D.B.A. EPSTEIN (ed)
113 Lectures on the asymptotic theory of ideals, D. REES
114 Lectures on Bochner-Riesz means, K.M. DAVIS & Y-C. CHANG
115 An introduction to independence for analysts, H.G. DALES & W.H. WOODIN
116 Representations of algebras, P.J. WEBB (ed)
117 Homotopy theory, E. REES & J.D.S. JONES (eds)
118 Skew linear groups, M. SHIRVANI & B. WEHRFRITZ
119 Triangulated categories in the representation theory of finite-dimensional algebras, D. HAPPEL
121 Proceedings of *Groups - St Andrews 1985*, E. ROBERTSON & C. CAMPBELL (eds)
122 Non-classical continuum mechanics, R.J. KNOPS & A.A. LACEY (eds)
124 Lie groupoids and Lie algebroids in differential geometry, K. MACKENZIE
125 Commutator theory for congruence modular varieties, R. FREESE & R. MCKENZIE
126 Van der Corput's method of exponential sums, S.W. GRAHAM & G. KOLESNIK
127 New directions in dynamical systems, T.J. BEDFORD & J.W. SWIFT (eds)
128 Descriptive set theory and the structure of sets of uniqueness, A.S. KECHRIS & A. LOUVEAU
129 The subgroup structure of the finite classical groups, P.B. KLEIDMAN & M.W.LIEBECK

130 Model theory and modules, M. PREST
131 Algebraic, extremal & metric combinatorics, M-M. DEZA, P. FRANKL & I.G. ROSENBERG (eds)
132 Whitehead groups of finite groups, ROBERT OLIVER
133 Linear algebraic monoids, MOHAN S. PUTCHA
134 Number theory and dynamical systems, M. DODSON & J. VICKERS (eds)
135 Operator algebras and applications, 1, D. EVANS & M. TAKESAKI (eds)
136 Operator algebras and applications, 2, D. EVANS & M. TAKESAKI (eds)
137 Analysis at Urbana, I, E. BERKSON, T. PECK, & J. UHL (eds)
138 Analysis at Urbana, II, E. BERKSON, T. PECK, & J. UHL (eds)
139 Advances in homotopy theory, S. SALAMON, B. STEER & W. SUTHERLAND (eds)
140 Geometric aspects of Banach spaces, E.M. PEINADOR and A. RODES (eds)
141 Surveys in combinatorics 1989, J. SIEMONS (ed)
142 The geometry of jet bundles, D.J. SAUNDERS
143 The ergodic theory of discrete groups, PETER J. NICHOLLS
144 Introduction to uniform spaces, I.M. JAMES
145 Homological questions in local algebra, JAN R. STROOKER
146 Cohen-Macaulay modules over Cohen-Macaulay rings, Y. YOSHINO
147 Continuous and discrete modules, S.H. MOHAMED & B.J. MÜLLER
148 Helices and vector bundles, A.N. RUDAKOV *et al*
149 Solitons nonlinear evolution equations & inverse scattering, M. ABLOWITZ & P. CLARKSON
150 Geometry of low-dimensional manifolds 1, S. DONALDSON & C.B. THOMAS (eds)
151 Geometry of low-dimensional manifolds 2, S. DONALDSON & C.B. THOMAS (eds)
152 Oligomorphic permutation groups, P. CAMERON
153 L-functions and arithmetic, J. COATES & M.J. TAYLOR (eds)
154 Number theory and cryptography, J. LOXTON (ed)
155 Classification theories of polarized varieties, TAKAO FUJITA
156 Twistors in mathematics and physics, T.N. BAILEY & R.J. BASTON (eds)
157 Analytic pro-*p* groups, J.D. DIXON, M.P.F. DU SAUTOY, A. MANN & D. SEGAL
158 Geometry of Banach spaces, P.F.X. MÜLLER & W. SCHACHERMAYER (eds)
159 Groups St Andrews 1989 volume 1, C.M. CAMPBELL & E.F. ROBERTSON (eds)
160 Groups St Andrews 1989 volume 2, C.M. CAMPBELL & E.F. ROBERTSON (eds)
161 Lectures on block theory, BURKHARD KÜLSHAMMER
162 Harmonic analysis and representation theory for groups acting on homogeneous trees, A. FIGA-TALAMANCA & C. NEBBIA
163 Topics in varieties of group representations, S.M. VOVSI
164 Quasi-symmetric designs, M.S. SHRIKANDE & S.S. SANE
165 Groups, combinatorics & geometry, M.W. LIEBECK & J. SAXL (eds)
166 Surveys in combinatorics, 1991, A.D. KEEDWELL (ed)
167 Stochastic analysis, M.T. BARLOW & N.H. BINGHAM (eds)
168 Representations of algebras, H. TACHIKAWA & S. BRENNER (eds)
169 Boolean function complexity, M.S. PATERSON (ed)
170 Manifolds with singularities and the Adams-Novikov spectral sequence, B. BOTVINNIK
171 Squares, A.R. RAJWADE
172 Algebraic varieties, GEORGE R. KEMPF
173 Discrete groups and geometry, W.J. HARVEY & C. MACLACHLAN (eds)
174 Lectures on mechanics, J.E. MARSDEN
175 Adams memorial symposium on algebraic topology 1, N. RAY & G. WALKER (eds)
176 Adams memorial symposium on algebraic topology 2, N. RAY & G. WALKER (eds)
177 Applications of categories in computer science, M.P. FOURMAN, P.T. JOHNSTONE, & A.M. PITTS (eds)
178 Lower K- and L-theory, A. RANICKI
179 Complex projective geometry, G. ELLINGSRUD, C. PESKINE, G. SACCHIERO & S.A. STRØMME (eds)
180 Lectures on ergodic theory and Pesin theory on compact manifolds, M. POLLICOTT
181 Geometric group theory I, G.A. NIBLO & M.A. ROLLER (eds)
182 Geometric group theory II, G.A. NIBLO & M.A. ROLLER (eds)
183 Shintani zeta functions, A. YUKIE
184 Arithmetical functions, W. SCHWARZ & J. SPILKER
185 Representations of solvable groups, O. MANZ & T.R. WOLF
186 Complexity: knots, colourings and counting, D.J.A. WELSH
187 Surveys in combinatorics, 1993, K. WALKER (ed)
190 Polynomial invariants of finite groups, D.J. BENSON
191 Finite geometry and combinatorics, F. DE CLERCK *et al*
192 Symplectic geometry, D. SALAMON (ed)
197 Two-dimensional homotopy and combinatorial group theory, C. HOG-ANGELONI W. METZLER & A.J. SIERADSKI (eds)

London Mathematical Society Lecture Note Series. 192

Symplectic Geometry

Edited by

Dietmar Salamon
University of Warwick

CAMBRIDGE UNIVERSITY PRESS
Cambridge, New York, Melbourne, Madrid, Cape Town, Singapore, São Paulo

Cambridge University Press
The Edinburgh Building, Cambridge CB2 2RU, UK

Published in the United States of America by Cambridge University Press, New York

www.cambridge.org
Information on this title: www.cambridge.org/9780521446990

First published 1993

A catalogue record for this publication is available from the British Library

ISBN-13 978-0-521-44699-0 paperback
ISBN-10 0-521-44699-6 paperback

Transferred to digital printing 2006

Contents

List of Participants

S. Angenent (Madison)
H. Braden (Edinburgh)
E. Ciriza (Trieste)
N. Chernov (Moscow)
S. Donaldson (Oxford)
S. Dostoglou (Santa Barbara)
J. Eells (Cambridge)
Y. Eliashberg (Stanford)
H. Geiges (Cambridge)
C. Golé (stony Brook)
N.J. Hitchin (Warwick)
H. Hofer (Bochum)
S.Y. Husseini (Madison)
L. Jeffrey (Austin)
J.D.S. Jones (Warwick)
B. Kasper (Stony Brook)
M. Kazarian (Moscow)
A. King (Liverpool)
K. MacKenzie (Sheffield)
I. Marshall
J.N. Mather (Princeton)
D. McDuff (Stony Brook)
M. Micallef (Warwick)
J.W. Norris (Birmingham)
P. Pansu (Orsay)
L. Polterovich (Tel Aviv)
J. Pöschel (ETH Zürich)
P.H. Rabinowitz (Madison)
J.H. Rawnsley (Warwick)
E. Rees (Edinburgh)
J. Reineck (Buffalo)
J.W. Robbin (Madison)
D. Salamon (Warwick)
R. Sjamaar (MIT)
L. Traynor (Berkeley)
C.B. Thomas (Cambridge)
C. Viterbo (Orsay)
J.G. Wolfson (Ann Arbor)
E. Zehnder (ETH Zürich)

1 Introduction

Recall that a *symplectic manifold* is a $2n$-dimensional smooth manifold M together with a closed nondegenerate 2-form ω. A *symplectomorphism* is a diffeomorphism of M which preserves ω and an n-dimensional submanifold $L \subset M$ is called *Lagrangian* if ω vanishes on TL. Such structures arise naturally from Hamiltonian dynamics and geometric optics and they have been studied for many decades. The past ten years have seen a number of important developments and major breakthroughs in symplectic geometry as well as the discovery of new links with other subjects such as dynamical systems, topology, Yang-Mills theory, theoretical physics, and singularity theory.

Many of these new developments have been motivated by Gromov's paper [3] on pseudoholomorphic curves in symplectic geometry. The role pseudoholomorphic curves play in Gromov's work is reminiscent of the role of self-dual Yang-Mills instantons in Donaldson's theory on smooth 4-manifolds. Gromov used pseudoholomorphic curves to prove a number of surprising and hitherto inaccessible results in symplectic geometry. For example he proved that there is no symplectic isotopy moving the unit ball in $\mathbf{R}^{2n}$ through a hole in a hypersurface whose radius is smaller than 1 (*a symplectic camel cannot pass through the eye of a needle*). The paper by McDuff and Traynor below gives a proof of this theorem which is based on Eliashberg's techniques of filling by pseudoholomorphic discs.

Moduli spaces of pseudoholomorphic curves also play an important role in McDuff's work on symplectic 4-manifolds. In her contribution below she proves a uniqueness theorem for symplectic structures on $\mathbf{C}P^2$ with one or two points blown up. This problem is related to the question of connectedness of the space of symplectic embeddings of two disjoint balls into $\mathbf{C}P^2$. The paper by Ciriza deals with the uniqueness of symplectic structures for Kähler manifolds of nonpositive sectional curvature.

Gromov also proved that for every embedded compact Lagrangian submanifold $L \subset \mathbf{R}^{2n}$ there exists a holomorphic disc with boundary on L (a kind of generalization of the Riemann mapping theorem). This can be interpreted as an obstruction to Lagrangian embeddings. Polterovich in his paper proves new such obstructions involving the Maslov class.

Another result by Gromov is his celebrated *squeezing theorem* which asserts that there is no symplectic embedding of the unit ball in $\mathbf{R}^{2n}$ into a cylinder $B^2(r) \times \mathbf{R}^{2n-2}$ of radius less than 1. As a result he proved that the group of symplectomorphisms is closed with respect to the C^0 topology. Hofer interpreted Gromov's squeezing theorem as an example of symplectic invariants which he termed *capacities*. He discovered a number of other capacities, for example the *displacement energy*. In their contribution Hofer and Eliashberg prove a new inequality for the displacement energy and use this to deduce C^1 properties of a hypersurface in $\mathbf{R}^{2n}$ from C^0 information.

This can be viewed as an example of symplectic rigidity.

Another example of symplectic rigidity is Arnold's conjecture about the fixed points of exact symplectomorphisms (time-1-maps of Hamiltonian flows) on compact symplectic manifolds. He conjectured that the number of fixed points of such a symplectomorphism is bounded below by the sum of the Betti numbers. For the torus this was proved by Conley and Zehnder [1] using Morse theory for the symplectic action functional on the loop space. Angenent in his paper uses the symplectic action functional to give a new interpretation of Melnikov's formula for transverse intersections of stable and unstable manifolds in area preserving diffeomorphisms. Zehnder's paper deals with stability problems for symplectomorphisms of $\mathbf{R}^{2n}$. For $n = 1$ this is related to the existence of quasi-periodic solutions which can be established by KAM theory.

An important breakthrough came when Floer combined the variational approach of Conley and Zehnder with Gromov's elliptic techniques and Witten's approach to Morse theory to prove the Arnold conjecture for monotone symplectic manifolds [2]. His work can be summarized as an infinite dimensional version of Morse theory for the symplectic action where the critical points are periodic orbits of Hamiltonian systems and connecting orbits are pseudoholomorphic curves. The resulting invariants are the Floer homology groups. A similar version of Floer homology as an invariant of homology-3-spheres is closely related to Donaldson's theory of smooth 4-manifolds. This amplifies the close relation of pseudo-holomorphic curves in symplectic manifolds with self-dual Yang-Mills equations in 4 dimensions.

The relation between symplectic geometry and gauge theory is fundamental in two of the papers. The moduli space of flat connections over a Riemann surface is a symplectic manifold on which the mapping class group acts by symplectomorphisms. The paper by Dostoglou and Salamon examines the Floer homology groups of these symplectomorphisms. An entirely different relation between contact geometry and gauge theory was discovered by Rumin and this is explained by Pansu in his contribution.

The paper by Donaldson describes new links between *complex-symplectic structures* on 4-dimensional cobordisms (to be thought of as a complexification of the diffeomorphism group of a 3-manifold) and Ashtekhar's formulation of the self-dual Einstein equations.

The papers by Kazarian and Lerman/Montgomery/Sjamaar deal with singularities in symplectic geometry. In the former the singularities arise from geometric optics while the latter deals with symplectic reduction in cases where the quotient is not a manifold.

The paper by Robbin and Salamon explains how the metaplectic representation can be obtained from Feynman path integrals in phase space with general quadratic Hamiltonians. This leads to a simple model of Segal's axioms for topological quantum field theory.

References

[1] C.C. Conley and E. Zehnder, The Birkhoff-Lewis fixed point theorem and a conjecture by V.I. Arnold, *Inv. Math.* **73** (1983), 33–49.

[2] A. Floer, Symplectic fixed points and holomorphic spheres, *Commun. Math. Phys.* **120** (1989), 575–611.

[3] M. Gromov, Pseudoholomorphic curves in symplectic manifolds, *Invent. Math.* **82** (1985), 307–347.

Acknowledgements

I would like to thank the Science and Engineering Research Council and the London Mathematical Society for their generous support of the Conference at Warwick. I also would like to thank all those who helped in running the conference, in particular Shaun Martin and Elaine Shiels.

About this volume

This volume is based on lectures given at a workshop and conference on symplectic geometry at the University of Warwick in August 1990. The area of symplectic geometry has developed rapidly in the past ten years with major new discoveries that were motivated by and have provided new links with many other subjects such as dynamical systems, topology, gauge theory, mathematical physics and singularity theory. The conference brought together a number of leading experts in these interacting areas of mathematics. The contributions to this volume reflect the richness of the subject and include expository papers as well as original research. They will be an essential source for all research mathematicians in symplectic geometry.

Short description

This volume contains expository and research papers by leading experts in symplectic geometry and topology. The contributions reflect the rapid developments in this area in the past ten years and the diversity of the subject. They illuminate the interactions with many other areas such as dynamical systems, topology, gauge theory, mathematical physics and singularity theory.

A Variational Interpretation of Melnikov's Function and Exponentionally Small Separatrix Splitting

Sigurd Angenent
Mathematics Department, UW Madison
2 February, 1993

§1. Introduction

This note is about the exponentially small separatrix splitting which occurs when one studies the separatrices of maps of "standard type"

$$\Phi(u, v) = (u + \varepsilon, v + \varepsilon f_0(u + \varepsilon v)),$$

where f_0 is an entire function, or when one considers the Poincaré–map associated with the ODE

$$u''(t) = F(t/\varepsilon, u(t)) \tag{1.1}$$

for small values of $\varepsilon > 0$, and nonlinearities F with $F(t + 1, u) = F(t, u)$ which are analytic in the u variable. We recall that the Poincaré–map Φ_ε is defined in terms of the first order system

$$u' = v, v' = F(t/\varepsilon, u)$$

which is equivalent to the second order ODE (1.1); Φ_ε sends $(u(0), v(0))$ to $(u(\varepsilon), v(\varepsilon))$, where $(u(t), v(t)), (0 \leq t \leq \varepsilon)$ is a solution of (1.1). For small

$\varepsilon > 0$ the theory of averaging tells us that we may regard (1.1) as a "small" perturbation of the averaged equation

$$u'' = F_0(u) \tag{1.2}$$

with $F_0(u) = \int_{\mathbf{T}} F(\tau, u)d\tau$.

If the Poincaré map associated with (1.2) has a hyperbolic fixed point with a homoclinic orbit, then one expects the same to be true[1] for the perturbed Poincaré map Φ_ε ($\varepsilon << 1$). One also expects the homoclinic orbit of perturbed map to come from a transverse intersection of the invariant manifolds through the hyperbolic fixed point. Melnikov's method allows one to verify this for smooth perturbations of (1.2) with fixed period, e.g. equations of the form $u'' = F_0(u) + \mu g(t, u)$. However, it has been observed that the method does not apply directly to (1.1). Holmes, Marsden and Scheurle[2] were the first to try to adjust Melnikov's method to the averaging situation. They gave an asymptotic expression for the separatrix splitting if F is of the form $F(\tau, u) = \sin u + \delta\varepsilon^p g(t)$, with g periodic, p sufficiently large, and δ a small parameter. This result was later improved by various authors, the best result to date being due to Delshams, Teresa and Seara[3].

In section 2 we give a variational interpretation of the Melnikov function, and in the subsequent sections show how this interpretation can be adapted to study the homoclinic orbits of (1.1). Like Holmes et.al. we only get an upper bound for the size of the splitting in the most general setting, while we only get transverse homoclinic intersections for a special nonlinearity, $F(\tau, u) = u - {}^3/_2 u^2 + \delta\varepsilon^{10} H'(\tau)$, with $H(\tau)$ periodic, and δ a small parameter. For this particular example the variational approach is an improvement on the results of Holmes et.al. but fails to give the result of Delshams et.al.

One advantage the variational point of view may have over others, is that it can easily be generalized, to find entire solutions of elliptic PDE's such as

$$\Delta u = F_0(u) + \mu g(x, u(x)), \qquad u(\infty) = 0, \tag{1.3}$$

where $g(x, u)$ is periodic in the x-variable; given a nontrivial solution $U(x)$ of the spatially homogeneous equation $\Delta u = F_0(u)$ which vanishes at $x = \infty$,

---

[1] See chapter 4 of [GH83] for a discussion of averaging.

[2] [HMS88]

[3] See [DTS91] and the references given there.

the analysis in section 2 allows one to find solutions of (1.3) close to some translate $U(x+\vartheta)$ of $U(x)$.

Although there is no obvious Poincaré–map in this situation, one can still show[4] that nondegenerate solutions of (1.3) generate many more solutions of (1.3), much in the same way that a transverse homoclinic point of the Poincaré–map generates an abundance of homoclinic orbits.

The two main examples we have in mind throughout the paper are a forced Duffing equation

$$u'' - F_0(u) = \delta g(t), \tag{1.4}$$

and a "kicked anharmonic oscillator"

$$u''(t) = \varepsilon \sum_{j\in\mathbf{Z}} \delta(t - j\varepsilon) \cdot F_0(u(t)) \tag{1.5}$$

with $F_0(u) = u - {}^3/_2 u^2$.

In the second example the equation is to be interpreted in the sense of distributions: a solution is a Lipschitz function whose second distributional derivative satisfies (1.5). In fact, solutions will be piecewise linear, and their values $u_j = u(j\varepsilon)$ satisfy the recurrence relation

$$u_{j+1} - 2u_j + u_{j-1} = \varepsilon^2 F_0(u_j). \tag{1.6}$$

One easily verifies that the Poincaré–map Φ_ε is given by the standard type map $(u, v) \mapsto (u + \varepsilon v, v + \varepsilon F_0(u + \varepsilon v))$.

In section 3 we introduce a large class of nonlinearities F which includes both of these examples.

§2. A variational account of the Melnikov function

We assume in this section that $\varepsilon = 1$, and that the nonlinearity F is of the form $F(t,u) = F_0(u) + \mu g(t,u)$, where g is some smooth function, μ is small, and F_0 satisfies

$$F_0(0) = 0, \quad F_0'(0) > 0. \tag{2.1}$$

This last condition implies that the origin is a hyperbolic fixed point for the local flow Ψ_t generated by the system $u' = v, v' = F_0(u)$.

---

[4] See [A86].

The potential energy associated with F_0 is given by

$$V_0(u) = -\int_0^u F_0(\omega)d\omega.$$

We shall assume that $V_0(u) < 0$ for some $u > 0$, and that $V_0'(\alpha) = -F_0(\alpha) < 0$ where α is the smallest positive root of $V_0(\alpha) = 0$. Under this assumption the stable and unstable manifolds W^u, W^s of the origin coincide; they are parametrized by $(U(t), U'(t))$ where U is the unique positive and even solution of

$$U'' = F_0(U), \qquad U(\pm\infty) = 0. \tag{2.2}$$

Consider the Poincaré map Φ_μ of the perturbed system $u' = v, v' = F(\mu, t, u)$. If μ is small Φ_μ will have a hyperbolic fixed point $\mathcal{O}_\mu$ near the origin, whose stable and unstable manifolds we denote by W_μ^s, W_μ^u. Since Φ_μ depends smoothly on μ, the fixed point $\mathcal{O}_\mu$ as well as the W_μ^s, W_μ^u vary smoothly with μ.

For most perturbations $g(t, u)$ the invariant manifolds W_μ^s, W_μ^u will not coincide when $\mu \neq 0$. Melnikov's method was designed to compute the separation between the invariant manifolds for small values of μ, and in particular, to find the transverse intersections in $W_\mu^s \cap W_\mu^u$.

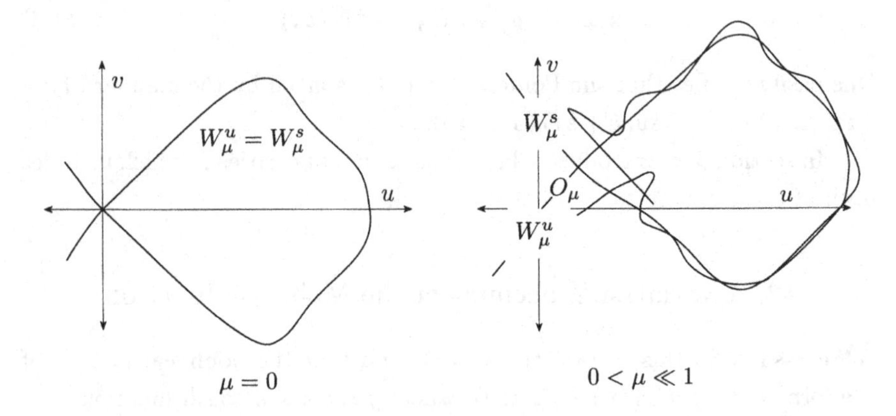

It is a commonplace[5] to remark that these transverse intersections are of interest since they are known to be a cause of complicated dynamics of the Poincaré map Φ_μ.

We shall now proceed to describe a variational method which produces a result equivalent to Melnikov's. To begin with we construct a periodic

[5] See [Mo73, HG84] and the references given there.

solution which corresponds to the hyperbolic fixed point $\mathcal{O}_\mu$ by applying the implicit function theorem to the map $\mathcal{F} : \mathbf{R} \times C^2(\mathbf{T}) \to C^0(\mathbf{T})$ given by

$$\mathcal{F}(\mu, p) = p'' - F_0(p) - \mu g(t, p).$$

Here $\mathbf{T} = \mathbf{R}/\mathbf{Z}$ and $C^k(\mathbf{T})$ is the space of k times continuously differentiable functions $u(t)$ with $u(t+1) \equiv u(t)$.

We have $\mathcal{F}(0,0) = 0$ while $\mathrm{d}_p\mathcal{F}(0,0)$, the derivative of $\mathcal{F}$ w.r.t. p, is given by $\mathrm{D}^2 - F_0'(0)$; since $F_0'(0) > 0$ the operator $\mathrm{d}_p\mathcal{F}(0,0)$ has a bounded inverse from $C^0(\mathbf{T})$ to $C^2(\mathbf{T})$, and we have a smooth branch of solutions $p(\mu, \cdot) \in C^2(\mathbf{T})$ of $\mathcal{F}(\mu, p) = 0$ with $p(0,t) \equiv 0$. The fixed point $\mathcal{O}_\mu$ is now given by $(p(\mu, 0), p'(\mu, 0))$.

Homoclinic orbits of Φ_μ correspondend to solutions $u(t)$ of $u'' = F(\mu, t, u)$ which are defined for all $t \in \mathbf{R}$, and which are asymptotic to the small solution $p(\mu, t)$ as $t \to \pm\infty$. To find such solutions we substitute $u(t) = v(t) + p(\mu, t)$ and obtain the following equation for v:

$$v'' = \hat{F}(\mu, t, v(t)), \qquad v(\pm\infty) = 0, \tag{2.3}$$

where

$$\begin{aligned} \hat{F}(\mu, t, v) &= F(\mu, t, p + v) - p'' \\ &= F_0(p + v) - F_0(p) + \mu \{g(t, p + v) - g(t, p)\} \end{aligned}$$

with $p = p(\mu, t)$, and $' = \partial/\partial t$. The corresponding potential energy is given by

$$\hat{V}(\mu, t, v) = \int_0^v \hat{F}(\mu, t, \omega) d\omega;$$

it satisfies $|\hat{V}(\mu, t, v)| \leq Cv^2$ for small v, and hence the functional

$$\mathcal{A}_\mu(v) = \mathcal{A}(\mu, v) = \int_{\mathbf{R}} \left({}^1\!/_2 v'(t)^2 - \hat{V}(\mu, t, v(t)) \right) dt \tag{2.4}$$

is well defined for $v \in H^1(\mathbf{R})$.

2.1. Lemma. *Critical points of $\mathcal{A}_\mu$ are exactly the solutions of (2.3), and hence they correspond to the homoclinic orbits of Φ_μ, i.e. to the intersections of W^s_μ and W^u_μ.*

For small μ a critical point of $\mathcal{A}_\mu$ is nondegenerate if and only if the corresponding intersection of W^s_μ and W^u_μ is transverse.

Proof. The first statement holds since (2.3) is the Euler-Lagrange equation for $\mathcal{A}_\mu$.

Concerning the connection between nondegeneracy and transversality we remark first of all that the Poincaré–maps Φ_μ and $\hat{\Phi}_\mu$, where the latter is derived from $w'' = \hat{F}(\mu, t, w)$, are conjugate. The conjugation is provided by the translation

$$\tau : (u_0, u_0') \mapsto (v_0, v_0') = (u_0 + p(\mu, 0), u_0' + p_t(\mu, 0)).$$

Thus if $P \in W_\mu^u \cap W_\mu^s$, then $\tau(P) \in \hat{W}_\mu^u \cap \hat{W}_\mu^s$, and $\hat{W}_\mu^u$ and $\hat{W}_\mu^s$ intersect transversally at $\tau(P)$ *iff* W_μ^u and W_μ^s do so at P. We may therefore consider $\hat{W}_\mu^u$ and $\hat{W}_\mu^s$ instead of W_μ^u and W_μ^s.

Let $v \in H^1$ be a critical point of $\mathcal{A}_\mu$. Then $v \in C^\infty$, and $P = (v(0), v'(0))$ is the corresponding intersection of $\hat{W}_\mu^u$ and $\hat{W}_\mu^s$. The second derivative of $\mathcal{A}_\mu$ at v is given by

$$\mathrm{d}^2\mathcal{A}_\mu(v) \cdot (\varphi, \psi) = \langle \mathrm{L}\varphi, \psi \rangle,$$

where $\mathrm{L} : H^1 \to H^{-1}$ is the differential operator

$$\mathrm{L} = -\mathrm{D}^2 + Q(t); \qquad Q(t) =_{\mathrm{def}} \frac{\partial \hat{F}}{\partial u}(\mu, t, v(t)),$$

and where $\langle \varphi, \psi \rangle = \int_{\mathbf{R}} \varphi\psi$.

When μ is small $p(\mu, t)$ is also small, so it follows from

$$\hat{F}_u(\mu, t, v) = F_0'(p(\mu, t) + v) + \mu g_u(t, p(\mu, t) + v),$$

$F_0'(0) > 0$, and $v(\pm\infty) = 0$ that for small μ

$$\liminf_{t \to \pm\infty} Q(t) > 0. \tag{2.5}$$

Hence L is Fredholm with index zero for small μ.

Indeed, $\mathrm{L}_0 = -\mathrm{D}^2 + Q_0(t)$ with $Q_0(t) = \hat{F}_u(\mu, t, 0)$ is invertible, since $\inf Q_0(t) > 0$; $\mathrm{L} - \mathrm{L}_0$ is given by multiplication with $Q(t) - Q_0(t)$, which vanishes at $t = \pm\infty$ and hence is a compact operator from H^1 to H^{-1}; so L is indeed Fredholm.

By definition the critical point v will be nondegenerate *iff* $\mathrm{L} = \mathrm{d}^2\mathcal{A}_\mu(v)$ is invertible, which, due to L's Fredholmness, will be the case *iff* L is injective. The nullspace of L consists of those $y \in H^1$ which satisfy

$$y'' = Q(t)y. \tag{2.6}$$

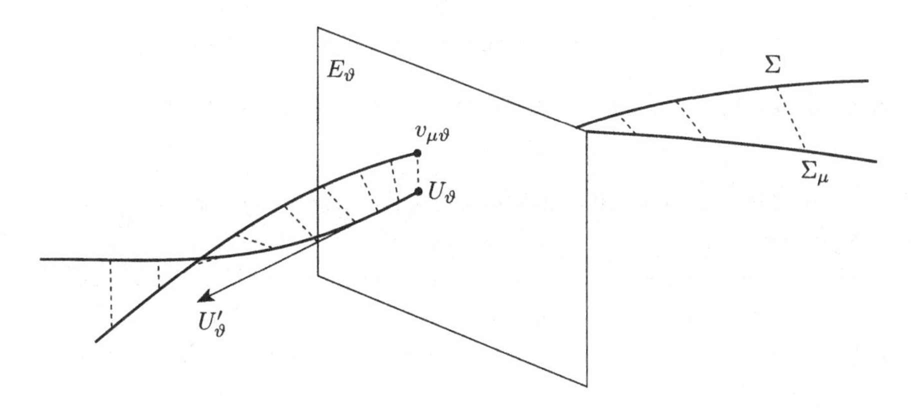

Figure 2—The curve of critical points, the transverse slice, and the perturbed curve.

At $t = \pm\infty$ $Q(t)$ is bounded away from zero, so solutions of (2.6) as well as their derivatives either grow or decay exponentially. Hence there are two solutions $y_\pm(t)$ of (2.6) with $y_+(+\infty) = 0$ and $y_-(-\infty) = 0$, respectively.

Since (2.6) is the variational equation of (2.3), the vectors $\xi_+ = (y_+(0), y'_+(0))$ and $\xi_- = (y_-(0), y'_-(0))$ span the tangent spaces $T_P W^s_\mu$ and $T_P W^u_\mu$, respectively.

If the intersection is not transverse, then $\xi_- = \lambda\xi_+$, and hence $y_- = \lambda y_+$; in this case y_+ not only vanishes at $t = \infty$, but also at $t = -\infty$. Condition (2.5) implies that bounded solutions of (2.6) decay exponentially as $t \to \pm\infty$, so y_+ belongs to the nullspace of L, and v is a degenerate fixed point.

Conversely, if v is degenerate, then the $y_\pm$ are multiples of each other, whence the $\xi_\pm$ are also multiples of each other; i.e. the intersection at P is nontransverse.

Q. E. D.

For $\mu = 0$ we have a curve $\Sigma \subset H^1(\mathbf{R})$ of critical points of $\mathcal{A}_\mu$, consisting of all translates $U_\vartheta(t) = U(\vartheta + t), \vartheta \in \mathbf{R}$ of U. All these critical points are degenerate, but they are nondegenerate in the direction transverse to the curve Σ. Indeed, define

$$E^r_\vartheta = \left\{ v \in H^r(\mathbf{R}) : \int_{\mathbf{R}} v(t) U_\vartheta{}'(t) dt = 0 \right\}, \qquad r = \pm 1.$$

Using $U_\vartheta(\pm\infty) = 0$ one easily verifies that $U_\vartheta \in E^1_\vartheta$. Since $U_\vartheta{}'$ spans the tangent space to the curve Σ, E^1_ϑ intersects Σ transversally at U_ϑ.

The second derivative of $\mathcal{A}_0 = \mathcal{A}(0,\cdot)$ at U_ϑ is given by $\mathrm{d}^2\mathcal{A}_0(U_\vartheta)\cdot(\varphi,\psi) = \langle \mathrm{A}_\vartheta\varphi, \psi\rangle$, where $\mathrm{A}_\vartheta : H^1 \to H^{-1}$ is the operator $\mathrm{A}_\vartheta = -\mathrm{D}^2 + F_0'(U_\vartheta(t))$.

2.2. Lemma. *A_ϑ is a Fredholm operator of index zero. Its kernel is spanned by $U_\vartheta{}'$, and its range is E_ϑ^{-1}.*

Proof. That A_ϑ is Fredholm follows from the same arguments we used in the previous lemma.

The kernel of A_ϑ clearly contains $U_\vartheta{}'$ – just differentiate (2.2) – and by considering the growth or decay of solutions of $y''(t) = F_0'(U_\vartheta(t))y(t)$ at $t = \pm\infty$ one finds that the kernel of A_ϑ can be at most one–dimensional. Hence it must be spanned by $U_\vartheta{}'$.

It follows that the range of A_ϑ has codimension one, while integration by parts shows that $\mathrm{A}_\vartheta(H^1) \subset E_\vartheta^{-1}$. Hence $\mathrm{A}_\vartheta(H^1) = E_\vartheta^{-1}$.

Q. E. D.

2.3. Corollary. *U_ϑ is a Morse critical point of $\mathcal{A}_0|E_\vartheta^1$.*

It follows that the critical point U_ϑ of $\mathcal{A}_0|E_\vartheta^1$ will persist under small perturbations, so that we obtain a smooth family of critical points $v_{\vartheta,\mu} \in E_\vartheta^1$ which is defined for small μ and $\vartheta \in \mathbf{R}$.

If $\mathcal{A}_\mu$ has a critical point v_* near any of the U_ϑ's then $v_* = v_{\vartheta,\mu}$ for some $\vartheta \in \mathbf{R}$, but the converse is not true: not every $v_{\vartheta,\mu}$ is a critical point of $\mathcal{A}_\mu$. To determine which of the $v_{\vartheta,\mu}$ are critical points we observe that, since $v_{\vartheta,\mu}$ is a critical point of $\mathcal{A}_\mu$ subject to the constraint $\langle U_\vartheta{}', v\rangle = 0$, we have

$$\mathrm{d}\mathcal{A}_\mu(v_{\vartheta,\mu}) = \lambda U_\vartheta{}',$$

where $\lambda = \lambda(\mu,\vartheta)$ is a Lagrange multiplier. This obviously implies:

2.4. Proposition. *$v_{\vartheta,\mu}$ is a critical point of $\mathcal{A}_\mu$ iff $\lambda(\mu,\vartheta) = 0$.*

Rather than considering $\lambda(\mu,\vartheta)$ we introduce

$$a(\mu,\vartheta) = \mathcal{A}_\mu(v_{\vartheta,\mu}),$$

and compute

$$\begin{aligned}\frac{\partial a}{\partial\vartheta} &= \left\langle \mathrm{d}\mathcal{A}_\mu(v_{\vartheta,\mu}), \frac{\partial v_{\vartheta,\mu}}{\partial\vartheta}\right\rangle \\ &= \lambda(\mu,\vartheta)\left\langle U_\vartheta{}', \frac{\partial v_{\vartheta,\mu}}{\partial\vartheta}\right\rangle.\end{aligned}$$

For $\mu = 0$ we have $v_{\vartheta,\mu} = U_\vartheta$, and hence $\frac{\partial v_{\vartheta,\mu}}{\partial \vartheta} = U_\vartheta{}'$, so that we find

$$\lambda(\mu, \vartheta) = \frac{\partial a}{\partial \vartheta}(\mu, \vartheta)\,(\Gamma + o(1))\,, \qquad (\mu \to 0),$$

with $\Gamma = \int_{\mathbf{R}} U'(t)^2 dt > 0$. Hence for small μ the zeroes of $\lambda(\mu, \cdot)$ and $a_\vartheta(\mu, \cdot)$ coincide.

The functional $\mathcal{A}(0, v)$ is invariant under translations of v, i.e. under the substitution $v(t) \mapsto v(t + \vartheta)$, so $a(0, \vartheta) = \mathcal{A}(0, U_\vartheta)$ does not depend on ϑ. Therefore $a_\vartheta(0, \vartheta) \equiv 0$, and $a_\vartheta(\mu, \vartheta)/\mu$ is a well defined smooth function for small μ; we have

$$a_\vartheta(\mu, \vartheta) = \mu \frac{\partial^2 a}{\partial \vartheta \partial \mu}(0, \vartheta) + O(\mu^2), \qquad (\mu \to 0).$$

The second derivative can be computed directly:

$$\begin{aligned}
\frac{\partial^2 a}{\partial \vartheta \partial \mu}(0, \vartheta) &= \frac{\partial}{\partial \vartheta}\left(\frac{\partial}{\partial \mu}\mathcal{A}_\mu(v_{\mu,\vartheta})\right) \\
&= \frac{\partial}{\partial \vartheta}\left(\frac{\partial \mathcal{A}}{\partial \mu}(0, U_\vartheta) + \mathrm{d}\mathcal{A}_0(U_\vartheta)\frac{\partial v_{\mu,\vartheta}}{\partial \mu}\right) \\
&= \frac{\partial}{\partial \vartheta}\left(\frac{\partial \mathcal{A}}{\partial \mu}(0, U_\vartheta)\right) \qquad (\text{since } \mathrm{d}\mathcal{A}_0(U_\vartheta) = 0) \\
&= -\frac{\partial}{\partial \vartheta}\int_{\mathbf{R}} \frac{\partial \hat{V}}{\partial \mu}(0, t, U_\vartheta(t))dt \qquad (\text{by } (2.4)) \\
&= \int_{\mathbf{R}} \frac{\partial \hat{F}}{\partial \mu}(0, t, U_\vartheta(t))U_\vartheta{}'(t)dt.
\end{aligned}$$

Using our definition of $\hat{F}$ we find that

$$\frac{\partial \hat{F}}{\partial \mu}(0, t, v) = (F_0'(v) - F_0'(0))\,q(t) + g(t, v) - g(t, 0),$$

where $q(t) = \frac{\partial p}{\partial \mu}(0, t)$ is obtained by solving

$$q''(t) - F_0'(0)q(t) = g(t, 0), \qquad q(t+1) \equiv q(t).$$

This identity implies that

$$\begin{aligned}
\int_{\mathbf{R}} g(t, 0)U_\vartheta{}'(t)dt &= \int_{\mathbf{R}} (U_\vartheta{}''' - F_0'(0)U_\vartheta{}')\,q(t)dt \\
&= \int_{\mathbf{R}} U_\vartheta{}'(t)\,(F_0'(U_\vartheta(t)) - F_0'(0))\,q(t)dt;
\end{aligned}$$

from which we get the following formula for $a_{\mu\vartheta}$

$$a_{\mu\vartheta}(0, \vartheta) = \int_{\mathbf{R}} g(t, U_\vartheta(t))U_\vartheta{}'(t)dt.$$

This last integral is precisely the Melnikov function. In general we can now easily prove:

2.5. Theorem (Melnikov). *We assume that the Melnikov function $\mathcal{M}(\vartheta) = a_{\mu\vartheta}(0, \vartheta)$ has a simple zero at $\vartheta = \vartheta_0$. Then there exists a smooth branch of nondegenerate critical points $w_\mu = v_{\mu,\vartheta(\mu)}$ of $\mathcal{A}_\mu$ with $w_0 = U_{\vartheta_0}$. In particular, for small enough μ the stable and unstable manifolds W^s_μ, W^u_μ have a transverse intersection at $P_\mu = (w_\mu(0) + p(\mu, 0), w'_\mu(0) + p_t(\mu, 0))$.*

Proof. We have just shown that $b(\mu, \vartheta) = a_\vartheta(\mu, \vartheta)/\mu$ is a smooth function, defined for small μ, with $b(0, \vartheta) = \mathcal{M}(\vartheta)$. The implicit function theorem provides us with a smooth branch of zeroes $\vartheta(\mu)$ of $b(\mu, \vartheta)$, hence of $a_\vartheta(\mu, \vartheta)$. Defining $w_\mu = v_{\mu,\vartheta(\mu)}$ as we did, we get a branch of critical points.

To show that the w_μ are nondegenerate we consider the splitting $H^1 = E^1_{\vartheta(\mu)} \oplus [U'_{\vartheta(\mu)}]$; w.r.t. this splitting we can represent the second derivative of $\mathcal{A}_\mu$ at w_μ by

$$\mathrm{d}^2\mathcal{A}_\mu(w_\mu) = \begin{pmatrix} \mathrm{A}_{\vartheta(\mu)}|E^1_{\vartheta(\mu)} + O(\mu) & O(\mu) \\ O(\mu) & a_{\vartheta\vartheta}(\mu, \vartheta(\mu)) \end{pmatrix}.$$

We have seen that $\mathrm{A}_{\vartheta(\mu)}|E^1_{\vartheta(\mu)}$ is invertible; we have also assumed that ϑ_0 is a simple zero of $\mathcal{M}$, i.e. $\mathcal{M}'(\vartheta_0) \neq 0$; using $a_{\vartheta\vartheta}(\mu, \vartheta(\mu)) = \mu\mathcal{M}'(\vartheta_0) + O(\mu^2)$ one can then show that $\mathrm{d}^2\mathcal{A}_\mu(w_\mu)$ is invertible, for small μ.

Q. E. D.

– *The size of the splitting* –

In the classical derivation of the Melnikov function one computes the distance $d(\mu, \vartheta)$ between W^u_μ and W^s_μ in some direction transverse to both invariant manifolds. The Melnikov function arises as the first coefficient of an expansion of $d(\mu, \vartheta)$ in powers of μ: $d(\mu, \vartheta) = \mu\mathcal{M}(\vartheta) + O(\mu^2)$.

One can show that $d(\vartheta, \mu) = \lambda(\vartheta, \mu) + O(\mu^2)$ which gives an approximate interpretation of $\lambda(\mu, \vartheta)$. Instead we will now show that one can give an exact measure of the size of the separation between W^u_μ and W^s_μ in terms of $a(\mu, \vartheta)$.

Homoclinic points lie on the stable manifold of $\mathcal{O}_\mu$ so they can be ordered linearly by their position on this invariant manifold. Their position on the unstable manifold defines another ordering of homoclinic points and these two orderings need not coincide, in general.

Our construction identifies certain homoclinic points with critical points of the function $a(\mu, \cdot)$ so that these particular homoclinics have a third ordering.

For small μ all three of these orderings coincide. That this is so becomes clear if one considers the curve $\gamma_\mu(\vartheta) = (v_{\mu,\vartheta}(0), v'_{\mu,\vartheta}(0))$ with $\vartheta \in \mathbf{R}$. When $\mu = 0$ we have $v_{\mu,\vartheta}(0) = U_\vartheta(0) = U(\vartheta)$, so that $\gamma_\mu(\vartheta) = (U(\vartheta), U'(\vartheta))$ parametrizes the stable and unstable manifolds of $\mathcal{O}_0$. Hence for small μ the curve parametrized by γ_μ will be C^1 close to both the stable and unstable manifolds of $\mathcal{O}_\mu$. Since the three orderings of critical points are determined by the order in which they occur on the curves γ_μ, W^s_μ and W^u_μ, these orderings coincide for small μ.

2.6. Proposition. *Let $P, Q \in W^u_\mu \cap W^s_\mu$ be two consecutive homoclinic points corresponding to two consecutive critical points $\vartheta_P < \vartheta_Q$ of $a(\mu, \cdot)$. Denote the region bounded by the segments of W^u_μ and W^s_μ that connect P and Q by Ω (some call such a region a "lobe").*

Then the area of the lobe Ω is given by

$$m(\Omega) = |\mathcal{A}_\mu(w_P) - \mathcal{A}_\mu(w_Q)| = |a(\mu, \vartheta_P) - a(\mu, \vartheta_Q)|.$$

This was shown by MacKay, Meiss and Percival[6].

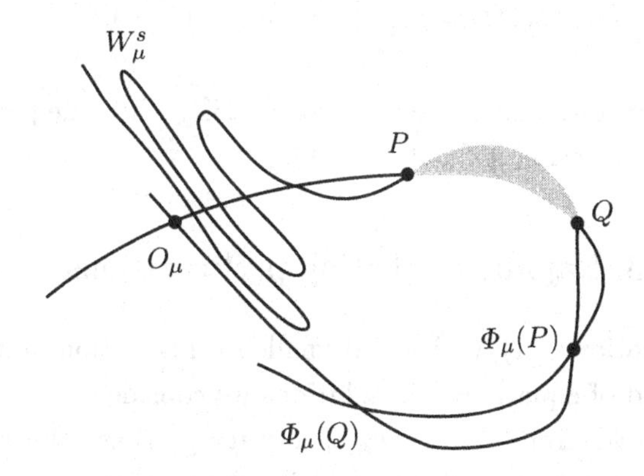

Fig 3—Two adjacent homoclinic points and the corresponding lobe

[6] See [MMS84] for a discussion of transport by area preserving maps. Their arguments are easily adapted to our setting.

The point of this observation was that it allows one to compute "the flux across a barrier," where the barrier is formed by the stable and unstable manifolds. More precisely, choose a homoclinic point P_μ corresponding to one of the critical points of $a(\mu, \cdot)$; Let $W^s_\mu[P_\mu, \mathcal{O}_\mu]$ and $W^u_\mu[\mathcal{O}_\mu, P_\mu]$ be the segments of the two invariant manifolds between $\mathcal{O}_\mu$ and P_μ; and define Ω to be the bounded region these two curve segments enclose. Then the flux across $\partial\Omega$ is by definition

$$\operatorname{flux}(\Phi_\mu, \Omega) = m\left(\Omega \Delta \Phi_\mu(\Omega)\right).$$

(We write $A\Delta B = (A \cap B^c) \cup (A^c \cap B)$.) Since one can decompose the difference $\Omega\Delta\Phi_\mu(\Omega)$ into a disjoint union of "lobes" one can actually compute the flux in terms of $a(\mu, \cdot)$: One finds that the flux is given by the sum of $|a(\mu, \vartheta_P) - a(\mu, \vartheta_Q)|$ over all consecutive critical points $\vartheta_P < \vartheta_Q$ in one period interval, i.e. the flux is equal to the total variation of $a(\mu, \cdot)$ over $\mathbf{T}$:

$$\operatorname{flux}(\Phi_\mu, \Omega) = \int_{\mathbf{T}} |a_\vartheta(\mu, \vartheta)| \, d\vartheta.$$

For small μ the flux is therefore given by

$$\operatorname{flux}(\Phi_\mu, \Omega) = \mu \int_{\mathbf{T}} |\mathcal{M}(\vartheta)| d\vartheta + O(\mu^2).$$

This completes our discussion of separatrix crossing when the perturbations are small and their period is fixed.

§3. Rapidly oscillating perturbations

We turn to equation (1.1), with $\varepsilon > 0$ small. In this section we will explain exactly what kind of equations and solutions we consider.

It is well known from the theory of averaging[7] that solutions of (1.1) are approximated by those of the averaged equation,

$$u''(t) = F_0(u(t)), \tag{3.1}$$

where

$$F_0(u) = \int_0^1 F(\tau, u) \, d\tau. \tag{3.2}$$

[7] See [GH, chapter 4].

Indeed, if we define

$$G(t, u) = \int_0^t (F(\tau, u) - F_0(u))\, d\tau,$$

then (1.1) can be rewritten as $u'' = F_0(u) + G_t(t/\varepsilon, u)$, and hence as

$$-u''(t) + F_0(u) - \varepsilon \left(\frac{dG(t/\varepsilon, u(t))}{dt} + G_u(t/\varepsilon, u(t))u'(t) \right) = 0. \qquad (3.3)$$

We shall regard (3.3) as the central equation in our discussion. This equation is a small perturbation of the averaged equation (3.1) and it is at least plausible that the deviation of solutions of (1.1) from those of the averaged equation is of order ε, on any bounded time interval; but to make a more precise statement we must first state our hypotheses.

We assume the autonomous term $F_0(u)$ is as in the previous section (i.e.(2.1) holds, and the corresponding potential energy is such that (2.2) again has a unique positive solution U); in addition to this we require F_0 to be an entire function of $u \in \mathbf{C}$.

Concerning $G(t, u)$ we assume that it is holomorphic in u, but only bounded measurable in the time variable. In other words, we assume that G may be written as

$$G(t, u) = \sum_{j=0}^{\infty} g_j(t) u^j, \qquad (3.4)$$

with $g_j \in L_\infty(\mathbf{T})$ and $\lim_{n\to\infty} \|g_n\|_{L_\infty}^{1/n} = 0$. We could also say that we require $u \mapsto G(\cdot, u)$ to be an entire function with values in $L_\infty(\mathbf{T})$.

Finally, we assume that $t \mapsto G(t, u)$ is right–continuous and vanishes at $t = 0$; i.e.

$$\lim_{t \downarrow 0} G(t, u) = 0 \qquad (3.5)$$

uniformly in u, on bounded subsets of $\mathbf{C}$.

Since we do not assume that G is differentiable w.r.t. the time variable, $G_t(t/\varepsilon, u(t))$ cannot be defined by simply substituting $u(t)$ in G_t. Hence we must define what it means for a function to be a solution of (1.1).

3.1. Definition. *A solution is an absolutely continuous function $u(t)$ which satisfies (3.3) in the sense of distributions.*

The right–continuity of G allows us to integrate (3.3), which leads to

$$u(t) = u_0 + tv_0 + \int_0^t (t-s)\left\{F_0(u(s)) - \varepsilon G_u(s/\varepsilon, u(s))u'(s)\right\} ds + \\ + \varepsilon \int_0^t G(s/\varepsilon, u(s))\, ds; \qquad (3.6)$$

$$u'(t) = v_0 + \int_0^t \left\{F_0(u(s)) - \varepsilon G_u(s/\varepsilon, u(s))u'(s)\right\} ds + \\ + \varepsilon G(s/\varepsilon, u(s)). \qquad (3.7)$$

The second identity shows that the derivative u' of any solution is bounded, i.e. solutions are Lipschitz continuous instead of merely absolutely continuous. The same identity also shows that $\lim_{t\downarrow 0} u'(t) = u'(0+)$ exists for any solution u, so that we can speak of the initial values $(u(0), u'(0+))$ of a solution.

By applying Picard–iteration and Gronwall's lemma to the two integral equations one can show the following.

3.2. Lemma. *For any $\varepsilon \geq 0$ and $u_0, v_0 \in \mathbf{C}$ (3.3) has a solution $u(\varepsilon, u_0, v_0; t)$ on a short enough time interval $-T_-(\varepsilon, u_0, v_0) < t < T_+(\varepsilon, u_0, v_0)$.*

The $T_\pm$ are lower semi continuous functions of ε, u_0, v_0, the solution $u(\varepsilon, u_0, v_0; t)$ depends continuously on $\varepsilon \geq 0$, and is holomorphic in u_0, v_0.

Thus, if the solution $u(t)$ of the averaged equation $u'' = F_0(u)$ with $u(0) = u_0, u'(0) = v_0$ exists on the closed interval $-t_- \leq t \leq t_+$, then for small enough $\varepsilon > 0$ (3.3) will have a solution on the same time interval, and $u(\varepsilon; t) \to u(t)$, $u'(\varepsilon; t) \to u'(t)$ uniformly on $[t_-, t_+]$.

In terms of the Poincaré–maps Φ_ε this may be stated as

$$\lim_{n\to\infty} \left(\Phi_{t/n}\right)^n (u_0, v_0) = \Psi_t(u_0, v_0),$$

where Ψ_t denotes the flow of the averaged system $u' = v, v' = F_0(u)$.

– *The two examples* –

If one assumes $G(t, u)$ does not depend on u, and puts $g(t) = G'(t)$ then one obtains the forced Duffing equation (1.2). Hence in terms of the first example we allow periodic forcing terms $g(t)$ which are distributional derivatives of bounded measurable functions with $\int_{\mathbf{T}} g(t)dt = 0$.

To obtain the second example one should choose $G(t, u) = Z(t)F_0(u)$ where

$$Z(t) = \lfloor t \rfloor - t = \max(n \in \mathbf{Z} : n \leq t) - t.$$

In this case our previous remarks show that the solutions of the difference equation (1.4) converge to the solution of the ODE $u'' = F_0(u)$ – this is ofcourse a well known fact from numerical mathematics.

§4. Separatrix splitting

In computing the splitting of the separatrices for small ε we try to follow the approach of section 2 as closely as possible. The main result we shall find in this section is:

4.1. Theorem. *Assume that U is analytic in $|\mathrm{Im} t| < \rho_0$. Then for any $\rho < \rho_0$ there exist $R(\rho), \delta(\rho) > 0$ such that the following holds for $\varepsilon[G]_R < \delta$.*

1. *Φ_ε has a hyperbolic fixed point $\mathcal{O}_\varepsilon$ near the origin; $\mathcal{O}_\varepsilon$ depends analytically on ε^2.*

2. *There is a homoclinic point $P_\varepsilon \in W^s_\varepsilon \cap W^u_\varepsilon$ such that*

$$\mathrm{flux}(\Phi_\varepsilon, \Omega_{\mathcal{O}_\varepsilon P_\varepsilon}) \le C_\rho e^{-2\pi\rho/\varepsilon},$$

where $\Omega_{\mathcal{O}_\varepsilon P_\varepsilon}$ is the domain enclosed by $W^s_\varepsilon[P_\varepsilon, \mathcal{O}_\varepsilon]$ and $W^u_\varepsilon[\mathcal{O}_\varepsilon, P_\varepsilon]$.

In what follows we will use the following seminorms to measure the size of F and G:

$$[G]_R = \mathrm{ess.sup}\left(|G(t,u)| : t \in \mathbf{T}, |u| \le R\right).$$

Since F and G are analytic in the u variable these seminorms also control the derivatives w.r.t. $u \in \mathbf{C}$ of F and G. Indeed, Cauchy's theorem implies that for almost every $t \in \mathbf{T}$

$$\left|\frac{\partial^k G}{\partial u^k}(t,u)\right| \le k![G]_{R+1}, \tag{4.1}$$

if $|u| \le R$.

We shall frequently use this observation in the following form. If we approximate F_0 or G by a Taylor series, then we can estimate the error in terms of the seminorms $[F_0]$ and $[G]$. Indeed,

$$G(t, u+v) = \sum_{k=0}^{n-1} \frac{\partial^k G}{\partial t^k}(t,u)v^k + R_n(u,v),$$

where

$$R_n(u,v) = v^n \int_0^1 \frac{(1-\lambda)^{n-1}}{(n-1)!} \frac{\partial^n G}{\partial u^n}(t, u+\lambda v)\, d\lambda,$$

so that it follows from (4.1) that

$$|R_n(u,v)| \le [G]_{|u|+|v|+1}|v|^n. \tag{4.2}$$

– Step 1–The hyperbolic fixed point –

As in section 2 we verify that the hyperbolic fixed point at the origin persists for small $\varepsilon > 0$, when the equation (1.1) is perturbed.

4.2. Lemma. *There is a $\delta_0 > 0$ such that (1.1) has an ε periodic solution $p_\varepsilon(t) = p(\varepsilon, t/\varepsilon)$ for any ε, G with $\varepsilon < \delta_0$ and $\varepsilon[G]_2 < \delta_0$. The solution $p(\varepsilon,t)$ is analytic in ε and may be written as $p(\varepsilon,t) = \sum_{j\ge1} p_j(t)\varepsilon^{2j}$. The $W^1_\infty(\mathbf{T})$ norm of the solution $p(\varepsilon,\cdot)$ is estimated by:*

$$\|p(\varepsilon,\cdot)\| \le C\varepsilon^2[G]_2.$$

In terms of the Poincaré map this means that Φ_ε has a fixed point

$$\mathcal{O}_\varepsilon = (p(\varepsilon,0), p'(\varepsilon,0)/\varepsilon),$$

for $0 < \varepsilon < \min(\delta_0, \delta_0/[G]_2)$. We postpone the proof of this lemma until section 6.

– Step 2–The modified equation –

After substituting $u(t) = v(t) + p(\varepsilon, t/\varepsilon)$ in (1.1) one gets the following equation for v:

$$v'' = \hat{F}(t/\varepsilon, v(t)) \tag{4.3}$$

where $\hat{F}(t,v) = F(t, p(\varepsilon,t)+v) - \varepsilon^{-2}p''(\varepsilon,t)$.

As long as it causes no confusion we will write $p(t)$ instead of $p(\varepsilon,t)$ and we will also suppress the ε dependence of $\hat{F}$.

The averaged equation is $v'' = \hat{F}_0(v)$, where

$$\begin{aligned}\hat{F}_0(v) &= \int_{\mathbf{T}} \{F(t, v+p(t)) - \varepsilon^{-2}p''(t)\}\, dt \\ &= \int_{\mathbf{T}} \{F_0(v+p(t)) - G_u(t, v+p(t))p'(t)\}\, dt.\end{aligned}$$

The average free part of $\hat{F}$ is given by $\hat{G}_t(t,v)$ where

$$\begin{aligned}\hat{G}(t,v) &= \int_0^t \left\{\hat{F}(\tau, v+p(\tau)) - \varepsilon^{-2}p''(\tau)\right\} d\tau - t\hat{F}_0(v) \\ &= \int_0^t \left\{F_0(v+p(\tau)) - \hat{F}_0(v) - G_u(\tau, v+p(\tau))p'(\tau)\right\} d\tau + \\ &\quad + \frac{p'(0)-p'(t)}{\varepsilon^2} + G(\tau, p(\tau)+v).\end{aligned}$$

We also define the modified potentials $\hat{V}_0(v) = \int_0^v \hat{F}_0(w)dw$.

Equation (4.3) is equivalent to (3.3), with F_0 and G replaced by $\hat{F}_0, \hat{G}$. Solutions of (4.3) are exactly the critical points of

$$\mathcal{A}_\varepsilon(v) = \int_{\mathbf{R}} \left\{ {}^1/_2 v'(t)^2 - \hat{V}_0(v(t)) - \varepsilon \hat{G}(t/\varepsilon, v(t))v'(t) \right\} dt.$$

The modified nonlinearities $\hat{F}$ and $\hat{G}$ are close to F, G when ε is small. How close they are is measured by the following lemma.

4.3. Lemma. *Let $\delta_0 > 0$ be as in lemma 4.1. Then for $\varepsilon < \delta_0, \varepsilon[G]_2 < \delta_0$ one has*

$$|\hat{F}_0(v) - F_0(v)| \leq c(1 + [G]_{R+2})[G]_2\varepsilon^2.$$
$$\left|\hat{G}(t,v) - G(t, v + p(t)) - \frac{p'(0) - p'(t)}{\varepsilon^2}\right| \leq c(1 + [G]_{R+2})[G]_2\varepsilon^2.$$
$$|\hat{G}(t,v)| \leq c[G]_{R+2}$$
$$|\hat{V}_0(v) - V_0(v)| \leq c(1 + [G]_{R+2})[G]_2\varepsilon^2$$

whenever $|v| \leq R$ and $|w| \leq 1$. Here c is a constant which only depends on R and $[F_0]_{R+2}$.

The first two inequalities are obtained from the definitions of $\hat{F}_0$ and $\hat{G}$ and our estimate in lemma 4.1, i.e. $|p| + |p'| \leq c[G]_2\varepsilon^2$. The third inequality then follows from the second, and one obtains the last one by integrating first estimate.

One can improve these inequalities a little by observing that we have defined $\hat{F}, \hat{G}$ and $\hat{V}$ so that

$$F_0(0) = \hat{F}_0(0) = \hat{G}(t, 0) = 0,$$
$$V_0(0) = V_0'(0) = \hat{V}_0(0) = \hat{V}_0'(0) = 0.$$

Since all these functions are analytic in the u variable this means that

$$|G(t,v)| \leq c[G]_{R+2} \cdot \left|\frac{v}{R}\right|,$$
$$|\hat{V}_0(v) - V_0(v)| \leq c(1 + [G]_{R+2})[G]_2\varepsilon^2 \cdot \left|\frac{v}{R}\right|^2$$

for $|v| \leq R$. This has the following implications for the functional $\mathcal{A}_\varepsilon$.

4.4. Lemma. *Assume again that $\varepsilon < \delta_0, \varepsilon[G]_2 < \delta_0$. Then $\mathcal{A}_\varepsilon$ is a holomorphic function on $H^1(\mathbf{R}, \mathbf{C})$, which satisfies*

$$|\mathcal{A}_\varepsilon(v) - \mathcal{A}_0(v)| \le c[G]_{R+2}\varepsilon,$$

for any $v \in H^1(\mathbf{R}, \mathbf{C})$ with $\|v\|_{H^1} \le R$. Moreover,

$$\left\|\mathrm{d}^k\mathcal{A}_\varepsilon(v) - \mathrm{d}^k\mathcal{A}_0(v)\right\| \le k!c[G]_{R+2}\varepsilon$$

holds whenever $\|v\|_{H^1} \le R - 1$.

Here and in the following sections we let $H^r = H^r(\mathbf{R}, \mathbf{C})$ stand for the Sobolev space of complex valued functions with r derivatives in L_2.

Proof. For any $v \in H^1(\mathbf{R})$ one has $\|v\|_{L_\infty} \le \|v\|_{H^1}$ so that $\|v\|_{H_1} \le R$ implies $\|v\|_{L_\infty} \le R$.

The first part of this lemma now follows by applying our estimates of $\hat{G}(t,v)$ and $|\hat{V}_0(v) - V_0(v)|$ to

$$\mathcal{A}_\varepsilon(v) - \mathcal{A}_0(v) = \int_{\mathbf{R}} \left\{V_0(v) - \hat{V}_0(v) - \varepsilon G(t/\varepsilon, v)v'\right\} dt.$$

The estimate for the derivatives then follows from Cauchy's formula and the analyticity of $\mathcal{A}_\varepsilon - \mathcal{A}_0$.

– *Step 3–The curve of critical points* –

As in section 2 the translates of U form a curve of critical points of $\mathcal{A}_0$. But in the present situation it turns out to be expedient to note that $U(t)$ is an analytic function defined on a neighborhood of the real axis: Since $U(t)$ is analytic in $S_{\rho_0} = \{t \in \mathbf{C} : |\mathrm{Im}t| < \rho_0\}$, every translate $U_\vartheta(t) = U(\vartheta + t)$ with $\vartheta \in S_{\rho_0}$ will also be a critical point of $\mathcal{A}_0$. The curve Σ of critical points is therefore a holomorphic curve in the *complex* Hilbert space $H^1(\mathbf{R}, \mathbf{C})$.

We define $E_\vartheta^{\pm1}$ to be the subspaces of $H^{\pm1}$ consisting of all w with $\int_{\mathbf{R}} wU_\vartheta{}' = 0$. The reader should note that for non-real values of ϑ this space is the L_2–orthogonal complement of the complex conjugate of $U_\vartheta{}'$ rather than $U_\vartheta{}'$ itself. With this definition the spaces $E_\vartheta^{\pm1}$ depend holomorphically on ϑ.

Analogously to lemma 2.2 one now shows

4.5. Lemma. $\mathrm{A}_\vartheta = \mathrm{d}^2\mathcal{A}_0(U_\vartheta) : H^1 \to H^{-1}$ *is Fredholm with index zero.*

$\mathrm{A}_\vartheta|E_\vartheta^1$ *is injective, the range of* A_ϑ *is* E_ϑ^{-1}*, its kernel is spanned by* $U_\vartheta{}'$.

U_ϑ is a nondegenerate critical point of $\mathcal{A}|E^1_\vartheta$.

The stability of nondegenerate critical points implies that any small perturbation of $\mathcal{A}_0|E^1_\vartheta$ will have a nondegenerate critical point near U_ϑ. In particular, $\mathcal{A}_\varepsilon|E^1_\vartheta$ will have such a critical point $v_{\vartheta,\varepsilon} \in E^1_\vartheta$ if ε is small enough.

4.6. Lemma. *For any $\rho < \rho_0$ there is a $\delta = \delta(\rho) > 0$ such that $\mathcal{A}_\varepsilon|E^1_\vartheta$ has a unique critical point $v_{\varepsilon,\vartheta} \in E^1_\vartheta$ with*

$$\|v_{\varepsilon,\vartheta} - U_\vartheta\|_{H^1} \leq c[G]_{R+2}\varepsilon$$

for any $\vartheta \in S_\rho$. Here $R = 1 + \sup_{\vartheta \in S_\rho} \|U_\vartheta\|_{H^1}$, and c is a constant that only depends on ρ and $[F_0]_{R+2}$.

The $v_{\varepsilon,\vartheta}$ are holomorphic in ϑ, and ε periodic in the sense that $v_{\varepsilon,\vartheta+\varepsilon}(t) = v_{\vartheta,\varepsilon}(t+\varepsilon)$.

As H^1 valued function of ϑ they are also C^1 close to U_ϑ, with

$$\left\| \frac{\partial v_{\varepsilon,\vartheta}}{\partial \vartheta} - U_\vartheta{}' \right\|_{H^1} \leq c[G]_{R+2}\varepsilon.$$

We shall obtain $v_{\vartheta,\varepsilon}$ from the Lagrange multiplier equation

$$d\mathcal{A}_\varepsilon(v) = \lambda U_\vartheta{}', \qquad \langle v, U_\vartheta{}' \rangle = 0 \tag{4.4}$$

in section 7.

Once again the $v_{\varepsilon,\vartheta}$ are not necessarily critical points of $\mathcal{A}_\varepsilon$, but any critical point of $\mathcal{A}_\varepsilon$ which is close to some U_ϑ must be a $v_{\varepsilon,\vartheta}$. At the same time, a $v_{\varepsilon,\vartheta}$ is a critical point *iff* the corresponding Lagrange multiplier $\lambda(\varepsilon,\vartheta)$, defined by (4.4), vanishes.

– *Step 5–The function $a(\varepsilon,\vartheta)$* –

Consider $a(\varepsilon,\vartheta) = \mathcal{A}_\varepsilon(U_\vartheta + w_{\varepsilon,\vartheta})$. Then $a(\varepsilon,\vartheta+\varepsilon) = a(\varepsilon,\vartheta)$ for all $\vartheta \in S_\rho$.

Just as in section 2 the critical points of $a(\varepsilon,\cdot)$ correspond precisely to the homoclinics of the Poincaré-map. Indeed, for $\vartheta \in S_\rho$ and $\varepsilon[G]_{R+2} \leq \delta(\rho)$ we have

$$\frac{\partial a}{\partial \vartheta}(\varepsilon,\vartheta) = \lambda(\varepsilon,\vartheta)\left\langle U_\vartheta{}', \frac{\partial v_{\varepsilon,\vartheta}}{\partial \vartheta} \right\rangle = \lambda(\varepsilon,\vartheta)\left(\Gamma + O(\varepsilon[G]_{R+2})\right), \tag{4.5}$$

where $\Gamma = \int_{\mathbf{R}} U'(t+\vartheta)^2 dt > 0$ does not depend on ϑ.

Unlike the case of the Melnikov function we cannot expand $a(\varepsilon, \vartheta)$ in powers of ε. For the moment we shall be satisfied with an upper estimate for the total variation of $a(\varepsilon, \cdot)$.

We begin with a crude estimate:

$$\begin{aligned}|a(\varepsilon,\vartheta) - \mathcal{A}(U_\vartheta)| &\le |\mathcal{A}_0(U_\vartheta + w_{\varepsilon,\vartheta}) - \mathcal{A}_0(U_\vartheta)| + |\mathcal{B}_\varepsilon(U_\vartheta + w_{\varepsilon,\vartheta})| \\ &\le \sup_{\|z\|\le R+2} |\mathcal{A}_0(z)| \cdot \|w_{\varepsilon,\vartheta}\|^2 + c[G]_{R+2}\varepsilon \\ &\le c[G]_{R+2}\varepsilon.\end{aligned}$$

Expand $a(\varepsilon, \vartheta)$ in a Fourier series: if

$$a(\varepsilon,\vartheta) = \sum_n a_n \exp(2\pi i n\vartheta/\varepsilon),$$

then

$$\begin{aligned}|a_n| &\le e^{-2\pi|n|\rho/\varepsilon}\frac{1}{\varepsilon}\int_0^\varepsilon |a(t \pm i\rho)|\,dt \\ &\le c\varepsilon[G]_{R+2}e^{-2\pi|n|\rho/\varepsilon}.\end{aligned}$$

Hence for real valued ϑ we get:

$$|a_\vartheta(\varepsilon,\vartheta)| \le \frac{2\pi}{\varepsilon}\sum_n n|a_n| \le c[G]_{R+2}e^{-2\pi\rho/\varepsilon}.$$

Define the region $\Omega_{\mathcal{O}_\varepsilon P_\varepsilon}$ as at the end of section 2: i.e. let P_ε be a homoclinic point associated with a zero of $a_\vartheta(\varepsilon, \cdot)$. Such a point certainly exists, since $a(\varepsilon, \vartheta)$ is a periodic function of ϑ– simply take a point $\vartheta \in \mathbf{R}$ where $a(\varepsilon, \cdot)$ is maximal, or minimal. Now let $\Omega_{\mathcal{O}_\varepsilon P_\varepsilon}$ be the bounded region enclosed by $W^s_\varepsilon[P_\varepsilon, \mathcal{O}_\varepsilon]$ and $W^u_\varepsilon[\mathcal{O}_\varepsilon, P_\varepsilon]$. Then we find that

$$\begin{aligned}\text{flux}(\Phi_\varepsilon, \Omega) &\le \int_0^\varepsilon |a_\vartheta(\varepsilon,\vartheta)|d\vartheta \\ &\le c[G]_{R+2}\varepsilon e^{-2\pi\rho/\varepsilon}.\end{aligned}$$

§5. Holomorphic Contraction Mapping Lemma

In what follows we shall frequently use the contraction mapping lemma to obtain solutions of differential equations. Since we are working with holomorphic maps of Banach spaces the usual contraction mapping lemma admits a stronger formulation. In practice this stronger version will not allow us to prove anything we couldn't do with the usual version, but it does save us a small amount of work: We no longer have to go through the calculations which show that the maps we consider are contractions.

5.1. HCM–lemma. *Let E be a complex Banach space, and let $f : B_r \to B_{\theta r}$ be a holomorphic mapping, where $B_\rho = \{x \in E : \|x\| < \rho\}$.*

If $\theta < {}^1/_2$, then $f|B_{\theta r}$ is a contraction, and hence has a unique fixed point in $B_{\theta r}$.

If F is another complex Banach space, and $f : F \times E \to E$ is holomorphic such that $f(\vartheta_0, \cdot)$ maps B_r into $B_{\theta r}$ for some $\theta < {}^1/_2$, $\vartheta_0 \in F$, then $f(\vartheta, \cdot)$ has a unique fixed point $x(\vartheta) \in B_{\theta r}$ for $\vartheta \in F$ close enough to ϑ_0. This fixed point depends holomorphically on ϑ, and

$$\|\mathrm{d}x(\vartheta_0)\| \leq \frac{1-\theta}{1-2\theta}\|\mathrm{d}_\vartheta f(\vartheta_0, x(\vartheta_0))\|.$$

Proof. Use Cauchy's inequality to estimate the derivative of f on $B_{\theta r}$ in terms of its sup norm on B_r:

$$\begin{aligned}\|\mathrm{d}f(x)\| &\leq \frac{1}{\rho}\sup\left(\|f(y)\| : \|y-x\| < \rho\right)\\ &\leq \frac{\sup\left(\|f(y)\| : y \in B_r\right)}{r - \|x\|}\\ &\leq \frac{\theta r}{r - \theta r} = \frac{\theta}{1-\theta}\\ &< 1,\end{aligned}$$

where we chose $\rho = r - \|x\|$, and used $\theta < {}^1/_2$. Thus f is a contraction on $B_{\theta r}$ which must have a fixed point.

By the implicit function theorem we see that the fixed point $x(\vartheta)$ depends smoothly (i.e. holomorphically) on parameters; the estimate for $\mathrm{d}x(\vartheta_0)$ follows directly after differentiating $x(\vartheta) = f(\vartheta, x(\vartheta))$.

Q. E. D.

5.2. HCM–lemma, special case. *If the holomorphic map $f : E \to E$ satisfies*

$$\|f(x)\| \leq a + b\|x\|^2,$$

with $ab < 1/16$, then f has a unique fixed point x_0 with $\|x\| \leq 2a$.

The special case follows from the HCM–lemma, after observing that $f(B_{4a}) \subset B_{4\theta a}$, where $\theta = (1 + 16ab)/4 < {}^1/_2$.

§6. Proof of lemma 4.1

We define $W^1_\infty(\mathbf{T})$ to be the space of all complex valued Lipschitz continuous functions on $\mathbf{T}$. By Rademacher's theorem $W^1_\infty(\mathbf{T})$ consists of those distributions on $\mathbf{T}$ whose first derivative can be represented by a bounded measurable function. $W^1_\infty(\mathbf{T})$ is a complex Banach space, with norm

$$\|f\|_{W^1_\infty} =_{\text{def}} \|f\|_{L_\infty} + \|f'\|_{L_\infty}.$$

We define $W^{-1}_\infty(\mathbf{T})$ to be the space of all distributions f which may be written as $f = g + h'$, for certain $g, h \in L_\infty(\mathbf{T})$. With the following norm:

$$\|f\|_{W^{-1}_\infty} =_{\text{def}} \inf\left(\|g\|_{L_\infty} + \|h\|_{L_\infty} : f = g + h'\right)$$

$W^{-1}_\infty(\mathbf{T})$ is a complex Banach space. One can identify this space as the dual of $W^1_1(\mathbf{T})$, the space of absolutely continuous functions on $\mathbf{T}$.

One substitutes $u(t) = p(t/\varepsilon)$ in (1.1), or (3.3) and finds the following equation for p:

$$\mathrm{D}^2 p(s) - \varepsilon^2 F(s, p(s)) = 0, \tag{6.1}$$

where the linear operator $\mathrm{D}^2 : W^1_\infty(\mathbf{T}) \to W^{-1}_\infty(\mathbf{T})$ is given by $\mathrm{D}^2 p(s) = p''(s)$. This operator is bounded, Fredholm; its kernel consists of all constant functions, its range is $X^- = \left\{q \in W^{-1}_\infty(\mathbf{T}) : \int_{\mathbf{T}} q = 0\right\}$. When restricted to $X^+ = X^- \cap W^1_\infty(\mathbf{T})$ D^2 is injective. We denote its inverse by $\mathrm{K} : X^- \to X^+$, and extend it to an operator $\mathrm{K} : W^{-1}_\infty(\mathbf{T}) \to X^+ \subset W^1_\infty(\mathbf{T})$ by defining $\mathrm{K}c = 0$ for any constant function c.

Write $p(s)$ as $p(s) = c + q(s)$, where c is constant and $\int_{\mathbf{T}} q(s)ds = 0$. Then $q'' = p''$, so that application of K to (6.1) leads to

$$q = -\varepsilon^2 \mathrm{K}\left\{F_0(c + q(s)) + G_t(s, c + q(s))\right\} =_{\text{def}} \varphi_\varepsilon(c, q). \tag{6.2}$$

Here φ_ε is a holomorphic map from $\mathbf{C} \times X^+$ to X^+. If $|c| \leq 1$, $\|q\| \leq 1$ then φ_ε can be estimated by

$$\|\varphi_\varepsilon(c, q)\| \leq C\varepsilon^2\left(|c| + \|q\| + [G]_2\right),$$

where $\|\cdot\|$ denotes the $W^1_\infty(\mathbf{T})$ norm. Applying the HCM–lemma, it follows that for small enough ε and any $|c| \leq 1$ there is a unique small solution $q_{\varepsilon,c}$ of $q = \varphi_\varepsilon(c, q)$. This solution is a holomorphic X^+–valued function of c and ε, and satisfies

$$\|q\| \leq C\varepsilon^2\left(|c| + [G]_2\right). \tag{6.3}$$

The function $c + q_{\varepsilon,c}$ will satisfy (6.1) upto a constant. Requiring this constant to vanish will give us one additional equation, which determines c. To get this equation we integrate (6.1), and write $F_0(u) = F_0'(0)u + F_2(u)u^2$. This leads to

$$c = \frac{-1}{F_0'(0)} \int_{\mathrm{T}} \left\{ F_2(p)p^2 - G_u(s,p)q' \right\} ds =_{\text{def}} \psi_\varepsilon(c).$$

Here $p = c + q_{\varepsilon,c}$, and we have replaced $G_t(s,p)$ by $(G(s,p))' - G_u(s,p)p'$. Using (6.3) we get the following estimate for ψ_ε:

$$|\psi_\varepsilon(c)| \le C|c|^2 + C\varepsilon^2[G]_2^2,$$

If $\varepsilon[G]_2$ is small then we can again apply the HCM–lemma and conclude that ψ_ε has a fixed point c_ε with

$$|c_\varepsilon| \le C\varepsilon^2[G]_2^2.$$

We obtain the desired solution by adding c and q: $p(\varepsilon,\cdot) = c_\varepsilon + q_{\varepsilon,c_\varepsilon}$. That $p(\varepsilon,\cdot)$ only contains even powers of ε follows from the uniqueness of the solution for fixed ε, and from the fact that the original equation (6.1) does not change if one changes the sign of ε.

§7. Proof of Lemma 4.5

Consider the Euler-Lagrange equation (4.4). In this equation we substitute $v = U_\vartheta + w$, we write $\mathcal{A}_\varepsilon(v) = \mathcal{A}_0(v) + \mathcal{B}_\varepsilon(v)$, and we expand $\mathrm{d}\mathcal{A}_0(v)$ in a Taylor series around U_ϑ:

$$\begin{aligned} \mathrm{d}\mathcal{A}_0(U_\vartheta + w) &= w'' + F_0(U_\vartheta + w) - F_0(U_\vartheta) \\ &= \left\{ -\mathrm{D}^2 + F_0'(U_\vartheta(t)) \right\} w + R_2(U_\vartheta, w) \\ &= \mathrm{A}_\vartheta w + R_2(U_\vartheta, w) \end{aligned}$$

where R_2 may be estimated by (4.2), i.e. if $|U_\vartheta(t)| \le R$ and $|w| \le 1$, then $|R_2(U_\vartheta(t), w)| \le [F_0]_{R+2}|w|^2$.

Thus $w = v - U_\vartheta$ satisfies

$$\mathrm{A}_\vartheta w + R_2(U_\vartheta, w) + \mathrm{d}\mathcal{B}_\varepsilon(U_\vartheta + w) = \lambda U_\vartheta{}', \qquad \left\langle w, U_\vartheta{}' \right\rangle = 0. \tag{7.1}$$

Let $\mathrm{T}_\vartheta : H^{-1} \to H^1$ be the pseudo inverse of A_ϑ; i.e.

$$\mathrm{T}_\vartheta | E_\vartheta^{-1} = \left(\mathrm{A}_\vartheta | E_\vartheta^1 \right)^{-1}, \quad \text{and} \quad \mathrm{T}_\vartheta U_\vartheta{}' = 0.$$

The range of T_ϑ is E^1_ϑ. One can compute T_ϑ from the resolvent of A_ϑ via the following formula[8]

$$\mathrm{T}_\vartheta = \frac{-1}{2\pi i}\oint_{|\zeta|=r}(\zeta - \mathrm{A}_\vartheta)^{-1}\frac{d\zeta}{\zeta}.$$

Here $r > 0$ should be chosen small enough, so that $\zeta - \mathrm{A}_\vartheta$ has a bounded inverse for any ζ with $0 < |\zeta| \le r$.

This expression for T_ϑ shows that T_ϑ depends holomorphically on $\vartheta \in S_{\rho_0}$ and that for any $0 < \rho < \rho_0$

$$\sup_{\vartheta\in S_\rho}\|\mathrm{T}_\vartheta\|_{H^{-1}\to H^1} = C_\rho < \infty.$$

Now we apply T_ϑ to both sides of (7.1) to get

$$\begin{aligned} w &= -\mathrm{T}_\vartheta\left\{R_2(U_\vartheta, w) + \mathrm{d}\mathcal{B}_\varepsilon(U_\vartheta + w)\right\} \\ &=_{\mathrm{def}} \varphi(\varepsilon, \vartheta; w). \end{aligned}$$

Clearly $\varphi(\varepsilon, \vartheta; \cdot)$ is a holomorphic map of H^1 to itself. If $\|U_\vartheta\| \le R-1$ and $\|w\| \le 1$ then we have

$$\begin{aligned} \|\varphi(\varepsilon, \vartheta; w)\| &\le \|\mathrm{T}_\vartheta\|\left\{\|R_2(U_\vartheta, w)\|_{H^{-1}} + \|\mathrm{d}\mathcal{B}_\varepsilon(U_\vartheta + w)\|_{H^{-1}}\right\} \\ &\le C_\rho\left\{[F_0]_{R+2}\|w\|^2 + c[G]_{R+2}\varepsilon\right\} \\ &\le c[G]_{R+2}\varepsilon + c\|w\|^2, \end{aligned}$$

where c only depends on ρ, R and $[F_0]_{R+2}$. By the HCM-lemma there is a $\delta_1 > 0$ such that $\varphi(\varepsilon, \vartheta; \cdot)$ has a unique fixed point $w_{\varepsilon,\vartheta}$ with

$$\|w_{\varepsilon,\vartheta}\| \le c[G]_{R+2}\varepsilon \tag{7.2}$$

as long as $[G]_{R+2}\varepsilon < \delta_1$.

It is clear from the definition of $\varphi(\varepsilon, \vartheta, w)$ that φ is holomorphic in ϑ, with

$$\frac{\partial\varphi}{\partial\vartheta} = -\frac{\partial\mathrm{T}_\vartheta}{\partial\vartheta}\{\cdots\} - \mathrm{T}_\vartheta\left\{\frac{\partial R_2(U_\vartheta, w)}{\partial U_\vartheta}U_\vartheta' + \mathrm{d}^2\mathcal{B}_\varepsilon(U_\vartheta + w)U_\vartheta'\right\}.$$

Thus

$$\left\|\frac{\partial\varphi}{\partial\vartheta}\right\| \le c\left([G]_{R+2}\varepsilon + \|w\|^2\right),$$

and the HCM–lemma implies that $w_{\varepsilon,\vartheta}$ is also holomorphic in ϑ, with $\|\partial_\vartheta w_{\varepsilon,\vartheta}\| \le c[G]_{R+2}\varepsilon$.

---

[8] See [DS88].

§8. The forced Duffing equation

So far we haven't found an explicit formula for $a(\varepsilon,\vartheta)$, for small ε. In this section we show how one can do this for the forced Duffing equation

$$u'' - u + {}^3/_2 u^2 = g(t/\varepsilon), \tag{8.1}$$

provided $g(t) = G'(t)$, and $\|G\|_{L_\infty}$ is small enough:

8.1. Theorem. *If $g(t) = \delta\varepsilon^{10} H'(t)$ where $\|H\| \in L_\infty(\mathbf{T})$ satisfies $\int_{\mathbf{T}} H = 0$, then*

$$\frac{\partial a}{\partial \vartheta} = \delta\varepsilon^8 e^{-2\pi^2/\varepsilon}\left\{-32\pi^3|\hat{H}_1|\sin\left(\frac{2\pi\vartheta}{\varepsilon} - \phi\right) + o(1)\right\}, \qquad (\delta,\varepsilon \to 0)$$

uniformly in $\vartheta \in \mathbf{R}$, where $\hat{H}_1$ is the first Fourier coefficient of $H(t)$, and $\phi = \arg \hat{H}_1$.

If $\hat{H}_1 \neq 0$ then, if ε and δ are small enough, a_ϑ has precisely two nondegenerate zeroes in any period interval, and hence the Poincaré–map $\Phi_{\delta,\varepsilon}$ has two transverse homoclinic orbits.

– *The small solution* –

Let $p(\varepsilon,\tau)$ be the small solution, i.e. $p(\varepsilon,\tau+1) \equiv p(\varepsilon,\tau)$ and $p(t) = p(\varepsilon, t/\varepsilon)$ satisfies (8.1). Since the perturbation $G(t,u) \equiv G(t)$ does not depend on u, all the seminorms $[G]_R = \|G\|_{L_\infty}$ are the same, and p satisfies $\|p\|_{L_\infty} \leq c\|G\|_{L_\infty}\varepsilon^2$, provided $\|G\|_{L_\infty}\varepsilon < \delta_0$. To calculate a_ϑ for small ε, it seems that we need to now the first order term in the expansion of p.

8.2. Lemma. $\|p + \varepsilon^2 \mathrm{K} g\|_{W^1_\infty} \leq C\varepsilon^4 \|g\|_{W^{-1}_\infty}$, *where the operator* K *is as in section 5.*

Since $\int_{\mathbf{T}} H = 0$, it follows from the way K was defined that $(\mathrm{K}H')' = H$, so this lemma implies that

$$\|p' + \delta\varepsilon^{12} H\|_{L_\infty} \leq C\delta\varepsilon^{14}. \tag{8.2}$$

Proof. As in section 6 we write $p = c + q$ with $\int_{\mathbf{T}} q = 0$ and c a constant. In view of (6.2),

$$q = -\varepsilon^2 \mathrm{K}\{g + F_0(p)\},$$

and the estimate for $\|p\|_{W^1_\infty}$, this implies

$$\|q + \varepsilon^2 \mathrm{K} g\|_{W^1_\infty} \leq C\varepsilon^4 \|g\|_{W^{-1}_\infty}.$$

For c we have

$$0 = \int_{\mathbf{T}} F_0(c+q)dt = F_0(c) + \int_{\mathbf{T}} (F_0(c+q) - F_0(c) - F_0'(c)q)\,dt,$$

which implies $|F_0(c)| \le C\|q\|^2$, and hence $|c| \le C\varepsilon^4 \|g\|^2_{W^{-1}_\infty}$. Adding our estimates for c and q yields the desired inequality.

Q. E. D.

– *The modified equation* –

Put $u = v + p$. Then u corresponds to a homoclinic orbit *iff*

$$v'' - v + {}^3/_2 v^2 + 3p(\varepsilon, t/\varepsilon)v = 0, \qquad v(\pm\infty) = 0. \tag{8.3}$$

The solutions of (8.3) are precisely the critical points of $\mathcal{A}_\varepsilon : H^1 \to \mathbf{C}$, where $\mathcal{A}_\varepsilon = \mathcal{A}_0 + \mathcal{B}_\varepsilon$, and

$$\mathcal{A}_0(v) = \int_{\mathbf{R}} \left({}^1/_2 v'^2 + {}^1/_2 v^2 - {}^1/_2 v^3\right) dt,$$

$$\mathcal{B}_\varepsilon(v) = -\int_{\mathbf{R}} {}^3/_2 p(\varepsilon, t/\varepsilon) v(t)^2\, dt.$$

The homoclinic orbit of the unperturbed equation is parametrized by $(U(t), U'(t))$, where

$$U(t) = \operatorname{sech}^2 \frac{t}{2} = -\sum_{k\in\mathbf{Z}} \frac{4}{(t-(2k+1)\pi i)^2}. \tag{8.4}$$

We consider the Euler–Lagrange equations $d\mathcal{A}_\varepsilon(v) = \lambda U_\vartheta{}'$, which upon substitution of $v = U_\vartheta + w$ may be written as

$$w'' - (1 - 3U_\vartheta)\,w + {}^3/_2 w^2 + 3pU_\vartheta + 3pw = \lambda U_\vartheta{}'. \tag{8.5}$$

If T_ϑ again denotes the pseudo inverse of the operator

$$\mathrm{A}_\vartheta = -\mathrm{D}^2 + F_0'(U_\vartheta) = -\mathrm{D}^2 + (1 - 3U_\vartheta),$$

then (8.5) is equivalent with

$$w = \mathrm{T}_\vartheta \left\{3pU_\vartheta + 3pw + {}^3/_2 w^2\right\}. \tag{8.6}$$

To analyze this equation we use our complete knowledge of the operator A_ϑ to first give an asymptotic description for T_ϑ as $\mathrm{Im}\vartheta \uparrow \pi$, and then use this asymptotics to compute $\lambda(\varepsilon, \vartheta)$ for $\mathrm{Im}\vartheta$ close to π.

– *The resolvent of* A_ϑ –

For $\omega \in \mathbf{C} - [1,\infty)$ we denote by $\sqrt{1-\omega}$ the square root with positive real part. If $\omega \in \mathbf{C} - [1,\infty)$ then the equation

$$-y'' + F_0'(U(t))y = \omega y \tag{8.7}$$

has two solutions $y_\pm(\omega, t)$, with

$$\begin{aligned} y_+(\omega,t) &= e^{-t\sqrt{1-\omega}}(1+o(1)), \qquad (t\to\infty),\\ y_-(\omega,t) &= e^{t\sqrt{1-\omega}}(1+o(1)), \qquad (t\to-\infty). \end{aligned}$$

Both solutions are holomorphic functions of $t \in \mathbf{C}$, with $|\mathrm{Im}t| < \pi$; they also depend analytically on ω. Their Wronskian

$$W(\omega) = \begin{vmatrix} y_-(\omega,t) & y_+(\omega,t) \\ y_-'(\omega,t) & y_+'(\omega,t) \end{vmatrix}$$

is therefore a holomorphic function on $\mathbf{C} - [1,\infty)$; $A_0 - \omega$ is invertible *iff* $W(\omega) \neq 0$. E.g. we know that A_0 itself is not invertible and, indeed, $y_\pm(0,t) = U'(t)$.

It is well known[9] that if $W(\omega) \neq 0$, then the inverse of $A_0 - \omega$ is an integral operator, with kernel

$$K(\omega; t, s) = \frac{y_+(\omega,t)y_-(\omega,s)}{W(\omega)} \tag{8.8}$$

if $t \geq s$; if $s > t$ then $K(\omega;t,s) = K(\omega;s,t)$.

As $y_\pm(\omega, t+\vartheta)$ are solutions of

$$-y'' + F_0'(U_\vartheta(t))y = \omega y,$$

the next lemma follows in the same way.

8.3. Lemma. *If* $W(\omega) \neq 0$ *then* $A_\vartheta - \omega : H^1 \to H^{-1}$ *is invertible for* $\vartheta \in S_\pi$. *Its inverse is a bounded integral operator on* $L_\infty(\mathbf{R},\mathbf{C})$ *with kernel* $K(\omega; t+\vartheta, s+\vartheta)$.

One can verify that the operator with kernel $K(\omega;t+\vartheta, s+\vartheta)$ is bounded on L_∞ by estimating its norm as follows:

$$\|(A_\vartheta - \omega)^{-1}\|_{L_\infty\to L_\infty} \leq \text{ess. sup}_{t\in\mathbf{R}} \int_{\mathbf{R}} |K(\omega; t+\vartheta, s+\vartheta)|ds.$$

[9] E.g. see chapter XIII of [DS88].

To analyze the kernel K as $\operatorname{Im} t, \operatorname{Im} s$ get close to π we look at the singular points of the differential equation which defines $y_\pm$. By substituting the expansion (8.4) in (8.7) we find that the points $t = (2k+1)\pi i$ are regular singular points for (8.7); a short computation shows that the charcteristic exponents at the singular points are -3 and $+4$. It follows that the solutions $y_\pm$ of (8.7) satisfy

$$|y_\pm(\omega, t)| \le C_\omega |t^2 + \pi^2|^{-3}$$

near the two singular points $\pm\pi i$.

Combining this with the exponential growth of $y_\pm(\omega, t)$ at $t = \mp\infty$ we get for all $t \in S_\pi$

$$|y_\pm(\omega, t)| \le C_\omega \left(1 + |t^2 + \pi^2|^{-3}\right) e^{\mp\kappa \operatorname{Re} t},$$

and thus

$$|K(\omega; t, s)| \le \frac{C_\omega}{|W(\omega)|} \left(1 + |t^2 + \pi^2|^{-3}\right) \left(1 + |s^2 + \pi^2|^{-3}\right) e^{-\kappa|t-s|},$$

where $\kappa = \operatorname{Re}\sqrt{1-\omega}$.

– *The kernel of the pseudo inverse* T_ϑ –

Recall that T_ϑ is obtained by integrating the resolvent around a small circle centered at the origin. Hence T_ϑ is also an integral operator with kernel $T(\vartheta + t, \vartheta + s)$ where

$$T(t, s) = \frac{-1}{2\pi i} \oint_{|\omega|=r} K(\omega; t, s) \frac{d\omega}{\omega},$$

and $r > 0$ should be chosen sufficiently small. Our estimate for K implies

$$|T(t, s)| \le C_r \left(1 + |t^2 + \pi^2|^{-3}\right) \left(1 + |s^2 + \pi^2|^{-3}\right) e^{-\sqrt{1-r}|t-s|} \qquad (8.9)$$

for all $s, t \in S_\pi$.

8.4. Lemma. *For any* σ *with* $0 < \operatorname{Im}\sigma < 2\pi/\varepsilon$

$$\|\mathrm{T}_{\pi i - \varepsilon\sigma}\|_{L_\infty \to L_\infty} \le \frac{C(\sigma)}{\varepsilon^5}.$$

Proof. It follows from (8.9) that the kernel of $\mathrm{T}_{\pi i - \varepsilon\sigma}$ is dominated by $|T(\pi i + t - \varepsilon\sigma, \pi i + s - \varepsilon\sigma)| \le \left(1 + |t - \varepsilon\sigma|^{-3}\right)\left(1 + |s - \varepsilon\sigma|^{-3}\right) e^{-\sqrt{1-r}|t-s|}$. Hence

$$\int_{\mathbf{R}} |T(\pi i + t - \varepsilon\sigma, \pi i + s - \varepsilon\sigma)| ds \le \frac{C(\sigma)}{\varepsilon^2}\left(1 + |t - \varepsilon\sigma|^{-3}\right) \le \frac{C}{\varepsilon^5},$$

from which the lemma follows.

Q. E. D.

– *The fixed point equation with* $\mathrm{Im}\vartheta \approx \pi$ –

We return to (8.6): let

$$\varphi_{\varepsilon,\vartheta}(w) = \mathrm{T}_\vartheta \left\{3pU_\vartheta + 3pw + {}^3/_2 w^2\right\}.$$

Then $\varphi_{\varepsilon,\vartheta} : L_\infty \to L_\infty$ is a holomorphic map.

For $\vartheta = \pi i - \varepsilon\sigma$, $\mathrm{Im}\sigma > 0$ we have:

$$\begin{aligned}\|\varphi_{\varepsilon,\vartheta}\| &\le \frac{C}{\varepsilon^5}\left\{3\|p\|\left(\|U_\vartheta\| + \|w\|\right) + {}^3/_2\|w\|^2\right\}\\ &\le \frac{C}{\varepsilon^5}\left\{\|p\|\cdot\|U_\vartheta\| + \|p\|^2 + \|w\|^2\right\}\\ &\le \frac{C}{\varepsilon^5}\left\{\|g\|_{W_\infty^{-1}} + \varepsilon^4\|g\|^2_{W_\infty^{-1}} + \|w\|^2\right\}.\end{aligned}$$

where we have used that $|U_\vartheta| \le C(\sigma)\varepsilon^{-2}$, $\|\cdot\|$ denotes the L_∞ norm, and C is a constant depending on σ.

By the HCM–lemma $\varphi_{\varepsilon,\vartheta}$ will have a unique small fixed point $w_{\varepsilon,\vartheta}$ provided $C\varepsilon^{-10}(\|g\| + \varepsilon^4\|g\|^2) < {}^1/_2$, i.e. if

$$\|g\|_{W_\infty^{-1}} \le C\delta\varepsilon^{10}$$

for some small enough $\delta > 0$. This is why we assumed that $g = \delta\varepsilon^{10}H'$, with $H \in L_\infty$ and $\delta > 0$ sufficiently small.

The fixed point w satisfies $\|w_{\varepsilon,\vartheta}\| \le C\delta\varepsilon^5$.

– *The functions* $\lambda(\varepsilon,\vartheta)$ *and* $a_\vartheta(\varepsilon,\vartheta)$ –

Multiplying (8.5) with $U_\vartheta{}'$ and integrating results in

$$\lambda(\varepsilon,\vartheta) = \frac{1}{\Gamma}\int_{\mathbf{R}} U_\vartheta{}'\left\{3pU_\vartheta + 3pw + {}^3/_2 w^2\right\} dt. \tag{8.10}$$

Since $\|p\| \le C\varepsilon^2\|g\|_{W_\infty^{-1}} \le C\delta\varepsilon^{12}$, we have

$$\begin{aligned}\int_{\mathbf{R}} U_\vartheta{}'\left\{3pw + {}^3/_2 w^2\right\} dt &\le C\int_{\mathbf{R}} |U_\vartheta{}'|(|p| + |w|)|w| dt\\ &\le C\delta^2\left(\varepsilon^{17} + \varepsilon^{10}\right)\int_{\mathbf{R}} |U_\vartheta{}'| dt\\ &\le C\delta^2\varepsilon^8.\end{aligned}$$

After integarting by parts in the first term of (8.10) we therefore get

$$\lambda(\varepsilon,\vartheta) = -\frac{3}{2\Gamma}\int_{\mathbf{R}} p'(\varepsilon,\tau)U(\vartheta + \varepsilon\tau)^2 d\tau + O\left(\delta^2\varepsilon^8\right). \tag{8.11}$$

By expanding $U(t)$ in a Laurent series around $t = \pi i$, and using the exponential decay of $U(t), t \in S_\pi$ as $\mathrm{Re}t \to \pm\infty$ one obtains

$$\left| U(\vartheta + \varepsilon\tau)^2 - \frac{16}{\varepsilon^4(\sigma - \tau)^4} \right| \leq \frac{C}{\varepsilon^2|\sigma - \tau|^2}, \qquad \forall \tau \in \mathbf{R}.$$

Combine this with lemma 8.2, and substitute in (8.11). The result is

$$\lambda(\varepsilon, \vartheta) = \frac{-24\delta\varepsilon^8}{\Gamma} \int_{\mathbf{R}} \frac{H(\tau)}{(\sigma - \tau)^4} d\tau + O\left(\delta\varepsilon^{10} + \delta^2\varepsilon^8\right)$$

From this we shall compute the Fourier expansion of $\lambda(\varepsilon, \vartheta)$. Put

$$\lambda(\varepsilon, \vartheta) = \sum_{n \in \mathbf{Z}} \lambda_n(\varepsilon) e^{2\pi i n \vartheta/\varepsilon}; \qquad H(\tau) = \sum_{n \in \mathbf{Z}} \hat{H}_n e^{2\pi i n \tau}.$$

Then contour integration shows

$$\int_{\mathbf{R}} \frac{H(\tau)}{(\sigma - \tau)^4} d\tau = -{}^4/{}_3\pi^3 i \sum_1^\infty n^3 \hat{H}_n e^{2\pi i \sigma}.$$

For $n \geq 1$ we then have

$$\begin{aligned} \lambda_{-n} &= \frac{1}{\varepsilon} \int_0^\varepsilon e^{2\pi i n \vartheta/\varepsilon} \lambda(\varepsilon, \vartheta) d\vartheta \\ &= -e^{-2\pi^2 n/\varepsilon} \int_i^{i+1} \lambda(\varepsilon, \pi i - \varepsilon\sigma) d\sigma \\ &= -\frac{32\pi^3 i}{\Gamma} e^{-2\pi^2 n/\varepsilon} \delta\varepsilon^8 \left(n^3 \hat{H}_n + O\left(\delta^2 + \delta\varepsilon^2\right)\right), \end{aligned}$$

where we have substituted $\vartheta = \pi i - \varepsilon\sigma$, and we have used the analyticity of $\lambda(\varepsilon, \cdot)$ to change the path of integration.

For $n > 0$ we have $\lambda_n(\varepsilon) = \overline{\lambda_{-n}(\varepsilon)}$, since λ is a real valued function for real ϑ. The terms in the Fourier series of λ with $n = 0, \pm 1$ dominate those with $|n| \geq 2$, so, writing $\hat{H}_1 = |\hat{H}_1| e^{i\phi}$, we get:

$$\lambda(\varepsilon, \vartheta) = \lambda_0(\varepsilon) - \frac{32\pi^3}{\Gamma} \delta\varepsilon^8 e^{-2\pi^2/\varepsilon} \left\{ |\hat{H}_1| \sin\left(\frac{2\pi i}{\varepsilon} - \phi\right) + O\left(\delta + \varepsilon^2\right) \right\}. \tag{8.12}$$

Finally, using

$$a_\vartheta = (\Gamma + o(1))\lambda(\varepsilon, \vartheta), \tag{8.13}$$

it follows from $\int_0^\varepsilon a_\vartheta(\varepsilon, \vartheta) d\vartheta = 0$ that $\lambda_0 = o\left(\delta\varepsilon^8 e^{-2\pi^2/\varepsilon}\right)$; combining (8.13) and (8.12) one then finds the desired expression for a_ϑ.

Q. E. D.

References

[A86] S. B. Angenent, *The Shadowing Lemma for Elliptic PDE,* Dynamics of Infinite Dimensional Systems, S. N. Chow & J. K. Hale, editors, Springer Verlag, NATO–ASI series F-**37**, (1988) pp. 7–22.

[DS88] N. Dunford and J. T. Schwarz, *Linear Operators, part II: Spectral Theory,* Wiley 1988.

[DTS91] A. Delshams, M. Teresa and M. Seara, *An asymptotic expresion for the splitting of separatrices of the rapidly forced pendulum,* University of Barcelona preprint, August 1991.

[GH83] J. Guckenheimer and P. Holmes, *Nonlinear Oscillations, Dynamical Systems and Bifurcations of Vector Fields,* Springer Verlag 1983.

[HMS88] P. Holmes, J. Marsden and J. Scheurle, *Exponentially small splittings of separatrices with applications to KAM theory and degenerate bifurcations,* Contemporary Mathematics **81** (1988) 213–244.

[MMS84] R. S. Mackay, J. D. Meiss, I. C. Percival, *Transport in Hamiltonian Systems,* Physica **13D** (1984) 55–81.

[Mo73] J. Moser, *Stable and random motions in dynamical systems,* Princeton University Press, 1973.

[N84] A. I. Neishtadt, *The separation of motions in systems with rapidly rotating phase,* J. Appl. Math. Mech. **48**(1984) 133–139.

GLOBAL DARBOUX THEOREMS AND A LINEARIZATION PROBLEM

Eleonora Ciriza
ICTP, Trieste, Italy
December 1990

1 Submanifolds of Kähler manifolds of non-positive curvature

After Gromov's discovery of the existence of exotic symplectic structures on $\mathbf{R}^{2n}$ one important problem has been the understanding of the standard symplectic structure itself. McDuff proved (in [8], [9]) a global version of the Darboux Theorem which states that

Theorem 1.1 *The Kähler form ω on a simply connected complete Kähler $2n$-dimensional manifold P of non-positive sectional curvature is diffeomorphic to the standard symplectic form ω_0 on $\mathbf{R}^{2n}$.*

This means in particular that the symplectic structure on a Hermitian symmetric space of non-compact type is standard. She also showed that

Theorem 1.2 *If L is a totally geodesic connected properly embedded Lagrangian submanifold of such a manifold P, then P is symplectomorphic to the cotangent bundle T^*L with its usual symplectic structure.*

Recall that a submanifold Q of P is said to be *symplectic* if ω restricts to a symplectic form on Q and is said to be *isotropic* if the restriction of ω to Q is identically zero. In the complex hyperbolic space $\mathbf{CH}^n$ of complex dimension n, the complex hyperbolic subspaces $\mathbf{CH}^i$, $0 \leq i \leq n$, are examples of totally geodesic symplectic submanifolds and the real hyperbolic subspaces $\mathbf{H}^{n-i}$, $0 \leq i \leq n$, are examples of totally geodesic isotropic submanifolds.

Throughout this section we assume that Q is a totally geodesic connected properly embedded submanifold of (P, ω). Then we have

Theorem 1.3 *The symplectomorphism constructed by McDuff takes a totally geodesic complex submanifold Q into a complex linear subspace of $\mathbf{R}^{2n}$.*

A simple example shows that this is no longer true when Q is isotropic. Nevertheless, we prove in [2]

Theorem 1.4 *If Q is isotropic of dimension k, then (P,Q,ω) is symplectomorphic to $(\mathbf{R}^{2n},\mathbf{R}^k,\omega_0)$, where $\mathbf{R}^k$ is an isotropic linear subspace of $\mathbf{R}^{2n}$.*

McDuff's symplectomorphism: Pick a point x_0 in $Q \subset P$ and let $\rho(x)$ be the distance from x to x_0. By using a Hessian comparison theorem for manifolds of nonpositive curvature McDuff shows that the 2-form $\omega_\rho = -d(Jd\rho^2)$ is symplectic and that $G_\rho \geq 4G$, where G_ρ is the Levi form $G_\rho(X,Y) = \omega_\rho(X,JY)$ and G is the original Kähler metric. Applying Moser's method (see [10], [14]) to the family of forms $\tau_t = t\omega+(1-t)\omega_\rho$, $0 \leq t \leq 1$, she obtains a sympletomorphism Φ_1 from (P,ω) to (P,ω_ρ). Then McDuff constructs a symplectomorphism Φ_2 from (P,ω_ρ) to $(\mathbf{R}^{2n},\omega_0)$. To do this, she shows that the Liouville vector field ξ_ρ defined by $\xi_\rho \lrcorner\, \omega_\rho = -Jd\rho^2$ is diffeomorphic to the radial vector field ξ_0 on $\mathbf{R}^{2n}$ given in polar coordinates by $\frac{r}{2}\frac{\partial}{\partial r}$. Further this diffeomorphism takes ω_ρ to a symplectic form which is linearly diffeomorphic to ω_0. So Φ_2 is the composite of a map which linearizes the Liouville vector field ξ_ρ with a map which adjusts the form.

Note Recall that the orthogonal space to the tangent space of Q with respect to a form Ω, at each point q in Q is defined to be

$$(T_qQ)^{\perp_\Omega} = \{v \in T_qP \ : \ \Omega(v,w) = 0 \ \forall w \in T_qQ\}.$$

In order to prove that the diffeomorphisms Φ_1 and Φ_2 preserve the submanifold Q we show that the ω- and ω_ρ-orthogonal spaces to the tangent space T_qQ are equal at each point q of Q, i.e. $(T_qQ)^{\perp_\omega} = (T_qQ)^{\perp_{\omega_\rho}}$. For details see [2] and [4].

Example: Consider $\mathbf{R}^4$ with the metric G given by the cartesian product of the Poincaré metric on $\mathbf{R}^2$ with the standard metric on $\mathbf{R}^2$. Construct $\Phi : (\mathbf{R}^4,\omega) \to (\mathbf{R}^4,\omega_0)$ as above with $x_0 = \{0\}$. Note that the rays through the origin are totally geodesic isotropic submanifolds of $(\mathbf{R}^4,G)$, however they are not mapped by Φ onto rays in $(\mathbf{R}^4,\omega_0)$. For the rays in $(\mathbf{R}^4,\omega_0)$ are the integral curves of the Liouville vector field ξ_0, therefore $\xi = \Phi^*\xi_0$ is a Liouville field on $(\mathbf{R}^4,\omega)$. But it is easy to see that in $(\mathbf{R}^4,\omega)$ no Liouville vector field points in the direction of the rays.

The isotropic case: In the example the symplectomorphism Φ^{-1} fails to preserve the property of being totally geodesic. Consequently, to deal with the case of a totally geodesic isotropic submanifold we consider the distance function to the submanifold instead of the distance to a point in the submanifold. Observe that Theorem 1.4 can be viewed as a natural extension of the extreme cases considered by McDuff: the dimension of Q is zero (Q is a point) or maximal (Q is Lagrangian). To obtain a Levi form ω_ρ which is connected to the geometry of Q one takes ρ to be the distance function from Q. This gives rise to a map Φ_1 which preserves Q. But the second step is harder. One has to show that the Liouville vector field ξ admits a C^1-linearizing conjugation along Q. In the case where Q is a point McDuff uses a Sternberg linearization theorem [13] and when Q is a Lagrangian submanifold she uses a linearization result of Nagano [11]. In the general case these theorems do not apply. However, in next section we show that although zero is an eigenvalue of the associated linear part of the vector field ξ and the non-zero eigenvalues are in resonance, ξ is C^k-conjugate, $k \geq 1$ to its linear part along Q.

2 The local structure of a Liouville vector field

A smooth vector field of a symplectic manifold (P, Ω) is called *Liouville* if its flow φ_t satisfies $\varphi_t^* \Omega = e^t \Omega$ (or equivalently when $\mathcal{L}_\xi \Omega = \Omega$, where $\mathcal{L}$ denotes the Lie derivative).

Assume that a Liouville vector field ξ of a $2n$-dimensional manifold (P, Ω) vanishes on a k-dimensional isotropic submanifold Q of P and that at every point q of Q $\xi(q) = 0$ and the eigenvalues of its 1-jet are 1, 1/2 and 0. This means that in some coordinates

$$J_q^1(\xi) = \sum_{i=1}^{k}(0x_i \frac{\partial}{\partial x_i} + 1y_i \frac{\partial}{\partial y_i}) + \sum_{r=k+1}^{n} (\frac{1}{2}x_r \frac{\partial}{\partial x_r} + \frac{1}{2}y_r \frac{\partial}{\partial y_r})$$

where

$$T_q Q = \{\cap_{j=k+1}^{n} Ker\, dx_j\} \cap \{\cap_{j=1}^{n} Ker\, dy_j\}.$$

The general linearization theorem would imply that the Liouville vector field ξ is C^0-conjugate to a linear map. But a smoother conjugacy is required in order to push forward the symplectic form. To prove that ξ is smoothly conjugate to its linear part, we use a linearization theorem due to G.Sell [12], which extends the linearization theorem of Sternberg [13] to the case of a vector field with resonant eigenvalues. We exploit the fact that the vector ξ is Liouville, together with the explicit algorithm that Sell's theorem provides to compute a lower bound for the order of smoothness of the conjugacy.

Resonance: *A collection of non-zero eigenvalues is resonant* if one of them is an integral linear combination (with nonnegative coefficients whose sum is at least two) of the others. i.e. Let $\lambda_1, ..., \lambda_N$ be a set of non-zero eigenvalues repeated with multiplicities and let $m = (m_1, ..., m_N)$ be nonnegative integers. Define $|m| = \sum m_i$ and $\gamma(\lambda_i, m) = \lambda_i - \sum m_r \lambda_r$. Then if a relation $\gamma(\lambda_i, m) = 0$ holds for $|m| \geq 2$ the eigenvalues are said to be in resonance, and $|m|$ is called the order of resonance. Thus the eigenvalues of our Liouville vector field ξ are resonant of order 2.

If $(z^1, ..., z^N)$ are coordinates with respect to the basis $(e_1, ..., e_N)$, let z^m stand for $z_1^{m_1}...z_N^{m_N}$. *The vector valued monomial* $z^m e_i$ *is resonant* if $\gamma(\lambda_i, m) = 0$ and $|m| \geq 2$.

Statement of Sell's theorem: Assume we have the equation $\dot{w} = A(q)w + F(w, q)$ where $A(q)$ is a matrix and $F(0, q) = 0$. When A is hyperbolic, let $\sum^+(A)$ denote those eigenvalues λ of A with $Re\lambda > 0$ and $\sum^-(A)$ those with $Re\lambda < 0$. If $\sum^i(A) \neq \emptyset$ where $i = +$ or $-$, *the spectral spread* is defined to be

$$\rho^i = \frac{max\{|Re\lambda| \ : \ \lambda \in \sum^i(A)\}}{min\{|Re\lambda| \ : \ \lambda \in \sum^i(A)\}}.$$

The r-smoothness of A is the largest integer $K \geq 0$ such that

1. $r - K\rho^- \geq 0$, if $\sum^+(A) = \emptyset$
2. $r - K\rho^+ \geq 0$, if $\sum^-(A) = \emptyset$
3. There exist positive numbers M, N with $r = M + N$, $M - K\rho^+ \geq 0$, $N - K\rho^- \geq 0$ if $\sum^+(A) \neq \emptyset$ and $\sum^-(A) \neq \emptyset$.

Now suppose that the following condition holds for some integer $r \geq 2$. (This is condition "B" in [12]).

$D^j F(q_0, q) = 0$ for $0 \leq j \leq r - 1$
and $Re\,\gamma(\lambda, m) \neq 0$ for all $\lambda \in \sum A(q) = \sum^+ A(q) \cup \sum^- A(q)$,
for all m with $|m| = r$ and for all $q \in \hat{V}$ neighborhood of q_0.

Then Sell's Theorem asserts that there is a C^K-smooth linearizing conjugation $w = z + \Phi(z, q)$ between $\dot{w} = A(\theta)w + F(w, \theta)$ and $\dot{z} = A(q)z$, where Φ varies smoothly in terms of the parameter q and is of class C^K in z, where K is the r-smoothness of $A(q_0)$.

In the case of the Liouville vector field ξ we have $Re\,\gamma(\lambda, m) = 0$ for some pair (λ, m) with $|m| = 2$. Hence Sell's theorem must be applied with $r \geq 3$. Therefore one has to find coordinates in which ξ has no quadratic terms. The first step is to construct a Darboux chart near each point in Q which is adapted to ξ.

Construction of an adapted Darboux chart: Because the Liouville vector field ξ vanishes on the isotropic submanifold Q of (P, Ω), Q is an invariant manifold of the field ξ which consist of fixed points. Since at any point q in Q, the tangent space to P at q can be written as $T_qP = T_qQ \oplus N_q$, where $N_q = (T_qQ)^{\perp_G}$ and the nonzero eigenvalues of ξ at a singular point $q \in Q$ are assumed to be 1 or 1/2 we have that $m(D\varphi_\tau|_{N_q^s}) = e^{1/2}$ and $\|D\varphi_\tau|_{T_qQ}\| = e^0$.

Where $m(L) = inf\{\|Lv\| : \|v\| = 1\}$ is the minimum norm of the linear map L and $\|L\| = sup\{\|Lv\| : \|v\| = 1\}$ is the norm of L.

It follows that Q is r-normally hyperbolic for all $r \in \mathbf{N}$ and all $q \in Q$. Therefore we can apply the Center Manifold Theorem ([7], [5]), which states that under these conditions

i) P is smoothly foliated by strong unstable φ_t-invariant submanifolds W_q which are transverse to Q, i.e. $P = \bigcup_{q \in Q} W_q$.

ii) each W_q is a C^r-manifold and the map $\pi : P \to Q$ given by $\pi(W_q) = q$ is C^r. Points of W_q are characterized by the fact that the distance from $\varphi_t(p)$ to $\varphi_t(q)$ goes to zero exponentially fast as t goes to $-\infty$.

Let us denote by W the foliation of P by the strong unstable manifolds W_q, where W_q is the leaf through a point q in Q. Since W is a coisotropic foliation $(TW)^\perp$ considered as a subbundle of TW is integrable. (See [14] p.11). Consequently $TW^\perp$ is the tangent bundle to an isotropic foliation of W, which we denote by $W^\perp$. Now let q be a point in the isotropic submanifold Q, and let U be a neighborhood of q in P sufficiently small so that the foliations of U defined by $W^\perp$ and W are simple, i.e. the set of the leaves of the foliation are smooth manifolds and the corresponding projections

$$\Phi : U \longrightarrow \frac{P \cap U}{W^\perp \cap U} = B_{W^\perp}$$

and

$$\Psi : U \longrightarrow \frac{P \cap U}{W \cap U} = B_W$$

are submersions. Then there is a unique Poisson structure on $B_{W^\perp}$ such that Φ is a Poisson morphism, whose rank at $\Phi(x)$ is equal to the rank of the 2-form $\Omega|_W$ induced by Ω on the leaf W_x through x, which equals $2n - 2k$, and there is a unique Poisson structure on B_W such that Ψ is a Poisson morphism, whose rank at $\Psi(x)$ equals the rank of $\Omega|_{W_x^\perp}$, which equals zero since $W_x^\perp$ is isotropic. (See [6]).

We use these unique Poisson structures on $B_{W^\perp}$ and B_W to construct Darboux coordinates adapted to the leaves of the foliations W and $W^\perp$. More precisely, we construct coordinates $(U; x_1, ..., x_n, y_1, ..., y_n)$ near a point q in Q such that

- $Q \cap U = \{y_i = 0\}_{i=1}^k \cap \{x_r = 0\}_{r=k+1}^n \cap \{y_r = 0\}_{r=k+1}^n$,
- the leaf through a point q in $Q \cap U$ is given by

$$W_q \cap U = \{p \in U \ : \ x_i(p) = x_i(q) \ i = 1, ..., k\},$$

and

- $\Omega = \sum_{i=1}^k dx_i \wedge dy_i + \sum_{r=k+1}^n 2dx_r \wedge dy_r.$

In these coordinates the eigenspace for $L(\xi)$ corresponding to the eigenvalue $\lambda = 1/2$ is exactly the space spanned by $\{\frac{\partial}{\partial x_r}, \frac{\partial}{\partial y_r}\}_{r=k+1}^{n}$ and the eigenspace corresponding to the eigenvalue $\lambda = 1$ is exactly the space spanned by $\{\frac{\partial}{\partial y_i}\}$. i.e. the linear part of ξ is

$$L(\xi) = \sum_{i=1}^{k} y_i \frac{\partial}{\partial y_i} + \sum_{r=k+1}^{n} \left(\frac{1}{2} x_r \frac{\partial}{\partial x_r} + \frac{1}{2} y_r \frac{\partial}{\partial y_r}\right).$$

Details can be found in [2] and [3].

In the situation we are considering, the eigenvalues of the Liouville vector field ξ (i.e. the eigenvalues of $A(q)$) at the singular point q satisfy the integral relations

$$\lambda_i = 1\lambda_r + 1\lambda_s$$

where $\lambda_i = 1$, $i = 1, ..., k$ $\lambda_r = \lambda_s = \frac{1}{2}$, $r, s = k+1, ..., n$. According to the definition the possible resonant monomials in each fiber are

$$(*) \qquad x_r x_s \frac{\partial}{\partial y_i}\ ,\ y_r y_s \frac{\partial}{\partial y_i}\ ,\ x_r y_s \frac{\partial}{\partial y_i}$$

for $i = 1, ..., k$; $r, s = k+1, ..., n$. Thus

$$\xi = L(\xi) + \sum_{r=k+1}^{n} E_r \frac{\partial}{\partial x_r} + \sum_{\ell=1}^{n} F_\ell \frac{\partial}{\partial y_\ell}$$

where for each $q = (x_1, ..., x_k)$ the functions $E_r(q, \cdot), F_\ell(q, \cdot)$ vanish to higher order than $(\sum_{r=k+1}^{n} |x_r|^2 + \sum_{l=1}^{n} |y_\ell|^2)^{1/2}$. Denote by ξ^q the restriction of ξ to W_q

Proposition 2.1 *For each q the Taylor expansion of ξ^q contains no resonant quadratic terms.*

<u>Proof:</u> The Taylor expansions of ξ^q fit together to give a Taylor expansion of the vector field ξ in terms of the coordinates y_i, x_s, y_s, $i = 1, ..., k$; $s = k+1, ..., n$, with coefficients which are functions of the x_i, $i = 1, ..., k$. It suffices to show that this expansion has no terms of the form (*). Notice that the vector field $\eta = \xi - L(\xi)$ is hamiltonian since

$$\mathcal{L}_\eta \Omega = d([\xi - L(\xi)] \lrcorner\, \Omega) = \Omega - \Omega = 0.$$

Let H be a hamiltonian function for η (i.e. $\eta \lrcorner\, \Omega = dH$). If η (therefore ξ) had any resonant monomial, then the usual Taylor expansion of the hamiltonian function H in terms of all the x_l, y_l would contain nonzero terms of the type:

$$x_i x_r x_s,\ x_i x_r y_s,\ x_i y_r y_s$$

for $1 \leq i \leq k$; $k+1 \leq s,r \leq n$. Consequently η would also contain terms of the type

$$x_i x_s \frac{\partial}{\partial x_r}, \; x_i x_s \frac{\partial}{\partial y_r}, \; x_i y_s \frac{\partial}{\partial x_r} \; or \; x_i y_s \frac{\partial}{\partial y_r}$$

which is impossible since the functions E_r and F_ℓ do not contain terms which depend linearly on x_s or y_s, $k+1 \leq s \leq n$. Observe that for each leaf W_q, the x_i are constants and so are coefficients.

□

By Poincaré-Dulac's theorem [1] p.184, the non-resonant quadratic terms can be annhilated by a polynomial change of variable, i.e.: the system $\dot{w} = \xi^q(w)$ can be reduce to the form $\dot{w} = A(q)w + G(w,q) + F(w,q)$ where $D^0F(q_0,q) = D^1F(q_0,q) = D^2F(q_0,q) = 0$ and G contains only the possible resonant terms. But by Proposition 2.1 $G(w,q)$ vanishes identically. Therefore we can apply Sell's algorithm for $r = 3$ and obtain that the linearizing conjugation is of class $K = 1$. Since the eigenvalues of $A(q_0)$ are 1 or 1/2, the spectral spread ρ^+ equals 2. Because $\sum^- A(q_0) = \emptyset$ the r-smoothness of $A(q_0)$ is the largest integer $K \geq 0$ such that $r - K2 \geq 0$. Thus for $K > 1$ in order to be able to apply Sell's algorithm we take $r = 2K$ and since the only resonances occur with $|m| = 2$ we can perform a further C^∞-substitution, as in the proof of Poincaré theorem [1] p.181, to reduce the system $\dot{w} = A(q)w + F(w,q)$ to the form $\dot{w} = A(q)w + o(|w|^r)$. Hence Sell's theorem guarantees that the linearizing conjugation Φ is of class C^K. This completes the proof of the following.

Theorem 2.2 *For $k \geq 1$, there is a C^k-smooth linearizing conjugation between the Liouville vector field ξ and its linear part on a neighborhood of each singular point q of ξ.*

This means that there is a local diffeomorphism on a neighborhood of q in P which carries the trajectories of the flow generated by the vector field ξ to the trajectories of the flow of its linear part, preserving the direction of motion.

This talk was given at the Symplectic Geometry Workshop held at the University of Warwick in August 1990. I would like to thank D. Salamon for his kind invitation. I am also grateful to Prof. D. McDuff for her constant advise, to Prof. J. Eells and to Prof. A. Verjovsky for their support and hospitality at the International Centre for Theoretical Physics, Trieste.

References

[1] V.I. Arnold: *Geometrical methods in the theory of ordinary differential equations.* A Series of Comprehensive Studies in Mathematics 250. Springer-Verlag. (1988)

[2] E. Ciriza: *The symplectic structure of submanifolds of Kähler manifolds of non-positive curvature.* Ph.D Thesis, State University of New York at Stony Brook. (1989)

[3] E. Ciriza: *The local structure of a Liouville vector field.* To appear in Amer. J. Math.

[4] E. Ciriza: *On special submanifolds in symplectic geometry.* To appear in Diff. Geom. and its Appl.

[5] M.W. Hirsch, C.C. Pugh, M. Shub: *Invariant manifolds.* Lecture Notes in Math. 583. Springer-Verlag.(1977)

[6] P. Libermann, C-M. Marle.: *Symplectic Geometry and Analytical Mechanics.* D.Reidel Publishing Company, Holland. (1987)

[7] C. Robinson: *Stable manifolds in hamiltonian systems.* Contemporary Mathematics Vol 81, (1988)

[8] D. McDuff: *The symplectic structures of Kähler manifolds of non-positive curvature.* J.Diff. Geometry 28. (1988)

[9] D. McDuff: *Symplectic structures on* $\mathbf{R}^{2n}$. Proceedings of the Seminaire Sud-Rhodanien (Lyon 1986). Travaux en Cours. Hermann, Paris.

[10] J. Moser: *On the volume elements on a manifold.* Trans. Amer. Math. Soc. 120. (1965)

[11] T. Nagano: *1-formes with their exterior derivative of maximal rank.* J.Diff. Geometry 2. (1968)

[12] G. Sell: *Smooth linearization near a fixed point.* Am. J. Math. 107 (1985).1035-1091

[13] S. Sternberg: *On contractions and a theorem of Poincaré.* Amer. J. Math. 79. (1957)

[14] A. Weinstein: *Lectures on symplectic manifolds.* CBMS regional conference series in Math 29, Am. Math. Soc. (1977)

Complex cobordism, Ashtekar's equations and diffeomorphisms

S. K. DONALDSON

1. INTRODUCTION

In this note we will try to bring together two sets of ideas: first, the ideas of Segal [7] on the *complexification* of diffeomorphism groups and second the Ashtekar formulation [1] of the self-dual Einstein equation in 4 dimensions. The account here is rather informal; our main purpose is to point out the connection between these ideas, rather than proving specific results. In particular, while one may hope that interesting analytical problems and-global questions may be involved in a later stage, we make no attempt to tackle such questions here. Constructions from Symplectic Geometry will be fundamental in our discussion.

We will first describe this background in more detail. Recall that if G is a Lie group with Lie algebra $\mathfrak{g}$, the complexification $\mathfrak{g}^c = \mathfrak{g} \otimes \mathbf{C}$ of the Lie algebra always exists, as a complex Lie algebra. We say that G has a complexification if there is a Lie group G^c with Lie algebra $\mathfrak{g}^c$ which contains G as a subgroup realising the standard embedding $\mathfrak{g} \subset \mathfrak{g}^c$. It is well known that if G is a compact group such a complexification exists, and if G is finite dimensional a complexification exists up to a covering. But when we go to infinite dimensional groups the picture changes, and a complexification will not exist in general, even if one goes to a covering group. In his approach to Conformal Field Theory, Segal defines a semi-group $\mathcal{A}$ which, he argues, should play the role of the complexification of the group $\mathrm{Diff}(S^1)$ of orientation-preserving diffeomorphisms of the circle. An element of $\mathcal{A}$ is a 2-dimensional manifold X, with boundary, diffeomorphic to the standard annulus and with the additional structure

(1) An almost complex structure, smooth up to the boundary
(2) Fixed parametrisations $j^+, j^- : S^1 \to \partial X$ of the two boundary components

The composition in the semigroup $\mathcal{A}$ is defined by gluing together annuli. This semigroup enjoys many of the properties one would hope for in a complexification of $\mathrm{Diff}(S^1)$. In particular, there is a notion of the "action" of $\mathcal{A}$ on the complexified Lie algebra, i.e. the sections of $TS^1 \otimes \mathbf{C}$ over S^1, which gives a substitute for the adjoint action of the complexified group on its Lie algebra. This is the aspect of Segal's construction which we will attempt to generalise. The "action" is defined as follows: if V is a holomorphic vector field over X the boundary-values of V, pulled back by j^+, j^- give sections V^+, V^- of the complexified tangent bundle of the circle, and we then write $X(V^-) = V^+$. It is easy to see that V^+ is uniquely determined by X and

V^-, although the action is not really defined everywhere—for a fixed X, the transformed element $X(V^-)$ will not exist for all V^-, although for a given real-analytic V^-, the transform $X(V^-)$ will exist for all sufficiently "thin" annuli X. Contemplating Segal's scheme, one is lead to wonder if there are other situations in which it is useful to define a similar "adjoint orbit" of a fictitious complexified diffeomorphism group. More generally, if G is a Lie group, possibly infinite dimensional, which acts on a vector space $\mathcal{U}$ one can ask for a definition which will generalise that of the orbit of the complexification G^c on $\mathcal{U} \otimes \mathbf{C}$. There is an obvious answer to this question. We say that elements $U^+, U^- \in \mathcal{U} \otimes \mathbf{C}$ are "G^c related" if there are

(1) An element g of G

(2) Smooth 1-parameter families $U_1(t), U_2(t) \in \mathcal{U}$ parametrised by an interval $(-a, a) \subset \mathbf{R}$ with $U_1(-a) + iU_2(-a) = U^-, U_1(a) + iU_2(a) = g(U^+)$

(3) A smooth 1-parameter family $\xi(t)$ in the Lie algebra of G such that

$$\frac{d}{dt}(U_1 + iU_2) = i\xi.(U_1 + iU_2). \tag{1}$$

(In (2) we use the fact that $g \in G$ acts on $\mathcal{U}$ and so on $\mathcal{U}^c$.) It is easy to see that in the finite dimensional case $U^{\pm}$ are G^c-related if and only if they lie in the same G^c orbit so this is a reasonable definition. It is also not hard to check that this definition agrees with the one given by Segal, in the case of the diffeomorphism group of the circle. More generally if G is a group of diffeomeorphism of a d-manifold M and $\mathcal{U}$ is some space of tensor fields over M then we can re-interpret the data above as a triple of tensor fields on the product space $M \times (-a, a)$, satisfying a differential equation corresponding to (1). It seems however that this interpretation will not be interesting in general. The key feature of Segal's case is that the data on the product space has an intrinsic interpretation (as a complex structure and holomorphic vector field), independent of the particular space–time description of the cylinder. This is a kind of extra $d+1$-dimensional "Lorentz invariance" of the equation (1), which does not hold in general. Our purpose here is to point out that there is another case where we have a similar intrinsic interpretation of the equation (1) in $(d+1)$ dimensions.

The other main motivation for our discussion comes from *Nahm's equations*. These are a system of ordinary differential equations for a triple $T_1(t), T_2(t), T_3(t)$ of elements of a Lie algebra $\mathfrak{g}$, depending on a real parameter t. The equations are

$$\frac{dT_1}{dt} = [T_2, T_3]$$
$$\frac{dT_2}{dt} = [T_3, T_1]$$
$$\frac{dT_3}{dt} = [T_1, T_2]$$

The first two of these equations can be combined into

$$\frac{d}{dt}(T_1 + iT_2) = [iT_3, T_1 + iT_2],$$

so a solution of Nahm's equations, over an interval $[-a, a]$, gives data of the kind considered above, with the adjoint representation of G and $U_1 = T_1, U_2 = T_2, \xi = T_3$. Let us say that a pair U^+, U^- in $\mathfrak{g}^c$ are "related by a Nahm solution" if $U^+ = T_1^+ + iT_2^+, U^- = T_1^- + iT_2^-$ and there is a solution of Nahm's equations over $[-a, a]$ with $T_1(\pm a) = T_1^\pm, T_2(\pm a) = T_2^\pm$. Thus if U^+, U^- are related by a Nahm solution they are also G^c related. If the group G is compact then the converse holds: a pair is related by a Nahm solution if and only if they are G^c related, i.e. if and only if they lie in the same orbit of the complexified group. This is proved in [2] for the unitary groups by setting up Nahm's equations as the Euler-Lagrange equations for a functional on paths on the space G^c/G, and the argument has been extended to other compact groups by P. Saskida [7]. This identification is important in the work of Kronheimer [5] on adjoint orbits. From this point of view it is natural to investigate the corresponding question for infinite-dimensional groups. The case of gauge groups leads to boundary value problems for Yang-Mills fields of the sort considered in [3]. Here we will look at diffeomorphism groups, the point being that we can make contact with the work of Ashtekar, who showed that Nahm's equations for the Lie algebra of the group of volume-preserving diffeomorphisms of a 3-manifold Y are a form of the self-dual Einstein equations on $Y \times \mathbf{R}$.(Ashtekar's ideas have been extended by Mason and Newman [6], and we will make use of some of their simplifications.)

2. Diffeomorphisms of a 3-manifold and complex cobordisms

By a *complex symplectic surface* Z we mean a differentiable 4-manifold endowed with a complex structure and a holomorphic symplectic form $\theta \in \Omega_Z^{2,0}$, i.e. θ is holomorphic and $\theta \wedge \overline{\theta} = 2\Omega$ is a volume form. If we write $\theta = \theta_1 + i\theta_2$, where θ_1, θ_2 are real 2-forms, the algebraic constraints imposed at each point are

$$\theta_1 \wedge \theta_1 = \theta_2 \wedge \theta_2 \neq 0 \quad , \quad \theta_1 \wedge \theta_2 = 0. \tag{2}$$

The 2-forms θ_1, θ_2 are closed, so they may be viewed as a pair of ordinary symplectic structures on the 4-manifold Z. Conversely any pair of closed 2-forms which satisfy the algebraic conditions (2) define a complex symplectic structure. To see this, observe that at each point $\theta_1 + i\theta_2$ is a null 2-form and hence, via the Plucker embedding, defines a subspace $T^{1,0} \subset TZ \otimes \mathbf{C}$, and so an almost-complex structure on Z. If N is the Nijenhius tensor of this structure, a tensor in $T^{1,0} \otimes (T^{0,2})^*$, then $d\theta$ has two components, one $\overline{\partial}\theta \in \Omega_Z^{2,1}$ and the other $N\theta \in \Omega_Z^{1,2}$. If $d\theta = 0$ both of these components vanish and this can only happen if N is zero and θ is holomorphic.

Now let Z be a complex symplectic surface with boundary, i.e. the forms are smooth up to the boundary, and suppose the boundary has a component Y. Suppose we are given a volume form α on Y.Then we can define a pair of vector fields v_1, v_2 on Y as follows. We restrict the 2-forms θ_1, θ_2 to Y and then compose with the isomorphism between 2-forms on Y and vector fields, given to us by the volume form. Thus

$$i_{v_1}(\alpha) = \theta_1|_Y, i_{v_2}\alpha = \theta_2|_Y.$$

Another way of saying this is that we choose a Hamiltonian H on Z, constant on the boundary component Y and with $dH \wedge \alpha$ equal to the volume form on Z, over the boundary. Then we let v_1, v_2 be the usual Hamiltonian vector fields generated by this function and the two symplectic structures θ_1, θ_2. Either way, it is clear that the vector fields are volume-preserving on Y, that is

$$L_{v_i}\alpha = 0,$$

where L denotes the Lie derivative. We can put the vector fields together to define a single section $v = v_1 + iv_2$ of the complexified tangent bundle $TY \otimes \mathbf{C}$. Suppose now that Z has two boundary components Y^+, Y^- with volume forms. We get complex vector fields $v^+ \in \Gamma(TY^+ \otimes \mathbf{C}), v^- \in \Gamma(TY^- \otimes \mathbf{C})$. We write, somewhat prematurely, $Z(v^-) = v^+$, and we say that v^+, v^- are *cobordism related.* We will argue that this notion is closely comparable to that of G^c related complex vector fields, introduced above, where G is the group of volume-preserving diffeomorphisms of Y, and we use the adjoint representation. It is possible that the two relations are identical if we consider cobordisms which are topological products. In one direction, suppose we start with a solution of equation (1) in this context. That is, we have a 3-manifold Y with a volume form α and volume-preserving vector fields v_1, v_2, v_3 on Y depending on a real parameter $t \in [-a, a]$, and such that

$$\frac{d}{dt}(v_1 + iv_2) = [iv_3, v_1 + iv_2]. \tag{3}$$

We suppose also that v_1, v_2, v_3 are linearly independent at each point. We will define a complex symplectic structure on $Y \times [-a, a]$ using this data. For each "time" t, let $\epsilon_1, \epsilon_2, \epsilon_3$ be the basis of 1-forms on Y dual to the vector fields v_1, v_2, v_3. Then we can write

$$\alpha = f\ \epsilon_1 \wedge \epsilon_2 \wedge \epsilon_3$$

for a real-valued function f on $Y \times [-a, a]$. We define 2-forms on $Y \times [-a, a]$ by

$$\theta_1 = f(dt \wedge \epsilon_1 + \epsilon_2 \wedge \epsilon_3)$$
$$\theta_2 = f(dt \wedge \epsilon_2 + \epsilon_3 \wedge \epsilon_1).$$

It is clear that, just as a matter of algebra,

$$\theta_1^2 = \theta_2^2 = f dt \wedge \alpha \ , \ \theta_1 \wedge \theta_2 = 0.$$

We claim that θ_1, θ_2 are closed on $Y \times [-a, a]$ if (and only if) the vector fields v_1, v_2 satisfy the equation (3). This is a calculation which is expedited by the use of the following identity. If ξ, η are vector fields on some manifold M and ϕ is a form on M then

$$d(i_\xi i_\eta \phi) = i_{[\xi,\eta]}\phi + i_\eta L_\xi \phi - i_\xi L_\eta \phi + i_\xi i_\eta d\phi.$$

This can be seen by applying the formula $L_X = di_X + i_X d$ twice. In our situation, we take the vector fields $\xi = v_2, \eta = v_3$ on Y, at a fixed time t, and let $\phi = \alpha$. Then, since the fields are volume-preserving, we get

$$d(i_{v_2} i_{v_3} \alpha) = i_{[v_2,v_3]}\alpha.$$

Now note that $i_{v_1}\alpha = f\epsilon_1 \wedge \epsilon_2$ and $i_{v_2} i_{v_3}\alpha = f\epsilon_1$. The form $i_{v_1}\alpha$ is closed since v_1 is volume-preserving. Let us write, temporarily, $\underline{d}$ for the exterior derivative on forms over the 4-manifold $Y \times \mathbf{R}$, and reserve d for the exterior derivative on forms over Y, with the time regarded as a parameter. In this notation we have

$$\underline{d}\psi = d\psi + dt \wedge \frac{d\psi}{dt}.$$

Now the exterior derivative of θ_1 is

$$\underline{d}\theta_1 = dt \wedge (d(f\epsilon_1) + \frac{d}{dt}(f\epsilon_2 \wedge \epsilon_3)\ ) + d(f\epsilon_2 \wedge \epsilon_3).$$

As we have noted above, $f\epsilon_2 \wedge \epsilon_3$ is closed on Y, so $\underline{d}\theta_1 = 0$ if and only if

$$d(f\epsilon_1) + \frac{d}{dt}(f\epsilon_2 \wedge \epsilon_3) = 0,$$

that is if and only if

$$i_{[v_2,v_3]}\alpha = \frac{d}{dt}(i_{v_1}\alpha).$$

But the volume form α is independent of time, so

$$\frac{d}{dt}(i_{v_1}\alpha) = i_{\frac{dv_1}{dt}}\alpha.$$

Thus θ_1 is closed if and only if $\frac{dv_1}{dt} = [v_2, v_3]$, which is the real part of equation (3). Similarly, θ_2 is closed if and only if $\frac{dv_2}{dt} = [v_3, v_1]$: thus a solution to the equation (3) gives a complex symplectic structure on the cylinder $Z = Y \times [-a, a]$. If we now apply our construction on the boundary we get precisely the complex vector fields $(v_1+iv_2)(\pm a)$ on the two boundary components, so $Z((v_1+iv_2)(-a)) = (v_1+iv_2)(a)$. Thus we see that G^c-related vector fields are also cobordism-related

We will now set up the converse problem. We would like to show that any complex symplectic structure on a cylinder arises from this construction. While we will not reach any conclusive results in this direction, we will explain the nature of the partial differential equation which is involved. Observe that in the set-up above the vector fields

$$v_1, v_2, v_3 \text{ and } v_4 = \frac{\partial}{\partial t}$$

on $Y \times [-a, a]$ each preserve the volume form $\alpha \wedge dt$. So the rescaled fields $f^{-1}v_i$ preserve the intrinsic volume form $\Omega = \theta_1^2 = \theta_2^2 = f\alpha \wedge dt$. Now let Z be a complex symplectic manifold, diffeomorphic to the cylinder. Suppose we choose a "time" function $\tau : Z \to [-a, a]$, without critical points, equal to $\pm a$ on the boundary components. This gives a foliation of Z into level sets $Y_t = \tau^{-1}(t)$. Using τ as a Hamiltonian function, for the two symplectic structures we get volume-preserving vector fields w_1, w_2 on Z, tangent to the leaves Y_t. Suppose we can find another vector field w_4 which is transverse to these leaves and such that

(1) w_4 and $w_3 = Iw_4$ are volume preserving on Z
(2) w_3 is tangent to the leaves Y_t, i.e $\nabla_{w_3}\tau = 0$.

Then the complex structure on Z arises from a solution of (3) by the construction above. To see this, we define $f = (\nabla_{v_4}\tau)^{-1}$ and set $v_i = fw_i$ for $i = 1, \ldots 4$, and $e_i = f^{1/2}w_i$. Then

$$\theta_1(e_1, e_4) = f\theta_1(w_1, w_4) = f\nabla_{v_4}\tau = 1,$$

and

$$\theta_2(e_1, e_4) = \theta_1(e_1, Ie_4) = 0.$$

So the pair of vectors e_1, e_4 make up a standard complex basis of the tangent space with respect to the complex symplectic form θ. Now the vector field v_4 satisfies $\nabla_{v_4}\tau = 0$, so we may integrate it to define a product structure on Z, i.e. to fix an identification $Z \cong Y \times [-a, a]$, such that the function τ is the time component, and $v_4 = \frac{\partial}{\partial \tau}$. Since the vectors e_i give a standard basis at each point there is a single algebraic model for the situation and we can write

$$\theta_1 = f(dt \wedge \epsilon_1 + \epsilon_2 \wedge \epsilon_3)\ ,\ \theta_2 = f(dt \wedge \epsilon_2 + \epsilon_3 \wedge \epsilon_1),$$

for certain time-dependent 1-forms ϵ_i on Y, dual to the vector fields v_1, v_2, v_3. We define a volume form α on Y by $\alpha \wedge dt = f\Omega$. Then the vector fields

v_i preserve the volume form α, since the fields w_i preserve Ω. Thus we have recovered the set-up we had before, and our calculation shows that the vector fields v_i satisfy the evolution equation (3), since the forms θ_i are closed.

We will now translate the conditions on the vector field w_4 into a more handy notation. We will first make a digression to point out a simple but special construction which can be made on any complex symplectic surface Z. We let $\mathcal{W}$ be the space of vector fields w on Z such that both w and Iw are volume-preserving. On any n-manifold the volume-preserving vector fields are naturally identified with the closed $(n-1)$-forms, so we can identify $\mathcal{W}$ with 3-forms λ on Z such that $d\lambda = d(I\lambda) = 0$. If we identify the real 3-forms λ with the $(2,1)$forms μ, by $\lambda = \mu + \overline{\mu}$, then we identify $\mathcal{W}$ with the space of $(2,1)$ forms μ with $\overline{\partial}\mu = 0$. Now use the symplectic form θ to write such a form as a product $\mu = \theta \wedge \beta$ where β is of type $(0,1)$. In this way we identify $\mathcal{W}$ with the space of $(0,1)$-forms β with $\overline{\partial}\beta = 0$. Now if H is any complex-valued smooth function on Z we construct such a form β by $\beta = \overline{\partial}H$. So we get a map

$$m : C^\infty(Z) \to \mathcal{W}$$

which is surjective if $H^1(Z : \mathcal{O}_Z) = 0$. This is obviously analogous to the Hamiltonian construction in symplectic geometry. We can take the analogy further by noting that $\mathcal{W}$ is naturally a complex Lie algebra. We define a bracket $[\ ,\ ]_c$ in $\mathcal{W}$ by

$$[X,Y]_c = [X,Y] + [IX,IY]$$

It is clear from the definition of $\mathcal{W}$ that $[\ ,\ ]_c$ does map $\mathcal{W} \times \mathcal{W}$ to $\mathcal{W}$. It satisfies the Jacobi identity because the complex structure is integrable. (The integrability condition precisely asserts that the real vector fields are a subalgebra of the complexified vector fields under the embeddding $X \to X + iIX$, and $[\ ,\ ]_c$ is induced from the complexification by this embedding.) Now we define a "Poissson bracket" $\{\ ,\ \}_c$ on the complex valued functions by

$$\{F,G\}_c\, \theta \wedge \overline{\theta} = \overline{\partial}F \wedge \overline{\partial}G \wedge \theta.$$

Some short calculations show that $\{\ ,\}_c$ makes $C^\infty(Z)$ into a complex Lie algebra, and that m is a Lie algebra homomorphism. Thus we get a kind of complexified Hamiltonian mechanics on a complex symplectic surface. (Which should not be confused with the more standard holomorphic complexification, in which holomorphic functions on Z generate holomorphic vector fields preserving θ.)

We now return from our digression to the question of finding the transverse vector field w_4 satisfying the conditions above. The first condition, that w_4 and Iw_4 are volume preserving, means that w_4 is identified with a $\overline{\partial}$-closed $(0,1)$-form β. The second condition becomes

$$\text{Im}\,(\beta \wedge \overline{\partial}\tau) = 0.$$

We must also consider boundary conditions, on the two boundary components $Y^{\pm}$. We suppose that we have given volume forms $\alpha^{\pm}$ on the boundaries, and we want our solution to be compatible with these. This is a condition on w_4 only, independent of τ, since the volume form induced on the leaves is $\alpha_t = i_{w_4}(\Omega)$. Thus β should satisfy an extra condition on the boundary. To see this condition, recall that the boundary of a complex manifold has a CR structure, a "complex tangent space" H which in this case is just the space spanned by v_1, v_2 with the complex structure induced from Z. Thus the volume forms on the boundary and the complex symplectic form give a trivialisation of the complex tangent bundle H. In particular, there is a canonical section $\overline{\epsilon} = \epsilon_1 - i\epsilon_2$ of $\overline{H}^*$ and the boundary condition is just that the restriction of β to H be equal to $\overline{\epsilon}$.

To sum up, suppose $v^{\pm}$ are cobordism-related vector fields on Y, so there is a complex symplectic cobordism Z with volume forms on the boundary realising these vector fields. The boundary data gives sections $\overline{\epsilon}^{\pm}$ of $\overline{H}^*$ over $Y^{\pm}$. If Z is a topological cylinder then to show that $V^{\pm}$ are G^c related one has to find a $\overline{\partial}$-closed $(0,1)$-form β and a foliating function τ such that that β restricts to $\overline{\epsilon}$ on the boundary and $\overline{\partial}\tau \wedge \beta \wedge \theta$ is real throughout Z. The question of the existence of β alone is cohomological in character: it is a question of extending the class of $\overline{\epsilon}$ in the first cohomology of the $\overline{\partial}_b$ complex on the boundary to a class in $H^1(Z, \mathcal{O}_Z)$: if one extension β_0 exists we can modify it to $\beta = \beta_0 + \overline{\partial}H$, for any function H which vanishes on the boundary. Suppose now that we fix a β_0 and also an arbitrary foliating function τ. Then we need to solve the remaining equation

$$\text{Im}(\overline{\partial}H \wedge \overline{\partial}\tau \wedge \theta) = -\text{Im}(\beta_0 \wedge \overline{\partial}\tau \wedge \theta),$$

with $H = 0$ on ∂Z. This is an underdetermined linear equation: there is one real equation for two real functions A, B with $H = A + iB$. The equation is essentially 3-dimensional, it can be written more explicitly using the vector fields w_1, w_2 along the leaves Y_t as

$$\nabla_{v_1} A + \nabla_{v_2} B = Q,$$

with Q given and A, B to be found. It seems likely that there always is a solution to this underdetermined equation and this, along with the extension of $\overline{\epsilon}$, would imply that the two complex vector fields are G^c-equivalent.

3. Nahm's equations, hyperkahler metrics and other topics.

We will now explain how one can formulate the analogue of the main result about Nahm's equations in this context. We will begin by recalling the result of Ashtekar. If we have three time-dependent volume-preserving vector fields on Y which satisfy Nahm's equations $\frac{dv_i}{dt} = [v_j, v_k]$ $(i,j,k$ cyclic) then v_1, v_2 satisfy (3), and hence define a complex symplectic structure on $Y \times (-a, a)$. But the symmetry in the equations means that the same goes for the other pairs v_2, v_3, v_3, v_1. Thus we get a family of complex symplectic structures on the product space and it is clear that these are algebraically compatible with a single Riemannian metric g. It follows straightaway that g is a hyperkahler metric, with the three Kahler forms $\theta_1, \theta_2, \theta_3$. The vector field w_4 now becomes the gradient vector field of the time function τ. This function must be harmonic, since w_4 is volume preserving. Conversely if we are given a hyperkahler manifold Z and a harmonic function τ on Z we use the gradient vector field grad τ, away from critical points, to define a product structure. Now suppose we have cobordism related complex vector fields v^+, v^- on Y, i.e. we have a complex symplectic manifold Z, which we suppose is diffeomorphic to a cylinder, with volume forms on the boundary components, and v^+, v^- are the vector fields induced on these two boundary components. We can ask if v^+, v^- are related by a solution of Nahm's equations, as in the finite dimensional case. We can verify this if we can find a suitable hyperkahler structure on the cylinder Z. What we need is a kahler metric g on Z, with the given volume form Ω, and a harmonic function τ (with respect to this metric) equal to $\pm a$ on the two boundary components. Of course this harmonic function is uniquely determined, if Z is compact by the metric and boundary data. The final condition we need is that the volume forms on the boundary agree with those induced by τ. In the notation we have used above, the volume form on Z is

$$\Omega = f^2(\epsilon_1 \wedge \epsilon_2 \wedge \epsilon_3 \wedge d\tau),$$

so $f^{1/2}\epsilon_i, f^{1/2}d\tau$ form a standard orthonormal frame of 1-forms at each point. Thus the Riemannian gradient of τ had norm

$$|\text{grad } \tau|^2 = f^{-1}$$

and the 3-volume form on Y is

$$\alpha = f\epsilon_1 \wedge \epsilon_2 \wedge \epsilon_3 = |\text{grad } \tau| \ \text{Vol}_3,$$

where Vol_3 is the Riemannian volume element on the 3- dimensional leaves. This is just the "flux" of the gradient vector field grad τ through the leaves, or in terms of differential forms $\alpha = *d\tau$. To sum up, the problem of solving Nahm's equations on Z comes down to the following. For any compact Riemannian manifold (X, g) with boundary, whose boundary has two components $\partial^{\pm}X$ we define a measure η_g on the boundary by the flux of the unique harmonic function on X with boundary values $\pm a$. Then for our given hypercomplex Z, with volume form α on the boundary, we seek a kahler metric g with the prescribed 4-volume element Ω in the interior and with $\eta_g = \alpha$ on the boundary.

It seems possible that this may be an interesting kind of boundary value problem to investigate: it is reasonable to hope for existence and uniqueness in the solutions since we can regard the equations in the interior as a second order elliptic equation for 2 real-valued functions (the function τ and a Kahler potential for g), and the boundary conditions represent two additional equations over the boundary, which is as one would want.

Finally we should mention another problem where similar constructions seem to be interesting. This is is the case where one takes a real 2-dimensional symplectic manifold (Σ, ω) and the group G to be the central extension of the exact area-preserving diffeomorphisms of Σ, with Lie algebra $C^\infty(\Sigma)$ under the Poisson bracket: compare [9], for example. One can look for the structure on a 3-dimensional cobordism U between two such surfaces $\Sigma^\pm$ which will correspond to the adjoint action of the missing complexified group. The appropriate structure should enable one to find:

(1) a complex 2-form Φ on U which is real on the boundary, giving symplectic forms $\omega^\pm \in \Omega^2(\Sigma^\pm)$

(2) a complex function $H = H_1 + iH_2 : U \to \mathbf{C}$,

such that

$$d\Phi = d(H\Phi) = 0 \text{ in } U.$$

Identifying the appropriate structure would give a notion of the "adjoint action" of U, taking $H|_{\Sigma^-}$ to $H|_{\Sigma^+}$. One can also look at Nahm's equations in this context, which essentially comes down to studying harmonic functions in $\mathbf{R}^3$, since the equations $\frac{dh_i}{dt} = \{h_j, h_k\}$ for a family of maps $\underline{h}(t) : \Sigma \to \mathbf{R}^3$ just assert that the images of Σ are the level sets of a harmonic function.

An interesting feature of this second example is that there are many invariants on the Lie algebra. In the ordinary Lie algebra of real-valued functions one knows that for each p

$$I_p(h) = \int_\Sigma h^p \omega$$

is preserved by the adjoint action. One would like there to be corresponding invariants in the complexification, and this would follow if one could find a definition with the properties mentioned above. For if we have such a structure detailed above on a cobordism U then it follows from Stokes' Theorem that

$$\int_{\Sigma^+} H^p \omega^+ = \int_{\Sigma^-} H^p \omega^-.$$

Something similar occurs for the higher dimensional case we have been studying in this paper. Suppose the 3-manifold Y is a real-homology 3-sphere. Then we can define a quadratic form Q on the Lie algebra of the group G of volume-preserving diffeomorphisms as follows: we identify the Lie algebra with the closed 2-forms, and for any such form F we choose a 1-form A such that $F = dA$. Then we set

$$Q(F) = \int_Y F \wedge A.$$

It is easy to see that this does not depend on the choice of A. We would like the complexification of this invariant quadratic form to be preserved by the "action" of a complex-symplectic cobordism Z, and this is the case as long as the symplectic form θ is zero in $H^2(Z, \mathbf{C})$. For then we can write $\theta_i = dA_i$ over Z, and

$$Q(v_i^+) - Q(v_i^-) = \int_{\partial Z} \theta_i \wedge A_i$$

which is

$$\int_Z \theta_i \wedge dA_i = \int_Z \theta_i^2 = \mathrm{Vol}(Z).$$

Thus $\mathrm{Re}(Q(v^+) - Q(V^-)) = 0$, since $\theta_1^2 = \theta_2^2$ on Z and similarly one sees that the imaginary part vanishes since $\theta_1 \wedge \theta_2 = 0$.

It is worth pointing out that the three diffeomorphism groups we have considered in this paper are related: a diffeomorphism of a circle induces an area-preserving diffeomorphism of the 2-dimensional cotangent bundle, and an exact area-preserving diffeomorphism of a surface lifts to a volume-preserving diffeomorphism of a circle bundle over the surface.

The author acknowledges with thanks the hospitality of the Department of Mathematics at the University of Maryland during the preparation of this paper.

References

1. A. Ashtekar, *New Hamiltonian formulation of general relativity*, Phys. Rev. D **36** (1987), 1587-1603.
2. S.K.Donaldson, *Nahm's equations and the classification of monopoles*, Commun. Math. Phys. **96** (1984), 387-407.
3. S.K. Donaldson, *Boundary value problems for Yang-Mills fields*, Jour. Geometry and Physics **8** (1992), 89-122.
4. N.J.Hitchin, *On the construction of monopoles*, Commun. Math. Phys. **89** (1983), 145-190.
5. P.B. Kronheimer, *A hyper-kahlerian structure on co-adjoint orbits of a complex semi-simple Lie group*, Journal London Math. Soc (1990), 193-209.
6. L.J. Mason and E.T.Newman, *A connection between the Einstein and Yang-Mills equations*, Commun. Math. Phys. **121** (1989), 659-668.
7. P. Saskida, Unpublished dissertation.
8. G.B.Segal, Unpublished manuscript..
9. R. S. Ward, *Infinite dimensional gauge groups and special nonlinear gravitons*, Jour. Geometry and Physics **8** (1992), 317-325.

The Mathematical Institute, 24-29 St. Giles, Oxford

group G of volume-preserving diffeomorphisms as follows: we identify the Lie algebra with the closed 2-forms, and for any such form B we choose a 1-form A such that $B = dA$. Then we set

$$Q(B) = \int_Z B \wedge A.$$

It is easy to see that this does not depend on the choice of A. We would like the complexification of this invariant quadratic form to be preserved by the "action" of a complex symplectic cobordism Z, and this is the case as long as the symplectic form θ is zero in $H^2(Z, \partial Z)$. For then we can write $\theta_1 = d\lambda_1$ on Z and

$$Q(\omega_1) - Q(\omega_1') = \int_{\partial Z} \theta_1 \wedge \theta_1,$$

which is

$$\int_{\partial Z} \theta_1 \wedge d\lambda_1 = \int_Z \theta_1^2 = \text{Vol}(Z).$$

Thus Re$(Q(\omega^+)) - Q(\omega^-) = 0$, since $\omega^+ = \theta_1$ on Z and similarly one sees that the imaginary part vanishes since $\theta_2 \wedge \theta_3 = 0$.

It is worth pointing out that the three diffeomorphism groups we have considered in this paper are related: a diffeomorphism of a circle induces an area-preserving diffeomorphism of the 2-dimensional cotangent bundle, and an area-preserving diffeomorphism of a surface lifts to a volume-preserving diffeomorphism of a circle bundle over the surface.

The author acknowledges with thanks the hospitality of the Department of Mathematics at the University of Maryland during the preparation of this paper.

REFERENCES

1. A. Ashtekar, New Hamiltonian formulation of general relativity, Phys. Rev. D 36 (1987), 1587–1602.
2. S.K. Donaldson, Nahm's equations and the classification of monopoles, Commun. Math. Phys. 96 (1984), 387–407.
3. S.K. Donaldson, Boundary value problems for Yang-Mills fields, Journal of Geometry and Physics 8 (1992), 89–122.
4. N.J. Hitchin, On the construction of monopoles, Commun. Math. Phys. 89 (1983), 145–190.
5. P.B. Kronheimer, A hyperkählerian structure on coadjoint orbits of a semisimple complex Lie group, Journal London Math. Soc (1990), 193–208.
6. L.J. Mason and E.T. Newman, A connection between the Einstein and Yang-Mills equations, Commun. Math. Phys. 121 (1989), 659–668.
7. [illegible]
8. [illegible]

[illegible]

Instanton homology and symplectic fixed points

Stamatis Dostoglou and Dietmar Salamon*
Mathematics Institute
University of Warwick
Coventry CV4 7AL
Great Britain

February 3, 1993

1 Introduction

A gradient flow of a Morse function on a compact Riemannian manifold is said to be of Morse-Smale type if the stable and unstable manifolds of any two critical points intersect transversally. For such a Morse-Smale gradient flow there is a chain complex generated by the critical points and graded by the Morse index. The boundary operator has as its (x, y)-entry the number of gradient flow lines running from x to y counted with appropriate signs whenever the difference of the Morse indices is 1. The homology of this chain complex agrees with the homology of the underlying manifold M and this can be used to prove the Morse inequalities [30] (see also [24]).

Around 1986 Floer generalized this idea and discovered a powerful new approach to infinite dimensional Morse theory now called Floer homology. He used this approach to prove the Arnold conjecture for monotone symplectic manifolds [12] and discovered a new invariant for homology 3-spheres called instanton homology [11]. This invariant can roughly be described as the homology of a chain complex generated by the irreducible representations of the fundamental group of the homology 3-sphere M in the Lie group $SU(2)$. These representations can be thought of as flat connections on the principal bundle $M \times SU(2)$ and they appear as the critical points of the Chern-Simons functional on the infinite dimensional configuration space of connections on this bundle modulo gauge equivalence. The gradient flow lines of the Chern-Simons functional are the self-dual Yang-Mills instantons on the 4-manifold

*This research has been partially supported by the SERC.

$M \times \mathbf{R}$ and they determine the boundary operator of instanton homology. The instanton homology groups are denoted by $HF_*^{\text{inst}}(M)$. They play an important role in Donaldson's theory of 4-manifolds. If X is a 4-manifold whose boundary $M = \partial X$ is a homology-3-sphere then the Donaldson-polynomials of X take their values in the instanton homology groups of M.

Via a Heegard splitting a homology-3-sphere can be represented as the union

$$M = M_0 \cup_\Sigma M_1$$

of two handle bodies whose common boundary $\Sigma = \partial M_0 = \partial M_1$ is a Riemann surface. The moduli space of irreducible flat connections on $\Sigma \times SU(2)$ is a finite dimensional symplectic manifold $\mathcal{M}$ (with singularities) and the subset of those which extend to a flat connection on M_i form a Lagrangian submanifold $\mathcal{L}_i \subset \mathcal{M}$ for $i = 0, 1$. The intersection points of these Lagrangian submanifolds correspond precisely to the flat connections on M, that is the critical points of the Chern-Simons functional. Now there is a Floer theory for Lagrangian intersections in which the critical points are the intersection points of the Lagrangian submanifolds and the connecting orbits are pseudo-holomorphic curves $u : [0,1] \times \mathbf{R} \to \mathcal{M}$ which satisfy $u(i,t) \in \mathcal{L}_i$ and converge to intersection points $x^\pm \in \mathcal{L}_0 \cap \mathcal{L}_1$ as t tends to $\pm\infty$. The associated Floer homology groups are denoted by

$$HF_*^{\text{symp}}(\mathcal{M}, \mathcal{L}_0, \mathcal{L}_1).$$

Atiyah [1] and Floer conjectured that the instanton homology of a homology-3-sphere M is isomorphic to the Floer homology of the triple $\mathcal{M}, \mathcal{L}_0, \mathcal{L}_1$ associated to a Heegard splitting as above

$$HF_*^{\text{inst}}(M) = HF_*^{\text{symp}}(\mathcal{M}, \mathcal{L}_0, \mathcal{L}_1).$$

In this paper we address a similar but somewhat simpler problem which was suggested to us by A. Floer. The instanton homology groups can also be constructed for 3-manifolds which are not homology-3-spheres if instead of the trivial $SU(2)$-bundle we take a nontrivial $SO(3)$-bundle Q over M. Then the integral first homology of M is necessarily nontrivial and we assume that there is no 2-torsion in H_1. This assumption guarantees that every flat connection on Q has a discrete isotropy subgroup and is therefore a regular point for the action of the identity component of the group of gauge transformations on the space of connections. The instanton homology groups of the bundle Q are denoted by $HF_*^{\text{inst}}(M;Q)$.

A special case is where the bundle $Q = P_f$ is the mapping cylinder of a nontrivial $SO(3)$-bundle $\pi : P \to \Sigma$ over a Riemann surface Σ for an automorphism $f : P \to P$. The underlying 3-manifold is the mapping cylinder $M = \Sigma_h$ of Σ for the diffeomorphism $h : \Sigma \to \Sigma$ induced by f. Then the flat connections on P_f correspond naturally to the fixed points of the symplectomorphism

$$\phi_f : \mathcal{M}(P) \to \mathcal{M}(P)$$

induced by f on the moduli space $\mathcal{M}(P)$ of flat connections on the bundle P. This moduli space is a compact symplectic manifold (without singularities) of dimension $6k-6$ where $k \geq 2$ is the genus of Σ. It is well known that this manifold is connected and simply connected [21] and in [2] Atiyah and Bott proved that $\pi_2(\mathcal{M}_F(P)) = \mathbf{Z}$. For any symplectomorphism $\phi : \mathcal{M} \to \mathcal{M}$ of such a symplectic manifold there are Floer homology groups $HF_*^{\mathrm{symp}}(\mathcal{M}, \phi)$. In this theory the critical points are the fixed points of ϕ and the conecting orbits are pseudoholomorphic curves $u : \mathbf{R}^2 \to \mathcal{M}$ which satisfy $u(s+1,t) = \phi(u(s,t))$ and converge to fixed points $x^\pm$ of ϕ as t tends to $\pm\infty$. The Euler characteristic of $HF_*^{\mathrm{symp}}(\mathcal{M}, \phi)$ is the Lefschetz number of ϕ. The main result of this paper is the following.

Theorem 1.1 *There is a natural isomorphism of Floer homologies*

$$HF_*^{\mathrm{symp}}(\mathcal{M}(P), \phi_f) = HF_*^{\mathrm{inst}}(\Sigma_h; P_f).$$

The proof consists of two steps. The first is an index theorem about the spectral flows of two associated families of self adjoint operators and states that the relative Morse indices agree. The second step is a characterization of holomorphic curves in the moduli space $\mathcal{M}(P)$ as a limit case of self-dual Yang-Mills instantons on the bundle $P_f \times \mathbf{R}$ over the 4-manifold $\Sigma_h \times \mathbf{R}$. In the present paper we outline the main ideas of the proof. Details of the analysis will appear elsewhere.

Thanks to P. Braam, S. Donaldson, J.D.S. Jones, and J. Robbin for helpful discussions.

2 Instanton homology

In this section we discuss Floer's instanton homology for nontrivial SO(3)-bundle. All the theorems are due to Floer [11], [13]. Let M be a compact connected oriented 3-dimensional manifold without boundary and $\pi : Q \to M$ be a principal bundle with structure group $G = \mathrm{SO}(3)$. We identify the space $\mathcal{A}(Q)$ of connections on Q with the space of smooth Lie algebra valued 1-forms $a \in \Omega^1(Q, \mathfrak{g})$ which are equivariant with respect to the adjoint action of G and canonical in the fibres:

$$a_{px}(vx) = x^{-1}a_p(v)x, \qquad a_p(p\xi) = \xi$$

for $v \in T_pQ$, $x \in G$, and $\xi \in \mathfrak{g}$. The space $\mathcal{G}(Q)$ of gauge transformations can be canonically identified with the space smooth maps $g : Q \to G$ which are equivariant under the action of G on itself through inner automorphisms:

$$g(px) = x^{-1}g(p)x.$$

Thus gauge transformations are sections of the adjoint bundle $G_Q = Q \times_{\mathrm{ad}} G$ which consists of equivalence classes of pairs $[p, A]$ where $p \in Q$ and $g \in G$ under the equivalence relation $[p, g] \equiv [px, x^{-1}gx]$ for $x \in G$. With these identifications $\mathcal{G}(Q)$ acts on $\mathcal{A}(Q)$ by the formula

$$g^*a = g^{-1}dg + g^{-1}ag.$$

Let $\mathcal{G}_0(Q)$ denote the component of the identity in $\mathcal{G}(Q)$ and consider the quotient

$$\mathcal{C}(Q) = \mathcal{A}(Q)/\mathcal{G}_0(Q).$$

This space is not an infinite dimensional manifold since the group $\mathcal{G}_0(Q)$ does not act freely on $\mathcal{A}(Q)$. However, almost every connection is a regular point of the action. More precisely, let

$$\mathcal{G}_a = \{g \in \mathcal{G}(Q) : g^*a = a\}$$

denote the isotropy subgroup of a connection $a \in \mathcal{A}(Q)$.

Lemma 2.1 *The isotropy subgroup of a connection $a \in \mathcal{A}(Q)$ is discrete if and only if*

$$\mathcal{G}_a \cap \mathcal{G}_0(Q) = \{1\}.$$

Connections with a discrete isotropy subgroup are called **regular**.

Proof: Suppose that $g \in \mathcal{G}_a \cap \mathcal{G}_0(Q)$ and $g \neq 1$. Then g lifts to a map $\tilde{g} : Q \to SU(2)$ and $\tilde{g}(p) \in SU(2)$ commutes with the holonomy subgroup $\tilde{H} \subset SU(2)$ of a at the point $p \in Q$ lifted to $SU(2)$. Since $g \neq 1$ the matrix $\tilde{g}(p)$ has two distinct eigenvalues and both eigenspaces are invariant under the holonomy group $\tilde{H}$. Thus there exists a circle of matrices in $SU(2)$ commuting with $\tilde{H}$ and therefore $\mathcal{G}_a$ is not discrete. □

The group of components of $\mathcal{G}(Q)$ acts on $\mathcal{C}(Q)$. These components are characterized by two invariants, the **degree**

$$\deg : \mathcal{G}(Q) \to \mathbf{Z}$$

and the **parity**

$$\eta : \mathcal{G}(Q) \to H^1(M; \mathbf{Z}_2).$$

The definition of the degree is based on the next lemma. The proof was personally communicated to the second author by John Jones.

Lemma 2.2 **(1)** *If $w_2(Q) = 0$ then the induced map $\pi_* : H_3(Q, \mathbf{Z}) \to H_3(M, \mathbf{Z})$ is onto.*

(2) *If $w_2(Q) \neq 0$ then the induced map $\pi^* : H^3(M, \mathbf{Z}_2) \to H^3(Q, \mathbf{Z}_2)$ is zero and $H_3(M, \mathbf{Z})/\mathrm{im}\,\pi_* = \mathbf{Z}_2$.*

(3) *For every $g \in \mathcal{G}(Q)$ the induced map $g_* : H_3(Q, \mathbf{Z}) \to H_3(\mathrm{SO}(3), \mathbf{Z})$ descends to a homomorphism $H_3(M, \mathbf{Z}) \to H_3(\mathrm{SO}(3), \mathbf{Z})$.*

Proof: Examine the spectral sequence of the bundle $Q \to M$ with integer coefficients to obtain

$$H_3(Q, \mathbf{Z})/\ker \pi_* \simeq \mathrm{im}\,\pi_* \simeq \ker\Big(d_2 : H_3(M, \mathbf{Z}) \to H_1(M, \mathbf{Z}_2)\Big).$$

Now we also have that the image of the fundamental class $[M]$ under $d_2 : H_3(M, \mathbf{Z}) \to H_1(M, \mathbf{Z}_2)$ is the Poincaré dual of $w_2(Q)$. Hence π_* is onto if and only if $w_2(Q) = 0$ and $\mathrm{im}\,\pi_* = \langle 2[M] \rangle$ otherwise. This proves statements (1) and (2).

Now let $\iota : \mathrm{SO}(3) \to Q$ be the inclusion of a fiber and denote the induced map on homology by $\iota_* : H_3(\mathrm{SO}(3), \mathbf{Z}) \to H_3(Q, \mathbf{Z})$. Then it follows again by examining the spectral sequence that ι_* is injective and

$$\mathrm{im}\,\iota_* \subset \ker \pi_*, \qquad \ker \pi_*/\mathrm{im}\,\iota_* \simeq H_2(M, \mathbf{Z}_2)$$

For every gauge transformation $g : Q \to \mathrm{SO}(3)$ denote $g_* : H_3(Q, \mathbf{Z}) \to H_3(\mathrm{SO}(3), \mathbf{Z})$. The composition $g \circ \iota$ is homotopic to a constant and hence we have $\mathrm{im}\,\iota_* \subset \ker g_*$. This implies that g_* descends to a homomorphism $H_3(Q, \mathbf{Z})/\mathrm{im}\,\iota_* \to H_3(\mathrm{SO}(3), \mathbf{Z}) = \mathbf{Z}$. Any such homomorphism must vanish on the subgroup $\ker \pi_*/\mathrm{im}\,\iota_* \simeq H_2(M, \mathbf{Z}_2)$. Thus we have proved that

$$\ker \pi_* \subset \ker g_*$$

for every gauge transformation $g \in \mathcal{G}(Q)$. In view of statement (2) it suffices to prove that

$$w_2(Q) \neq 0 \qquad \Longrightarrow \qquad \mathrm{im}\,g_* \subset \langle 2[\mathrm{SO}(3)] \rangle$$

for every gauge transformation $g \in \mathcal{G}(Q)$. We prove instead that the dual homomorphism $g^* : H^3(\mathrm{SO}(3), \mathbf{Z}_2) \to H^3(Q, \mathbf{Z}_2)$ is zero. To see this consider the generator $\alpha \in H^1(\mathrm{SO}(3), \mathbf{Z}_2)$. Since $\iota^* g^* = 0$ it follows that $g^*\alpha = \pi^*\beta$ for some $\beta \in H^1(M, \mathbf{Z}_2)$. Hence $g^*(\alpha^3) = \pi^*(\beta^3) = 0$ where the last assertion follows from statement (2). This proves the lemma. □

For $g \in \mathcal{G}(Q)$ the induced homomorphism $H_3(M, \mathbf{Z}) = \mathbf{Z} \to H_3(G, \mathbf{Z}) = \mathbf{Z}$ is determined by an integer $\deg(g)$ called the **degree** of g. Alternatively, the degree can be defined as the intersection number of the submanifolds $\mathrm{graph}\,(g) = \{[p, g(p)] : p \in Q\}$ and $\mathrm{graph}\,(\mathbb{1})$ of the adjoint bundle G_Q.

The homomorphism $g_* : \pi_1(Q) \to \pi_1(G) = \mathbf{Z}_2$ descends to a homomorphism $\eta(g) : \pi_1(M) \to \mathbf{Z}_2$, called the **parity** of g. It is the obstruction for g to lift to a map $\tilde{g} : Q \to \mathrm{SU}(2)$. A gauge transformation is called **even** if $\eta(g) = 0$. Every even gauge transformation is of even degree but not vice versa. Moreover, the map $\eta : \mathcal{G}(Q) \to H^1(M, \mathbf{Z}_2)$ is always onto. Throughout we shall assume the following hypothesis.

Hypothesis (H1) *Every cohomology class $\eta \in H^1(M; \mathbf{Z}_2)$ can be represented by finitely many embedded oriented Riemann surfaces. Moreover $w_2(Q) \neq 0$.*

If M is orientable then every one dimensional integral cohomology class can be represented by finitely many embedded oriented Riemann surfaces. So the first part of hypothesis (H1) will be satisfied whenever M is orientable and there is no 2-torsion in H_1. Also note that (H1) implies the following weaker hypothesis.

Hypothesis (H2) *There exists an embedding $\iota : \Sigma \to M$ of a Riemann surface such that $\iota^* Q$ is the nontrivial $SO(3)$-bundle over Σ.*

Some important consequences of hypotheses (H1) and (H2) are summarized in the next lemma. The proof will be given in the appendix.

Lemma 2.3 **(1)** *Two gauge transformations are homotopic if and only if they have the same degree and the same parity.*

(2) *Assume (H1). Then for every $g \in \mathcal{G}(Q)$*

$$\deg(g) \equiv w_2(Q) \cdot \eta(g) \pmod 2. \tag{1}$$

Conversely, for every integer k and every $\eta \in H^1(M; \mathbf{Z}_2)$ with $k \equiv w_2(Q) \cdot \eta \pmod 2$ there exists a gauge transformation with $\deg(g) = k$ and $\eta(g) = \eta$.

(3) *Assume (H2). Then there exists a gauge transformation g of degree* 1.

Remark 2.4 **(i)** If $Q = \mathbf{R}P^3 \times SO(3)$ then there exists a gauge transformation of degree 1. However, for such a gauge transformation equation (1) is violated since $w_2(Q) = 0$. If M is a homology 3-sphere and $Q = M \times \mathrm{SO}(3)$ is the product bundle then equation (1) is trivially satisfied but there is no gauge transformation of degree 1.

(ii) Another interesting example is the (unique) nontrivial SO(3)-bundle $\pi : Q \to \mathbf{R}P^3$. This bundle does not satisfy hypothesis (H2). It can be represented in the form $Q = S^3 \times_{\mathbf{Z}_2} \mathrm{SO}(3)$ with $(x, A) \sim (-x, RA)$ where $R \in \mathrm{SO}(3)$ is a reflection, i.e. $R^2 = \mathbb{1}$. A gauge transformation of this bundle is a smooth map $g : S^3 \to \mathrm{SO}(3)$ such that $g(-x) = Rg(x)R$. Any such map lifts to a smooth map $\tilde{g} : S^3 \to \mathrm{SU}(2)$ and the degree of g in the above sense is the degree of the lift $\tilde{g}$ (or half the degree of g as a map between oriented 3-manifolds). The lift $\tilde{g}$ is necessarily of even degree. So in this case there is no gauge transformation of degree 1. Hence statements (2) and (3) of Lemma 2.3 are both violated.

The space $\mathcal{A}(Q)$ of connections on Q is an affine space whose associated vector space is $\Omega^1(\mathfrak{g}_Q)$. Here $\mathfrak{g}_Q = Q \times_G \mathfrak{g}$ is the vector bundle over M associated to Q via the adjoint representation of G on its Lie algebra. We think of an infinitesimal connection $\alpha \in \Omega^1(\mathfrak{g}_Q)$ as an invariant and horizontal 1-form on Q. The Lie algebra of $\mathcal{G}(Q)$ is the space of invariant $\mathfrak{g}$-valued functions on Q and so may be identified with $\Omega^0(\mathfrak{g}_Q)$. The infinitesimal action of $\mathcal{G}(Q)$ is given by the covariant derivative

$$d_a : \Omega^0(\mathfrak{g}_Q) \to \Omega^1(\mathfrak{g}_Q), \qquad d_a\eta = d\eta + [a \wedge \eta].$$

Thus the tangent space to the configuration space $\mathcal{C}(Q)$ at a regular connection a is the quotient $\Omega^1(\mathfrak{g}_Q)/\operatorname{im} d_a$.

The curvature of a connection is the 2-form

$$F_a = da + \frac{1}{2}[a \wedge a] \in \Omega^2(M; \mathfrak{g}_Q)$$

and determines a natural 1-form on the space of connections via the linear functional

$$\alpha \quad \mapsto \quad \int_M \langle F_a \wedge \alpha \rangle$$

on $T_a\mathcal{A}(Q) = \Omega^1(\mathfrak{g}_Q)$. Here $\langle \, , \rangle$ denotes the invariant inner product on the Lie algebra $\mathfrak{g}$ given by minus the Killing form or 4 times the trace. We shall denote this 1-form by F. The Bianchi identity asserts that F is closed. Since the affine space $\mathcal{A}(Q)$ is contractible this implies that F is the differential of a function. Integrating F along a path which starts at a fixed flat connection $a_0 \in \mathcal{A}(Q)$ we obtain the Chern-Simons functional

$$\mathcal{CS}(a_0 + \alpha) = \frac{1}{2}\int_M \left(\langle d_{a_0}\alpha \wedge \alpha \rangle + \frac{1}{3}\langle [\alpha \wedge \alpha] \wedge \alpha \rangle \right)$$

for $\alpha \in \Omega^1(\mathfrak{g}_Q)$. One can check directly that

$$d\mathcal{CS}(a)\alpha = \int_M \langle F_a \wedge \alpha \rangle$$

for $a \in \mathcal{A}(Q)$ and $\alpha \in \Omega^1(\mathfrak{g}_Q)$. Thus the flat connections on Q appear as the critical points of the Chern-Simons functional. Since the 1-form $d\mathcal{CS} = F$ is invariant and horizontal it follows that the difference $\mathcal{CS}(g^*a) - \mathcal{CS}(a)$ is independent of the connection a and is locally independent of the gauge transformation g. So it depends only on the component of $\mathcal{G}(Q)$ and it turns out that

$$\mathcal{CS}(a) - \mathcal{CS}(g^*a) = 8\pi^2 \deg(g). \tag{2}$$

Hence as a function on the quotient $\mathcal{A}(Q)/\mathcal{G}(Q)$ the Chern-Simons functional takes values in S^1. Note that this function is only well defined up to an additive constant which we have chosen such that $\mathcal{CS}(a_0) = 0$.

Denote the space of flat connections by

$$\mathcal{A}_{\text{flat}}(Q) = \{a \in \mathcal{A}(Q) : F_a = 0\}.$$

For every flat connection $a \in \mathcal{A}_{\text{flat}}(Q)$ there is a chain complex

$$\Omega^0(\mathfrak{g}_Q) \xrightarrow{d_A} \Omega^1(\mathfrak{g}_Q) \xrightarrow{d_A} \Omega^2(\mathfrak{g}_Q) \xrightarrow{d_A} \Omega^3(\mathfrak{g}_Q).$$

with associated cohomology groups $H^j_a(M)$. These cohomology groups can be identified with the spaces of harmonic forms

$$H^j_a(M) = \ker d_a \cap \ker d_a^*$$

where d_a^* denotes the L^2-adjoint of $d_a : \Omega^{j-1} \to \Omega^j$ with respect to a Riemannian metric on M. A flat connection is called **nondegenerate** if $H^1_a(M) = 0$. This is consistent with the Chern-Simons point of view since the Hessian of $\mathcal{CS}$ at a critical point $a \in \mathcal{A}_{\text{flat}}(Q)$ is given by $*d_a$ and should be viewed as an operator on the quotient $\Omega^1(\mathfrak{g}_Q)/\text{im}d_a$. Here $* : \Omega^j \to \Omega^{3-j}$ denotes the Hodge-$*$-operator with respect to the Riemannian metric on M and the invariant inner product on $\mathfrak{g}$. So the flat connection a is nondegenerate if and only if the Hessian of $\mathcal{CS}$ at a is invertible. Note that a flat connection a is both regular and nondegenerate if and only if the extended Hessian

$$D_a = \begin{pmatrix} *d_a & d_a \\ d_a^* & 0 \end{pmatrix}$$

is nonsingular. Here D_a is a selfadjoint operator on $\Omega^1(\mathfrak{g}_Q) \oplus \Omega^0(\mathfrak{g}_Q)$.

Lemma 2.5 *If (H2) is satisfied then every flat connection on Q is regular.*

Proof: By (H2) there exists an embedding $\iota : \Sigma \to M$ of an oriented Riemann surface such that $w_2(\iota^*Q) \neq 0$. Let $a \in \mathcal{A}_{\text{flat}}(Q)$ and $g \in \mathcal{G}_a \cap \mathcal{G}_0(Q)$. Then $\iota^*a \in \mathcal{A}_{\text{flat}}(\iota^*Q)$ and $\iota^*g \in \mathcal{G}_{\iota^*a} \cap \mathcal{G}_0(\iota^*Q)$. By Lemma 4.1 below $\iota^*g = \mathbb{1}$ and hence $g = \mathbb{1}$. □

Remark 2.6 It is easy to construct regular flat connections with $\mathcal{G}_a \neq \{\mathbb{1}\}$. The holonomy of such a connection is conjugate to the abelian subgroup of diagonal matrices in $SO(3)$ with diagonal entries ± 1

Via the holonomy the flat connections on Q correspond naturally to representations

$$\rho : \pi_1(M) \to \text{SO}(3).$$

The second Stiefel-Whitney class $w_2(Q) \in H^2(M;\mathbf{Z}_2)$ appears as the cohomology class

$$w_\rho \in H^2(\pi_1(M), \mathbf{Z}_2)$$

associated to ρ as follows. Choose any map $\tilde{\rho} : \pi_1(M) \to \mathrm{SU}(2)$ which lifts ρ and define $w_\rho(\gamma_1, \gamma_2) = \tilde{\rho}(\gamma_1\gamma_2)\tilde{\rho}(\gamma_2)^{-1}\tilde{\rho}(\gamma_1)^{-1} = \pm 1$. Then w_ρ is a cocycle:

$$\partial w_\rho(\gamma_1, \gamma_2, \gamma_3) = w_\rho(\gamma_2, \gamma_3)w_\rho(\gamma_1\gamma_2, \gamma_3)w_\rho(\gamma_1, \gamma_2\gamma_3)w_\rho(\gamma_1, \gamma_2) = 1.$$

The coboundaries are functions of the form

$$\partial f(\gamma_1, \gamma_2) = f(\gamma_1)f(\gamma_1\gamma_2)f(\gamma_2)$$

for some map $f : \pi_1(M) \to \{\pm 1\}$. Hence w_ρ is a coboundary if and only if the lift $\tilde{\rho} : \pi_1(M) \to \mathrm{SU}(2)$ can be chosen to be a homomorphism. The cohomology class of w_ρ is independent of the choice of the lift $\tilde{\rho}$. It determines the second Stiefel-Whitney class of the bundle Q via the natural homomorphism $\iota : H^2(\pi_1(M), \mathbf{Z}_2) \to H^2(M, \mathbf{Z}_2)$.

The Casson invariant $\lambda(M;Q)$ of the Manifold M (with respect to the bundle Q) can roughly be defined as "half the number of representations $\rho : \pi_1(M) \to SO(3)$ with $\iota(w_\rho) = w_2(Q)$" or "half the number of zeros of F". Here the flat connections are to be counted modulo even gauge equivalence and with appropriate signs. This is analogous to the Euler number of a vector field on a finite dimensional manifold. As in finite dimensions we will in general have to perturb the 1-form F to ensure finitely many nondegenerate zeros. In [11] Floer discovered a refinement of the Casson invariant which is called instanton homology. These homology groups result from Floer's new approach to infinite dimensional Morse theory applied to the Chern-Simons functional. The Casson invariant appears as the Euler characteristic of instanton homology.

The L^2-gradient of the Chern-Simons functional with respect to the Riemannian metric on M and the invariant inner product on $\mathfrak{g}$ is given by $\mathrm{grad}\,\mathcal{CS}(a) = *F_a \in \Omega^1(\mathfrak{g}_Q)$. Thus a gradient flow line of the Chern-Simons functional is a smooth 1-parameter family of connections $a(t) \in \mathcal{A}(Q)$ satisfying the nonlinear partial differential equation

$$\dot{a} + *F_a = 0. \tag{3}$$

The path $a(t)$ of connections on Q can also be viewed as a connection on the bundle $Q \times \mathbf{R}$ over the 4-manifold $M \times \mathbf{R}$. In this interpretation (3) is precisely the self-duality equation with respect to the product metric on $M \times \mathbf{R}$. Moreover, the Yang-Mills functional agrees with the flow energy

$$\mathcal{YM}(a) = \tfrac{1}{2}\int_{-\infty}^{\infty} \left(\|\dot{a}\|^2 + \|F_a\|^2\right) dt.$$

The key obstacle for Morse theory in this context is that equation (3) does not define a well posed initial value problem and the Morse index of every critical point is infinite. In [11] Floer overcame this difficulty by studying only the space of bounded solutions of (3) and constructing a chain complex as was done by Witten [30] in finite dimensional Morse theory (see also [24]). In order to describe how this works we make another assumption on the bundle Q.

Hypothesis (H3) Every flat connection on Q is nondegenerate.

This condition will in general not be satisfied. If there are degenerate flat connections then we perturb the Chern-Simons functional in order to ensure nondegenerate critical points for the perturbed functional.

If (H3) is satisfied then for every smooth solution $a(t) \in \mathcal{A}(Q)$ of (3) with finite Yang-Mills action there exist flat connections $a^{\pm} \in \mathcal{A}_{\text{flat}}(Q)$ such that $a(t)$ converges exponentially with all derivatives to $a^{\pm}$ as t tends to $\pm\infty$ (see for example [18]). Conversely, if $a(t)$ is a solution of (3) for which these limits exist then the Yang-Mills action is finite:

$$\mathcal{YM}(a) = \mathcal{CS}(a^-) - \mathcal{CS}(a^+).$$

Fix two flat connections $a^{\pm} \in \mathcal{A}_{\text{flat}}(Q)$ and consider the space of those solutions $a(t) \in \mathcal{A}(Q)$ of (3) which also satisfy

$$\lim_{t \to \pm\infty} a(t) = g_{\pm}^* a^{\pm}, \qquad g_{\pm} \in \mathcal{G}_0(Q). \tag{4}$$

These solutions are usually termed **instantons** or **connecting orbits** from a^- to a^+. The moduli space of these instantons is denoted by

$$\mathcal{M}(a^-, a^+) = \frac{\{a(t) \in \mathcal{A}(Q) : (3), (4)\}}{\mathcal{G}_0(Q)}.$$

The next theorem due to Floer [11] summarizes some key properties of these moduli spaces.

Theorem 2.7 *Assume (H2) and (H3). For a generic metric on M the moduli space $\mathcal{M}(a^-, a^+)$ is a finite dimensional oriented paracompact manifold for every pair of flat connections $a^{\pm} \in \mathcal{A}_{\text{flat}}(Q)$. There exists a function $\mu : \mathcal{A}_{\text{flat}}(Q) \to \mathbf{Z}$ such that*

$$\dim \mathcal{M}(a^-, a^+) = \mu(a^-) - \mu(a^+).$$

This function μ satisfies

$$\mu(a) - \mu(g^* a) = 4 \deg(g)$$

for $a \in \mathcal{A}_{\text{flat}}(Q)$ and $g \in \mathcal{G}(Q)$.

In our context the integer $\mu(a)$ plays the same role as the Morse index does in finite dimensional Morse theory. The number $\mu(a^-)-\mu(a^+)$ is given by the spectral flow [3] of the operator family $D_{a(t)}$ as the connection $a(t) \in \mathcal{A}(Q)$ runs from a^- to a^+. So the function $\mu : \mathcal{A}_{\text{flat}}(Q) \to \mathbf{Z}$ is only defined up to an additive constant. We will choose this constant such that $\mu(a_0) = 0$ where $a_0 \in \mathcal{A}_{\text{flat}}(Q)$ is a fixed flat connection.

To construct the instanton homology groups we can now proceed as in finite dimensional Morse theory [30]. The key idea is to construct a chain complex over the flat connections and to use the instantons to construct a boundary operator. For simplicity of the exposition we restrict ourselves to coefficients in $\mathbf{Z}_2$.

Let C be the vector space over $\mathbf{Z}_2$ generated by the flat connections modulo gauge equivalence. For now we divide only by the component of the identity $\mathcal{G}_0(Q)$. This vector space is graded by μ. It follows from Uhlenbeck's compactness theorem [28] that the space

$$C_k = \bigoplus_{\substack{a\in\mathcal{A}_{\text{flat}}(Q)/\mathcal{G}_0(Q)\\ \mu(a)=k}} \mathbf{Z}_2\langle a\rangle$$

is finite dimensional for every integer k. It follows also from Uhlenbeck's compactness theorem that the moduli space $\mathcal{M}(a^-, a^+)$ consists of finitely many instantons (modulo time shift) whenever

$$\mu(a^-) - \mu(a^+) = 1.$$

Let $n_2(a^-, a^+)$ denote the number of such instantons, counted modulo 2. These numbers determine a linear map $\partial : C_{k+1} \to C_k$ defined by

$$\partial\langle b\rangle = \sum_{\mu(a)=k} n_2(b,a)\langle a\rangle$$

for $b \in \mathcal{A}_{\text{flat}}(Q)$ with $\mu(b) = k+1$. In [11] Floer proved that (C, ∂) is a chain complex and that its homology is an invariant of the bundle $Q \to M$.

Theorem 2.8 *Assume (H2) and (H3).*

(1) *The above map $\partial : C \to C$ satisfies $\partial^2 = 0$. The associated homology groups*

$$HF_k^{\text{inst}}(M;Q) = \frac{\ker \partial_{k-1}}{\text{im}\partial_k}$$

are called the **Floer homology groups** *of the pair (M, Q).*

(2) *The Floer homology groups $HF_k^{\text{inst}}(M;Q)$ are independent of the metric on M used to construct them.*

To prove $\partial^2 = 0$ we must show that

$$\sum_{\substack{b\in\mathcal{A}_{\mathrm{flat}}(Q)/\mathcal{G}_0(Q)\\ \mu(b)=k}} n_2(c,b)n_2(b,a) \quad \in 2\mathbf{Z}$$

whenever $\mu(a) = k-1$ and $\mu(c) = k+1$. This involves a glueing argument for pairs of instantons running from c to b and from b to a. Such a pair gives rise to a (unique) 1-parameter family of instantons running from c to a. Now the space of instantons running from c to a (modulo time shift) is a 1-dimensional manifold and has therefore an even number of ends. This implies $\partial^2 = 0$. That any two instanton homology groups corresponding to different metrics are naturally isomorphic can be proved along similar lines.

Remark 2.9 (i) The Floer homology groups $HF_*^{\mathrm{inst}}(M;Q)$ are new invariants of the 3-manifold M which cannot be derived from the classical invariants of differential topology.

(ii) The Floer homology groups are graded modulo 4. From Lemma 2.3 it follows that there exists a gauge transformation $g \in \mathcal{G}(Q)$ of degree 1 and by Theorem 2.7 the map $a \mapsto g^*a$ induces an isomorphism

$$HF_{k+4}^{\mathrm{inst}}(M;Q) = HF_k^{\mathrm{inst}}(M;Q).$$

Note however that the even gauge transformations $g \in \mathcal{G}^{\mathrm{ev}}(Q)$ only give rise to a grading modulo 8. So the Euler characteristic of $HF_*^{\mathrm{inst}}(M;Q)$ is the number

$$\begin{aligned}\chi(HF_*^{\mathrm{inst}}(M;Q)) &= \sum_{k=0}^{3}(-1)^k \dim HF_k^{\mathrm{inst}}(M;Q)\\ &= \frac{1}{2}\sum_{a\in\mathcal{A}_{\mathrm{flat}}(Q)/\mathcal{G}^{\mathrm{ev}}(Q)} (-1)^{\mu(a)}.\end{aligned}$$

This is the **Casson invariant** $\lambda(M;Q)$.

(iii) By Lemma 2.3 the group of components of the space of degree-0 gauge transformations $\{g \in \mathcal{G}(Q) : \deg(g) = 0\}$ is isomorphic to the finite group

$$\Gamma = \{\eta(g) : g \in \mathcal{G}(Q), \deg(g) = 0\} \subset H^1(M;\mathbf{Z}_2).$$

This group acts on $HF_k^{\mathrm{inst}}(M;Q)$ for every k through permutations of the canonical basis. By Remark 2.6 the group Γ does not act freely. The above definition of the Casson invariant ignores this action of Γ.

(iv) The same construction works over the integers. For this we must assign a number $+1$ or -1 to each instanton running from a^- to a^+ whenever $\mu(a^-) - \mu(a^+) = 1$. This involves a consistent choice of orientations for the Moduli spaces $\mathcal{M}(a^-, a^+)$.

(v) If Hypothesis (H3) is violated then the Floer homology groups can still be defined. The construction then requires a suitable perturbation of the Chern-Simons functional. This will be discussed in Section 7.

(vi) It represents a much more serious problem if Hypothesis (H2) fails. This is because of the presence of flat connections which are not regular. Such connections cannot be removed by a gauge invariant perturbation.

(vii) If Q is the trivial $SO(3)$-bundle over a homology-3-sphere M then the only flat connection which is not regular is the trivial connection. In this case the difficulty mentioned in (vi) does not arise. This is in fact the context of Floer's original work on instanton homology [11].

3 Floer homology for symplectic fixed points

In [12] Floer developed a similar theory for fixed points of symplectomorphisms. In his original work Floer assumed that the symplectomorphism ϕ is exact and he proved that the Floer homology groups $HF_*^{\mathrm{symp}}(\mathcal{M},\phi)$ are isomorphic to the homology of the underlying symplectic manifold $(\mathcal{M},\omega)$. Like Floer we assume that $\mathcal{M}$ is compact and monotone. This means that the first Chern class c_1 of $T\mathcal{M}$ agrees over $\pi_2(\mathcal{M})$ with the cohomology class of the symplectic form ω. Unlike Floer we assume in addition that $\mathcal{M}$ is simply connected but we do not require the symplectomorphism $\phi : \mathcal{M} \to \mathcal{M}$ to be isotopic to the identity. Then there are Floer homology groups of ϕ whose Euler characteristic is the Lefschetz number. Here is how this works.

The fixed points of ϕ can be represented as the critical points of a function on the space of smooth paths

$$\Omega_\phi = \{\gamma : \mathbf{R} \to \mathcal{M} : \gamma(s+1) = \phi(\gamma(s))\}.$$

The tangent space to Ω_ϕ at γ is the space of vector fields $\xi(s) \in T_{\gamma(s)}\mathcal{M}$ along γ such that $\xi(s+1) = d\phi(\gamma(s))\xi(s)$. The space Ω_ϕ carries a natural 1-form

$$T_\gamma\Omega_\phi \to \mathbf{R} : \xi \mapsto -\int_\gamma \iota(\xi)\omega.$$

This 1-form is closed but not exact since Ω_ϕ is not simply connected. Since $\mathcal{M}$ is simply connected the fundamental group of Ω_ϕ is $\pi_1(\Omega_\phi) = \pi_2(\mathcal{M})$. The universal cover $\tilde{\Omega}_\phi$ can be explicitly represented as the space of homotopy classes of smooth maps $u : \mathbf{R} \times I \to \mathcal{M}$ such that

$$u(s,0) = x_0, \qquad u(s+1,t) = \phi(u(s,t)), \qquad u(s,1) = \gamma(s).$$

Here $x_0 = \phi(x_0)$ is a reference point chosen for convenience of the notation. The homotopy class $[u]$ is to be understood subject to the boundary condition at $t = 0$ and $t = 1$. The second homotopy group $\pi_2(\mathcal{M})$ acts on $\tilde{\Omega}_\phi$ by taking connected sums. A sphere in $\mathcal{M}$ can be represented by a function $v : \mathbf{R} \times I \to \mathcal{M}$ such that $v(s,0) = v(s,1) = v(0,t) = x_0$ and $v(s+1,t) = v(s,t)$ for $s,t \in \mathbf{R}$. The connected sum of $[u] \in \widetilde{\Omega}_\phi$ and $[v] \in \pi_2(\mathcal{M})$ is given by $u\#v(s,t) = v(2t,s)$ for $t \leq 1/2$ and $u\#v(s,t) = u(2t-1,s)$ for $t \geq 1/2$.

The pullback of the above 1-form on Ω_ϕ is the differential of the function

$$\tilde{a}_\phi : \widetilde{\Omega}_\phi \to \mathbf{R}, \qquad \tilde{a}_\phi(u) = \int u^*\omega$$

called the **symplectic action.** This function satisfies

$$\tilde{a}_\phi(u\#v) = \tilde{a}_\phi(u) + \int_{S^2} v^*\omega$$

for every sphere $v : S^2 \to \mathcal{M}$. Since $\mathcal{M}$ is monotone the symplectic form ω is integral and hence $\tilde{a}_\phi$ descends to a map $a_\phi : \Omega_\phi \to \mathbf{R}/\mathbf{Z}$. By construction the differential of a_ϕ is the above 1-form on Ω_ϕ. So the critical points of a_ϕ are the constant paths in Ω_ϕ and hence the fixed points of ϕ.

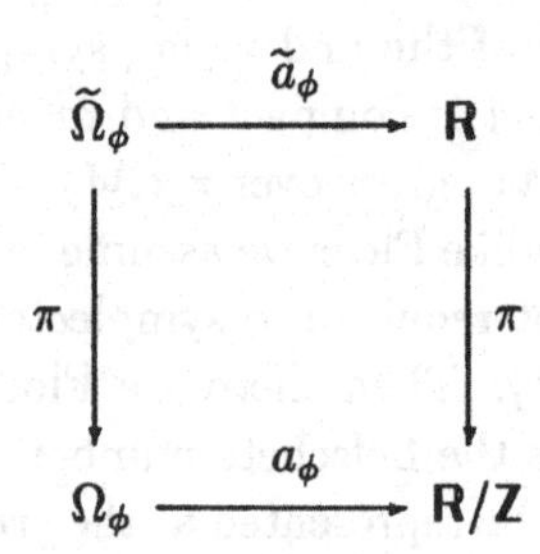

An almost complex structure $J : T\mathcal{M} \to T\mathcal{M}$ is said to be **compatible** with ω if the bilinear form

$$\langle \xi, \eta \rangle = \omega(\xi, J\eta)$$

defines a Riemannian metric on $\mathcal{M}$. The space $\mathcal{J}(\mathcal{M},\omega)$ of such almost complex structures is contractible. A symplectomorphism ϕ acts on $\mathcal{J}(\mathcal{M},\omega)$ by pullback $J \mapsto \phi^*J$. To construct a metric on Ω_ϕ choose a smooth family $J_s \in \mathcal{J}(\mathcal{M},\omega)$ such that

$$J_s = \phi^* J_{s+1}. \tag{5}$$

This condition guarantees that for any two vectorfields $\xi, \eta \in T_\gamma\Omega_\phi$ the expression $\langle \xi(s), \eta(s) \rangle_s = \omega(\xi(s), J_s(\gamma(s))\eta(s))$ is of period 1 in s. Hence define the inner product of ξ and η by

$$\langle \xi, \eta \rangle = \int_0^1 \langle \xi(s), \eta(s) \rangle_s \, ds.$$

The gradient of a_ϕ with respect to this metric on Ω_ϕ is given by $\operatorname{grad} a_\phi(\gamma) = J_s(\gamma)\dot{\gamma}$. Hence a gradient flow line of a_ϕ is a smooth map $u : \mathbf{R}^2 \to \mathcal{M}$ satisfying the nonlinear partial differential equation

$$\frac{\partial u}{\partial t} + J_s(u)\frac{\partial u}{\partial s} = 0 \tag{6}$$

and the periodicity condition

$$u(s+1,t) = \phi(u(s,t)). \tag{7}$$

Condition (5) guarantees that whenever $u(s,t)$ is a solution of (6) then so is $v(s,t) = \phi(u(s-1,t))$. Condition (7) requires that these two solutions agree.

As in section 2 equations (6) and (7) do not define a well posed Cauchy problem and the Morse index of any critical point is infinite. However, the solutions of (6) are precisely Gromov's pseudoholomorphic curves [16]. It follows from Gromov's compactness that every solution of (6) and (7) with finite energy

$$E(u) = \frac{1}{2}\int_{-\infty}^{\infty}\int_0^1 \left(|\partial_s u|_s^2 + |\partial_t u|_s^2\right) dsdt < \infty$$

has limits

$$\lim_{t\to\pm\infty} u(s,t) = x^\pm = \phi(x^\pm). \tag{8}$$

(See for example [24], [31].) Conversely, any solution of (6), (7), and (8) has finite energy. Given any two fixed points $x^\pm$ let

$$\mathcal{M}(x^-,x^+)$$

denote the space of these solutions.

For any function $u : \mathbf{R}^2 \to \mathcal{M}$ which satisfies (7) and (8) we introduce the Maslov index as follows. Let $\Phi(s,t) : \mathbf{R}^{2n} \to T_{u(s,t)}\mathcal{M}$ be a trivialization of $u^*T\mathcal{M}$ as a symplectic vector bundle such that

$$\Phi(s+1,t) = d\phi(u(s,t))\Phi(s,t).$$

Consider the paths of symplectic matrices $\Psi^\pm(s) = \Phi(s,\pm\infty)^{-1}\Phi(0,\pm\infty)$. These satisfy $\Psi^\pm(0) = \mathbb{1}$ and $\Psi^\pm(1)$ is conjugate to $d\phi(x^\pm)$. In particular 1 is not an eigenvalue of $\Psi^\pm(1)$. The homotopy class of such a path is determined by its Maslov index $\mu(\Psi^\pm) \in \mathbf{Z}$ introduced by Conley and Zehnder [5]. Roughly speaking the Maslov index counts the number of times s such that 1 is an eigenvalue of the symplectic matrix $\Psi^\pm(s)$. The integer

$$\mu(u) = \mu(\Psi^-) - \mu(\Psi^+)$$

is independent of the choice of the trivialization and is called the **Maslov index** of u. This number depends only on the homotopy class of u. For a detailed account of the Maslov index and its role in Floer homology we refer to [25]. The next theorem is due to Floer [12].

Lemma 3.1 *If u satisfies (7) and (8) then*

$$\mu(u\#v) = \mu(u) + 2\int_{S^2} v^* c_1$$

for any sphere $v : S^2 \to \mathcal{M}$.

Theorem 3.2 *For a generic family of almost complex structures satisfying (5) the space $\mathcal{M}(x^-, x^+)$ is a finite dimensional manifold for every pair of fixed points $x^\pm = \phi(x^\pm)$. The dimension of $\mathcal{M}(x^-, x^+)$ is given by the Maslov index*

$$\dim_u \mathcal{M}(x^-, x^+) = \mu(u)$$

locally near $u \in \mathcal{M}(x^-, x^+)$.

By Lemma 3.1 the Maslov index induces a map

$$\mu : \mathrm{Fix}(\phi) \to \mathbf{Z}_{2N}$$

such that $\mu(u) = \mu(x^-) - \mu(x^+) \pmod{2N}$ for every solution u of (7) and (8). Here the integer N is defined by $c_1(\pi_2(\mathcal{M})) = N\mathbf{Z}$. This function $\mu : \mathrm{Fix}(\phi) \to \mathbf{Z}_{2N}$ is only defined up to an additive constant. We may choose this constant such that

$$(-1)^{\mu(x)} = \mathrm{sign}\det(\mathbb{1} - d\phi(x)) \tag{9}$$

for every fixed point $x = \phi(x)$. The manifold $\mathcal{M}(x^-, x^+)$ is not connected and the dimension depends on the component. It follows from Theorem 3.2 that the dimension is well defined modulo $2N$

$$\dim \mathcal{M}(x^-, x^+) = \mu(x^-) - \mu(x^+) \pmod{2N}.$$

The Floer homology groups of ϕ can now be constructed as follows. Let C be the vector space over $\mathbf{Z}_2$ freely generated by the fixed points of ϕ. This vector space is graded modulo $2N$ by the Maslov index. It is convenient to define

$$C_k = \bigoplus_{\substack{x=\phi(x) \\ \mu(x)=k \pmod{2N}}} \mathbf{Z}_2\langle x\rangle$$

for every integer k keeping in mind that $C_{k+2N} = C_k$. Since $\mathcal{M}$ is monotone it follows from Gromov's compactness for pseudoholomorphic curves that the 1-dimensional part of the space $\mathcal{M}(x^-, x^+)$ consists of finitely many connecting orbits (modulo time shift) whenever $\mu(x^-) - \mu(x^+) = 1 \pmod{2N}$. Let $n_2(x^-, x^+)$ be the number of these connecting orbits modulo 2. This gives a linear map $\partial : C_{k+1} \to C_k$ via the formula

$$\partial\langle y\rangle = \sum_{\substack{x=\phi(x) \\ \mu(x)=k \pmod{2N}}} n_2(y, x)\langle x\rangle$$

for $y = \phi(y)$ with $\mu(y) = k + 1 \pmod{2N}$. In [12] Floer proved that (C, ∂) is a chain complex.

Theorem 3.3 *Assume that $\mathcal{M}$ is simply connected and monotone.*

(1) *The above map $\partial : C \to C$ satisfies $\partial^2 = 0$. The associated homology groups*

$$HF_k^{\mathrm{symp}}(\mathcal{M};\phi) = \frac{\ker \partial_{k-1}}{\mathrm{im}\partial_k}$$

are called the **Floer homology groups** *of the pair $(\mathcal{M},\phi)$.*

(2) *The Floer homology groups $HF_k^{\mathrm{symp}}(\mathcal{M},\phi)$ are independent of the almost complex structure J used to construct them; they depend on ϕ only up to symplectic isotopy.*

(3) *If ϕ is isotopic to the identity in the class of symplectomorphisms then the Floer homology groups of ϕ are naturally isomorphic to the homology of the manifold $\mathcal{M}$*

$$HF_k^{\mathrm{symp}}(\mathcal{M};\phi) = \bigoplus_{j=k(\mathrm{mod}\ 2N)} H_j(\mathcal{M};\mathbf{Z}_2).$$

Proof of (2): To prove that $HF_*^{\mathrm{symp}}(\mathcal{M};\phi)$ depends only on the isotopy class of ϕ we must generalize the above construction. Let

$$\phi_s = \psi_s^{-1} \circ \phi$$

be an isotopy of symplectomorphisms from $\phi_0 = \phi$ to ϕ_1. Since $\mathcal{M}$ is simply connected there exists a family of Hamiltonian vector fields $X_s : \mathcal{M} \to T\mathcal{M}$ such that

$$\frac{d}{ds}\psi_s = X_s \circ \psi_s.$$

This means that the 1-form obtained by contracting X_s with the symplectic form ω is exact

$$\iota(X_s)\omega = dH_s.$$

The isotopy can be chosen such that

$$\psi_{s+1} \circ \phi_1 = \phi_0 \circ \psi_s$$

or equivalently

$$\phi^* X_{s+1} = X_s, \qquad H_{s+1} \circ \phi = H_s. \tag{10}$$

Now replace equation (6) by

$$\frac{\partial u}{\partial t} + J_s(u)\frac{\partial u}{\partial s} - \nabla H_s(u) = 0. \tag{11}$$

Here $\nabla H_s = J_s X_s$ is the gradient of the Hamiltonian function $H_s : \mathcal{M} \to \mathbf{R}$ with respect to the metric induced by J_s. It follows from (5) that whenever $u(s,t)$ is a solution of (11) then so is $v(s,t) = \phi(u(s-1,t))$. Hence the periodicity condition (7) is consistent with (11). One can prove by exactly the same arguments as in [12] that $HF_*^{\text{symp}}(\mathcal{M};\phi)$ is isomorphic to the Floer homology constructed with (11) instead of (6). We denote these Floer homology groups by $HF_*^{\text{symp}}(\mathcal{M};\phi,H)$.

The stationary points of equation (11) (that is $\partial u/\partial t = 0$) with the boundary condition (7) are the paths $\gamma \in \Omega_\phi$ for which $\gamma(s) = \psi_s(\gamma(0))$. These paths are in one-to-one correspondence with the fixed points of $\phi_1 = \psi_1^{-1}\phi$. Now let $u : \mathbf{R}^2 \to \mathcal{M}$ be any solution of (11) and (7) and define

$$v(s,t) = \psi_s^{-1}(u(s,t)), \qquad I_s = \psi_s^* J_s.$$

Then ${\phi_1}^* I_{s+1} = I_s$ and $v(s,t)$ satisfies the partial differential equation

$$\frac{\partial v}{\partial t} + I_s(v)\frac{\partial v}{\partial s} = 0 \tag{12}$$

and the periodicity condition

$$v(s+1,t) = \phi_1(v(s,t)). \tag{13}$$

Hence there is a one-to-one correspondence between the solutions of (11) and (7) on the one hand and the solutions of (12) and (13) on the other hand. This shows that $HF_*^{\text{symp}}(\mathcal{M};\phi,H)$ is isomorphic to $HF_*^{\text{symp}}(\mathcal{M};\phi_1)$.

□

Remark 3.4 (i) A similar construction works for monotone symplectic manifolds $\mathcal{M}$ which are not simply connected. In this case Ω_ϕ will no longer be connected and there are Floer homology groups for every homotopy class of paths. Moreover, the fundamental group of Ω_ϕ will no longer be isomorphic to $\pi_2(\mathcal{M})$ and a_ϕ may not take values in S^1. If this is the case then the Floer homology groups will be modules over a suitable Novikov ring as in [17]. Finally, not every isotopy of symplectomorphisms corresponds to a time dependent Hamiltonian vector field but the Floer homology groups will only be invariant under Hamiltonian isotopy.

(ii) In [17] the construction of the Floer homology groups has been generalized to some classes of non-monotone symplectic manifolds. The results in [17] include the case where the first Chern class c_1 vanishes over $\pi_2(\mathcal{M})$ but ω does not. In this case the Floer homology groups are modules over the Novikov ring associated to the ordering on $\pi_2(\mathcal{M})$ determined by ω.

(iii) Theorem 3.3 implies the Arnold conjecture for simply connected monotone symplectic manifolds $\mathcal{M}$: If ϕ is the time-1-map of a time dependent Hamiltonian vector field with nondegenerate fixed points then the number of fixed points of ϕ can be estimated below by the sum of the Betti numbers. In [12] Floer proved this result without assuming $\mathcal{M}$ to be simply connected.

(iv) It follows from (9) that the Euler characteristic of $HF_*^{\mathrm{symp}}(\mathcal{M};\phi)$ is the Lefschetz number of ϕ

$$\begin{aligned}\chi(HF_k^{\mathrm{symp}}(\mathcal{M};\phi)) &= \sum_{k=0}^{2N-1}(-1)^k \dim HF_k^{\mathrm{symp}}(\mathcal{M};\phi) \\ &= \sum_{x=\phi(x)} \operatorname{sign}\det(\mathbb{1}-d\phi(x)) \\ &= L(\phi).\end{aligned}$$

4 Flat connections over a Riemann surface

Let $\pi : P \to \Sigma$ be principal $SO(3)$-bundle over a compact oriented Riemann surface Σ of genus k. Up to isomorphism there are only two such bundles characterized by the second Stiefel-Whitney class. We assume $w_2(P) = 1$ so the bundle is nontrivial. As in section 1 let $\mathcal{A}(P)$ denote the space of connections on P and $\mathcal{G}(P)$ denote the group of gauge transformations. The component of the identity

$$\mathcal{G}_0(P) = \{g \in \mathcal{G}(P) : g \sim 1\}$$

can be characterized as the subgroup of even gauge transformation that is those which lift to $SU(2)$. Alternatively $\mathcal{G}_0(P)$ can be described as the kernel of the epimorphism $\eta : \mathcal{G}(P) \to H^1(\Sigma;\mathbf{Z}_2)$ which, as in section 1, assigns to each gauge transformation its parity $\eta(g)$. Here η induces an isomorphism from the group of components of $\mathcal{G}(P)$ to $H^1(\Sigma;\mathbf{Z}_2) \approx {\mathbf{Z}_2}^{2k}$

Recall from [2] that the affine space $\mathcal{A}(P)$ is an infinite dimensional symplectic manifold with symplectic form

$$\omega_A(a,b) = \int_\Sigma \langle a \wedge b \rangle \tag{14}$$

for $a, b \in T_A\mathcal{A}(P) = \Omega^1(\mathfrak{g}_P)$. The gauge group $\mathcal{G}_0(P)$ acts on this manifold by symplectomorphisms. The Lie algebra $\mathrm{Lie}(\mathcal{G}_0(P)) = \Omega^0(\mathfrak{g}_P)$ acts by Hamiltonian vector fields $\mathcal{A}(P) \to \Omega^1(\mathfrak{g}_P) : A \mapsto d_A\xi$ where $\xi \in \Omega^0(\mathfrak{g}_Q)$. The associated Hamiltonian functions are $\mathcal{A}(P) \to \mathbf{R} : A \mapsto \int_\Sigma \langle F_A \wedge \xi \rangle$. Thus the curvature

$$\mathcal{A}(P) \mapsto \Omega^2(\mathfrak{g}_P) : A \mapsto F_A$$

is the moment map and the corresponding Marsden-Weinstein quotient is the moduli space

$$\mathcal{M}(P) = \mathcal{A}_{\mathrm{flat}}(P)/\mathcal{G}_0(P).$$

of flat connections modulo even gauge equivalence.

Since P is the nontrivial SO(3)-bundle the space $\mathcal{M}(P)$ is a compact manifold of dimension $6k-6$ provided that $k \geq 2$. Moreover, every conformal structure on Σ induces a Kähler structure on $\mathcal{M}(P)$. To see this consider the DeRham complex

$$\Omega^0(\mathfrak{g}_P) \xrightarrow{d_A} \Omega^1(\mathfrak{g}_P) \xrightarrow{d_A} \Omega^2(\mathfrak{g}_P)$$

twisted by a flat connection A. It follows from Lemma 4.1 below that $d_A : \Omega^0 \to \Omega^1$ is injective and hence $H^0_A(\Sigma) = 0$. The first cohomology $H^1_A(\Sigma) = \ker d_A \cap d_A^* \subset \Omega^1(\mathfrak{g}_P)$ appears as the tangent space of the manifold $\mathcal{M}(P)$ at A. A conformal structure on Σ determines a Hodge-$*$-operator

$$H^1_A(\Sigma) \to H^1_A(\Sigma) : a \mapsto *a$$

for $A \in \mathcal{A}_{\mathrm{flat}}(P)$. These operators form an integrable complex structure on $\mathcal{M}(P)$ which is compatible with the symplectic form (14) The associated Kähler metric is given by

$$\langle a, b \rangle = \int_\Sigma \langle a \wedge *b \rangle$$

for $a, b \in H^1_A(\Sigma)$.

Via the holonomy the space $\mathcal{M}(P)$ can be identified with the space of odd representations of the fundamental group of P in SU(2):

$$\mathcal{M}(P) = \frac{\mathrm{Hom}^{\mathrm{odd}}(\pi_1(P), \mathrm{SU}(2))}{\mathrm{SU}(2)}.$$

More precisely, since $\mathfrak{g} = \mathfrak{so}(3) = \mathfrak{su}(2)$ the holonomy of a flat connection $A \in \mathcal{A}_{\mathrm{flat}}(P)$ at a point $p_0 \in P$ determines a homomorphism $\rho_A : \pi_1(P) \to \mathrm{SU}(2)$ whose image is denoted by $H_A(p_0) = \rho_A(\pi_1(P))$. Since P is the nontrivial bundle its fundamental group is given by $2k+1$ generators

$$\alpha_1, \ldots \alpha_k, \beta_1, \ldots, \beta_k, \varepsilon$$

with relations

$$\prod_{j=1}^k [\alpha_j, \beta_j] = \varepsilon, \quad [\alpha_j, \varepsilon] = 1, \quad [\beta_j, \varepsilon] = 1, \quad \varepsilon^2 = 1.$$

A homomorphism $\rho : \pi_1(P) \to \mathrm{SU}(2)$ is called **odd** if $\rho(\varepsilon) = -1$. Any such homomorphism is given by a $2k$-tuple of matrices $U_j, V_j \in \mathrm{SU}(2)$ such that

$$\prod_{j=1}^k [U_j, V_j] = -1$$

The space of conjugacy classes of such homomorphisms is easily seen to be a compact manifold of dimension $6k-6$. In particular, every flat connection on P is regular.

Lemma 4.1 *If $A \in \mathcal{A}_{\text{flat}}(P)$ then $\mathcal{G}_A \cap \mathcal{G}_0(P) = \{1\}$.*

Proof: Let $g \in \mathcal{G}_A \cap \mathcal{G}_0(P)$ and let $\tilde{g} : P \to \mathrm{SU}(2)$ be a lift of g. Then $\tilde{g}(p_0)$ commutes with $H_A(p_0)$. By the above discussion $H_A(p_0)$ is not an abelian subgroup of SU(2). Hence $\tilde{g}(p_0) = \pm 1\!\!1$ and $g(p_0) = 1\!\!1$. □

The topology of the moduli space $\mathcal{M}(P)$ has been studied extensively by Atiyah and Bott [2] and Newsteadt [21]. We recall those results which are of interest to us.

Theorem 4.2 *Assume $k \geq 2$. The moduli space $\mathcal{M}(P)$ is connected and simply connected and $\pi_2(\mathcal{M}(P)) = \mathbf{Z}$.*

It is easy to see that $\mathcal{M}(P)$ is connected for $k \geq 1$ and simply connected for $k \geq 2$ [21]. Define

$$R_k : \mathrm{SU}(2)^{2k} \to \mathrm{SU}(2) : (U_1, \ldots, U_k, V_1, \ldots, V_k) \mapsto \prod_{j=1}^{k} [U_j, V_j].$$

A $2k$-tuple in $X_k = \mathrm{SU}(2)^{2k}$ is a singular point for R_k if and only if all $2k$ matrices commute. Hence the set S_k of singular points is a $(2k+2)$-dimensional stratified subvariety of $Z_k = {R_k}^{-1}(1\!\!1)$. Hence there is a fibration $R_k : X_k \setminus Z_k \to \mathrm{SU}(2) \setminus \{1\!\!1\}$ with fibre $F_k = {R_k}^{-1}(-1\!\!1)$ over a contractible base. The space $X_k \setminus Z_k$ is connected for $k \geq 1$ and simply connected for $k \geq 2$ and so is F_k. Now SO(3) acts freely on F_k and the quotient is $\mathcal{M}(P)$. The homotopy exact sequence of the fibration

$$\mathrm{SO}(3) \hookrightarrow F_k \to \mathcal{M}(P)$$

shows that $\mathcal{M}(P)$ is connected and simply connected. Using the linking number of a sphere in $X_k \setminus Z_k$ with the codimension-3 submanifold $Z_k \setminus S_k$ we see that $\pi_2(F_k) = \pi_2(X_k \setminus Z_k) = \mathbf{Z}$. But in this case the fundamental group of SO(3) enters the exact sequence

$$0 \to \mathbf{Z} \to \pi_2(\mathcal{M}(P)) \to \mathbf{Z}_2 \to 0$$

and we can only deduce that either $\pi_2(\mathcal{M}(P)) = \mathbf{Z}$ or $\pi_2(\mathcal{M}(P)) = \mathbf{Z} \oplus \mathbf{Z}_2$. We must rule out the latter case. This requires the approach of Atiyah and Bott [2] using infinite dimensional Morse theory.

Theorem 4.3 *Assume $k \geq 2$. Then the (infinite dimensional) space $\mathcal{A}_{\text{flat}}(P)$ of flat connections on P is simply connected and $\pi_2(\mathcal{A}_{\text{flat}}(P)) = 0$.*

The proof of this result is based on Morse theory for the Yang-Mills functional

$$\mathcal{YM}(A) = \int_\Sigma \|F_A\|^2$$

on the space of connections $\mathcal{A}(P)$. The idea is to extend a loop $A_1 : S^1 \to \mathcal{A}_{\text{flat}}(P)$ to a map $A : D \to \mathcal{A}(P)$ on the unit disc. Then use the gradient flow of the Yang-Mills functional to 'push this extension down' to the set $\mathcal{A}_{\text{flat}}(P)$ of absolute minima. This requires that all the non-minimal critical manifolds have Morse index at least 3 so that the extension A in general position does not intersect their stable manifolds. The same consideration gives $\pi_2(\mathcal{A}_{\text{flat}}(P)) = 0$ if these Morse indices are at least 4. The details of this argument are carried out in [6]. In [2] Atiyah and Bott used an alternative stratification of the space $\mathcal{A}(P)$ to prove Theorem 4.3.

Now the homotopy exact sequence of the fibration $\mathcal{G}_0 \hookrightarrow \mathcal{A}_{\text{flat}}(P) \to \mathcal{M}(P)$ shows that

$$\pi_2(\mathcal{M}(P)) = \pi_1(\mathcal{G}_0(P)) = \mathbf{Z}.$$

The last identity follows from Lemma 1.2 applied to the bundle $Q = P \times S^1$ over the 3-manifold $M = \Sigma \times S^1$. The isomorphism $\pi_1(\mathcal{G}_0(P)) = \mathbf{Z}$ is given by the degree of a loop $g(s) \in \mathcal{G}_0(P)$ regarded as a gauge transformation of $P \times S^1$.

We shall now give a more explicit description of the isomorphism from $\pi_1(\mathcal{G}_0(P))$ to $\pi_2(\mathcal{M}(P))$. Let $A_0 \in \mathcal{A}_{\text{flat}}(P)$ be a flat connection and let $g(\theta) \in \mathcal{G}_0(P)$ be a loop of gauge transformations such that $g(0) = g(1) = \mathbb{1}$. By Theorem 4.3 there exists a map $A : D \to \mathcal{A}_{\text{flat}}(P)$ on the unit disc such that

$$A(e^{2\pi i\theta}) = A_1(\theta) = g(\theta)^* A_0. \tag{15}$$

This map represents a sphere in the moduli space $\mathcal{M}(P)$ since the boundary of D is mapped to a point. An easy calculation shows that the integral of the symplectic form over this sphere is given by

$$\begin{aligned} \int_D \int_\Sigma \left\langle \frac{\partial A}{\partial s} \wedge \frac{\partial A}{\partial t} \right\rangle dsdt &= \frac{1}{2} \int_0^1 \int_\Sigma \left\langle \frac{dA_1}{d\theta} \wedge (A_1 - A_0) \right\rangle d\theta \\ &= 8\pi^2 \deg(g). \end{aligned} \tag{16}$$

The last identity follows from equation (2) applied to the Chern-Simons functional on the bundle $Q = P \times S^1$.

Remark 4.4 (i) The group $\pi_0(\mathcal{G}(P)) = \mathcal{G}(P)/\mathcal{G}_0(P) = \mathbf{Z}_2{}^{2k}$ of components of the gauge group acts on the moduli space by symplectomorphism $\mathcal{M}(P) \to \mathcal{M}(P) : [A] \mapsto [g^*A]$. This action is not free: it is easy to construct a flat connection whose isotropy subgroup $\mathcal{G}_A$ is nontrivial but discrete.

(ii) The tangent bundle of $\mathcal{M}(P)$ is a complex vector bundle and therefore has Chern classes. It follows from our theorem about the spectral flow [7] that the integral of the first Chern class $c_1 \in H^2(\mathcal{M}(P);\mathbf{Z})$ over the sphere $A : D \to \mathcal{A}_{\text{flat}}(P)$ satisfying (15) is given by $2\deg(g)$. This was already known to Atiyah and Bott [2].

5 Mapping cylinders

Continue the notation of the previous section. Any diffeomorphism $h : \Sigma \to \Sigma$ lifts to an automorphism $f : P \to P$ since h^*P is isomorphic to P. We assume throughout that h is orientation preserving. The automorphism f induces a symplectomorphism

$$\phi_f : \mathcal{M}(P) \to \mathcal{M}(P)$$

defined by $\phi_f([A]) = [f^*A]$. In the context of representations of the fundamental group this symplectomorphism is given by $\rho \mapsto \rho \circ f_*$. Hence the symplectomorphism ϕ_f depends only on the homotopy class of f. A fixed point of ϕ_f is an equivalence class of a flat connection $A_0 \in \mathcal{A}_{\text{flat}}(P)$ such that $g^*f^*A_0 = A_0$ for some gauge transformation $g \in \mathcal{G}_0(P)$. This can be written as

$$f_g^*A_0 = A_0$$

where the automorphism $f_g \in \text{Aut}(P)$ is defined by

$$f_g(p) = f(p)g(p)$$

for $p \in P$. The differential of ϕ_f at the fixed point A_0 is the linear map

$$f_g^* : H^1_{A_0}(\Sigma) \to H^1_{A_0}(\Sigma).$$

Here we identify $H^1_{A_0}$ with the quotient $\ker d_{A_0}/\operatorname{im} d_{A_0}$ rather than the space of harmonic forms which will in general not be invariant under f_g^*.

Remark 5.1 For every $g \in \mathcal{G}(P)$ and every $A \in \mathcal{A}(P)$

$$f^*g^*A = (g \circ f)^*f^*A.$$

Hence the symplectomorphism $\mathcal{M}(P) \to \mathcal{M}(P) : [A] \mapsto [g^*A]$ commutes with ϕ_f whenever g is homotopic to $g \circ f$ or equivalently the parity $\eta(g)$ is in the kernel of the homomorphism $\mathbb{1} - h^* : H^1(\Sigma, \mathbf{Z}^2) \to H^1(\Sigma, \mathbf{Z}^2)$. Define

$$\mathcal{G}_f(P) = \{g \in \mathcal{G}(P) : g \sim g \circ f\}, \qquad \Gamma_f = \mathcal{G}_f(P)/\mathcal{G}_0(P).$$

Then $\Gamma_f \simeq \ker(\mathbb{1} - h^*)$ is a finite group which acts on $\mathcal{M}(P)$ by symplectomorphisms which commute with ϕ_f. In particular, Γ_f acts on the fixed points of ϕ_f. More explicitly, if $[A_0] \in \text{Fix}(\phi_f)$ and $g_0 \in \mathcal{G}_f(P)$ then $[A_0'] = [g_0^*A_0] \in \text{Fix}(\phi_f)$. To see this choose $g \in \mathcal{G}_0(P)$ such that $g^*f^*A_0 = A_0$ and define $g' = (g_0 \circ f)^{-1}gg_0 \in \mathcal{G}_0(P)$. Then $g'^*f^*A_0' = A_0'$.

The automorphism f also determines a principal SO(3)-bundle

$$P_f \to \Sigma_h.$$

Here $Q = P_f$ denotes the mapping cylinder of P for the automorphism f. That is the set of equivalence classes of pairs $[p, s] \in P \times \mathbf{R}$ under the equivalence relation generated by $[p, s+1] \equiv [f(p), s]$. Likewise the 3-manifold $M = \Sigma_h$ denotes the mapping cylinder of the Riemann surface Σ for the diffeomorphism h. This bundle satisfies hypothesis (H1) of section 1. A connection $a \in \mathcal{A}(P_f)$ is a 1-form $a = A + \Phi\, ds$ where $A(s) \in \mathcal{A}(P)$, $\Phi(s) \in \Omega^0(\mathfrak{g}_P)$ for $s \in \mathbf{R}$ and

$$A(s+1) = f^*A(s), \qquad \Phi(s+1) = \Phi(s) \circ f. \tag{17}$$

Now the group $\mathcal{G}(P_f)$ of gauge transformations of P_f consists of smooth 1-parameter families of gauge transformations $g(s) \in \mathcal{G}(P)$ such that

$$g(s+1) = g(s) \circ f.$$

Such a gauge transformation acts on a connection $a = A + \Phi\, ds \in \mathcal{A}(P_f)$ by

$$g^*a = g^*A + \left(g^{-1}\dot{g} + g^{-1}\Phi g\right) ds.$$

Here we use the notation g^* ambiguously: g^*a denotes the action of $g \in \mathcal{G}(P_f)$ on $a \in \mathcal{A}(P_f)$ whereas g^*A denotes the pointwise action of $g(s) \in \mathcal{G}(P)$ on $A(s) \in \mathcal{A}(P)$.

Remark 5.2 Consider the normal subgroups

$$\mathcal{G}_0(P_f) \subset \mathcal{G}_\Sigma(P_f) \subset \mathcal{G}(P_f)$$

where $\mathcal{G}_\Sigma(P_f) = \{g \in \mathcal{G}(P_f) : g(s) \in \mathcal{G}_0(\Sigma)\ \forall\, s\}$ and $\mathcal{G}_0(P_f)$ is the component of $\mathbb{1}$. Then

$$\mathcal{G}(P_f)/\mathcal{G}_\Sigma(P_f) \simeq \Gamma_f, \qquad \mathcal{G}_\Sigma(P_f)/\mathcal{G}_0(P_f) \simeq \mathbf{Z}.$$

The second isomorphism is given by the degree while the first follows from the fact that $g_0 \in \mathcal{G}(P)$ extends to a gauge transformation of P_f if and only if $g_0 \in \mathcal{G}_f(P)$. The first isomorphism shows that Γ_f is the group of components of gauge transformations of P_f of degree zero

The path space Ω_{ϕ_f} can be naturally identified with a subquotient of the space of connections on P_f. Consider the subspace

$$\mathcal{A}_\Sigma(P_f) = \{A + \Phi\, ds \in \mathcal{A}(P_f) : F_A = 0,\ d_A^*\,(dA/ds - d_A\Phi) = 0\}.$$

If $A + \Phi\, ds \in \mathcal{A}_\Sigma(P_f)$ then Φ is uniquely determined by A. In fact $dA/ds - d_A\Phi$ represents the projection of dA/ds onto the space $H_A^1(\Sigma) = T_{[A]}\mathcal{M}(P)$ of harmonic forms.

Proposition 5.3 *There are natural bijections*

$$\Omega_{\phi_f} \simeq \mathcal{A}_\Sigma(P_f)/\mathcal{G}_\Sigma(P_f), \qquad \tilde{\Omega}_{\phi_f} \simeq \mathcal{A}_\Sigma(P_f)/\mathcal{G}_0(P_f).$$

In particular, Ω_{ϕ_f} is connected and

$$\pi_1(\Omega_{\phi_f}) \simeq \mathcal{G}_\Sigma(P_f)/\mathcal{G}_0(P_f) \simeq \pi_2(\mathcal{M}(P)).$$

Proof: A point in the space Ω_{ϕ_f} is a smooth path $\gamma : \mathbf{R} \to \mathcal{M}(P)$ such that $\gamma(s+1) = \phi_f(\gamma(s))$. Such a path lifts to a smooth map $A : \mathbf{R} \to \mathcal{A}_{\mathrm{flat}}(P)$ satisfying

$$A(s+1) = h(s)^* f^* A(s).$$

for some smooth map $h : \mathbf{R} \to \mathcal{G}_0(P)$. Two such pairs (A, h) and (A', h') represent the same path $\gamma : \mathbf{R} \to \mathcal{M}(P)$ if and only if there exists a smooth map $g : \mathbf{R} \to \mathcal{G}_0(P)$ such that

$$A'(s) = g(s)^* A(s), \qquad h'(s) = (g(s) \circ f)^{-1} h(s) g(s+1).$$

Every path γ can be represented by a pair (A, h) with $h(s) \equiv 1$. To see this choose, for any given pair (A, h), a path $g : [0,1] \to \mathcal{G}_0(P)$ such that $g(s) = 1$ for s near 0 and $g(s) = h(s)^{-1}$ for s near 1. Then $g(s)$ extends to a unique function $\mathbf{R} \to \mathcal{G}_0(P)$ satisfying

$$h(s) g(s+1) = g(s) \circ f.$$

This proves the first statement.

To prove the second statement recall from Theorem 4.3 that $\mathcal{A}_{\mathrm{flat}}(P)$ is connected and simply connected and $\pi_2(\mathcal{A}_{\mathrm{flat}}(P)) = 0$. Hence $\mathcal{A}_\Sigma(P_f)$ is connected and simply connected. By the homotopy exact sequence

$$\pi_1(\mathcal{A}_\Sigma(P_f)) \to \pi_1(\mathcal{A}_\Sigma(P_f)/\mathcal{G}_0(P_f)) \to \pi_0(\mathcal{G}_0(P_f))$$

the quotient $\mathcal{A}_\Sigma(P_f)/\mathcal{G}_0(P_f)$ is simply connected. Hence this quotient is the universal cover of $\mathcal{A}_\Sigma(P_f)/\mathcal{G}_\Sigma(P_f) \simeq \Omega_{\phi_f}$. □

The Chern-Simons functional on the space of connections $\mathcal{A}(P_f)$ is given by the formula

$$\mathcal{CS}(A + \Phi\, ds) = \frac{1}{2} \int_0^1 \int_\Sigma \left(\langle \dot{A} \wedge (A - A_0) \rangle + \langle F_A \wedge \Phi \rangle \right) ds$$

Here A_0 is any flat connection on P such that $f^* A_0 = A_0$. A simple calculation shows that the restriction of $\mathcal{CS}$ to $\mathcal{A}_\Sigma(P_f)$ induces the symplectic action functional on $\tilde{\Omega}_{\phi_f} = \mathcal{A}_\Sigma(P_f)/\mathcal{G}_0(P_f)$. We shall in fact prove that the critical points of the Chern-Simons functional, that is the flat connections on P_f, agree with the critical points of the symplectic action, that is the fixed points of ϕ_f. To see this note that the curvature of the connection $a = A + \Phi\, ds$ is the 2-form

$$F_{A+\Phi\, ds} = F_A + \left(d_A\Phi - \dot{A}\right) \wedge ds$$

Thus the flat connections on P_f are smooth families of flat connections $A(s) \in \mathcal{A}_{\mathrm{flat}}(P)$ such that $dA/ds \in \operatorname{im} d_A$ and (17) is satisfied.

Proposition 5.4 *The map $\mathcal{A}(P_f) \to \mathcal{A}(P) : A + \Phi\, ds \mapsto A(0)$ induces a bijection*

$$\mathcal{A}_{\mathrm{flat}}(P_f)/\mathcal{G}_\Sigma(P_f) \simeq \mathrm{Fix}(\phi_f).$$

In particular $\mathcal{A}_{\mathrm{flat}}(P_f)/\mathcal{G}(P_f) \simeq \mathrm{Fix}(\phi_f)/\Gamma_f$.

Proof: First assume that $a = A + \Phi\, ds \in \mathcal{A}_{\mathrm{flat}}(P_f)$. Then $A(s) \in \mathcal{A}_{\mathrm{flat}}(P)$ for every s and $\dot{A} = d_A\Phi$. Let $g(s) \in \mathcal{G}(P)$ be the unique solution of the ordinary differential equation $\dot{g} + \Phi g = 0$ with $g(0) = \mathbb{1}$. Then $g(s)^*A(s) \equiv A_0$ and hence it follows from (17) that $g(1)^*f^*A_0 = g(1)^*A(1) = A_0$. By construction, $g(1) \in \mathcal{G}_0(P)$ and hence A_0 represents a fixed point of ϕ_f. This proves that there is a well defined map $\mathcal{A}_{\mathrm{flat}}(P_f)/\mathcal{G}_\Sigma(P_f) \to \mathrm{Fix}(\phi_f)$ given by $A + \Phi\, ds \mapsto A(0)$.

We prove that this map is onto. Suppose that $A_0 \in \mathcal{A}_{\mathrm{flat}}(P)$ represents a fixed point of ϕ_f and let $g_1 \in \mathcal{G}_0(P)$ such that $g_1^*f^*A_0 = A_0$. Choose a smooth 1-parameter family of connections $g(s) \in \mathcal{G}_0(P)$ such that $g(0) = 1$, $g(1) = g_1$ and $g(s+1) = (g(s) \circ f)g_1$ for every s. Let $A(s) \in \mathcal{A}_{\mathrm{flat}}(P)$ be defined by $g(s)^*A(s) = A_0$ and $\Phi(s) = -\dot{g}(s)g(s)^{-1}$. Then $a = A + \Phi\, ds$ is the required flat connection on P_f.

We prove that the map is injective. Let $a, a' \in \mathcal{A}_{\mathrm{flat}}(P_f)$ such that $A'(0) = g_0^*A(0)$ for some $g_0 \in \mathcal{G}_0(P)$. Define $g(s) \in \mathcal{G}_0(P)$ to be the unique solution of the ordinary differential equation $\dot{g} = g\Phi' - \Phi g, \qquad g(0) = g_0$. Then

$$\frac{d}{ds}g^*A = g^{-1}\dot{A}g + d_{g^*A}(g^{-1}\dot{g}) = g^{-1}d_A\Phi g + d_{g^*A}(\Phi' - g^{-1}\Phi g) = d_{g^*A}\Phi'.$$

Here we have used $\dot{A} = d_A\Phi$. Since $A'(0) = g(0)^*A(0)$ and $\dot{A}' = d_{A'}\Phi'$ it follows that $g(s)^*A(s) = A'(s)$ for every s. Moreover it follows from (17) that $g(s+1)^{-1}g(s) \circ f \in \mathcal{G}_{A(s+1)} \cap \mathcal{G}_0(P)$. By Lemma 4.1 this implies $g(s+1) = g(s) \circ f$. Hence $g(s)$ defines a gauge transformation of P_f and $a' = g^*a$. $\square$

Proposition 5.5 *A flat connection $A + \Phi\, ds \in \mathcal{A}_{\mathrm{flat}}(P_f)$ is nondegenerate as a critical point of the Chern-Simons functional if and only if $[A(0)]$ is a nondegenerate fixed point of ϕ_f.*

Proof: Let $a = A + \Phi\,ds \in \mathcal{A}_{\mathrm{flat}}(P_f)$ be a flat connection and define $A_0 = A(0) \in \mathcal{A}_{\mathrm{flat}}(P)$. Then $f_g^* A_0 = A_0$ where $g \in \mathcal{G}_0(P)$ is defined by $g = g(1)$ for the unique solution $g(s) \in \mathcal{G}_0(P)$ of $\dot{g}(s) + \Phi(s)g(s) = 0$ with $g(0) = \mathbb{1}$. We must show that

$$\mathbb{1} - f_g^* : H^1_{A_0}(\Sigma) \to H^1_{A_0}(\Sigma)$$

is an isomorphism if and only if $H^1_{A+\Phi\,ds}(\Sigma_h) = 0$. The latter means that whenever $\alpha + \phi\,ds \in \Omega^1(\mathfrak{g}_{P_f})$ is an infinitesimal connection such that

$$d_A\alpha = 0, \qquad d_A\phi = \dot{\alpha} + [\Phi \wedge \alpha] \tag{18}$$

then there exists a $\xi \in \Omega^0(\mathfrak{g}_{P_f})$ such that

$$\alpha = d_A\xi, \qquad \phi = \dot{\xi} + [\Phi \wedge \xi]. \tag{19}$$

Replacing a by g^*a and f by $f_{g(1)}$ we may assume without loss of generality that $\Phi(s) \equiv 0$ and $A(s) \equiv A_0$.

Assume first that $\mathbb{1} - f^*$ is an isomorphism of $H^1_{A_0}$. Let $\alpha(s) \in \Omega^1(\mathfrak{g}_P)$ and $\phi \in \Omega^0(\mathfrak{g}_P)$ satisfy (18) and the boundary conditions $\alpha(s+1) = f^*\alpha(s)$, $\phi(s+1) = \phi(s) \circ f$. Then

$$f^*\alpha(0) - \alpha(0) = d_{A_0}\int_0^1 \phi(s)\,ds, \qquad d_{A_0}\alpha(0) = 0.$$

Since $\mathbb{1} - f^* : H^1_{A_0} \to H^1_{A_0}$ is injective there exists a $\xi_0 \in \Omega^0(\mathfrak{g}_P)$ such that $\alpha(0) = d_{A_0}\xi_0$. Define

$$\xi(s) = \xi_0 + \int_0^s \phi(\theta)\,d\theta.$$

Then $\phi = \dot{\xi}$, hence $\dot{\alpha} = d_{A_0}\dot{\xi}$, and hence $\alpha = d_{A_0}\xi$. The latter identity implies that $d_{A_0}\xi(s+1) = f^*d_{A_0}\xi(s)$ and hence $\xi(s+1) = \xi(s) \circ f$.

Conversely, suppose that $\mathbb{1} - f^* : H^1_{A_0} \to H^1_{A_0}$ is not injective. Then there exist $\alpha_0 \in \Omega^1(\mathfrak{g}_P)$ and $\xi_0 \in \Omega^0(\mathfrak{g}_P)$ such that

$$f^*\alpha_0 - \alpha_0 = d_{A_0}\xi_0, \qquad d_{A_0}\alpha_0 = 0, \qquad \alpha_0 \notin \mathrm{im}\,d_{A_0}.$$

Choose any function $\phi(s) \in \Omega^0(\mathfrak{g}_P)$ satisfying

$$\xi_0 = \int_0^1 \phi(s)\,ds, \qquad \phi(s+1) = \phi(s) \circ f.$$

For example take a cutoff function $\beta : [0,1] \to \mathbf{R}$ of mean value 1 which vanishes near $s = 0$ and $s = 1$ and define $\phi(s+j) = \beta(s)\xi_0 \circ f^j$ for $0 \le s \le 1$ and $j \in \mathbf{Z}$. Let $\alpha(s) \in \Omega^1(\mathfrak{g}_P)$ be the unique solution of $\dot{\alpha} = d_{A_0}\phi$ with $\alpha(0) = \alpha_0$. Then α and ϕ satisfy (18) but are not of the form (19) since $\alpha_0 \notin \mathrm{im}\,d_{A_0}$. This shows that $H^1_{A+\Phi\,ds} \neq 0$. □

Remark 5.6 It is an open question whether every symplectomorphism $\phi : \mathcal{M}(P) \to \mathcal{M}(P)$ is isotopic (within the group of symplectomorphisms) to one of the form ϕ_f for $f \in \mathrm{Aut}(P)$. It is also an open question whether f_0 and f_1 are isotopic whenever ϕ_{f_0} and ϕ_{f_1} are isotopic.

6 Instantons and holomorphic curves

The proof of Theorem 1.1 is based on a comparison between holomorphic curves in $\mathcal{M}(P)$ and self-dual instantons on the 4-manifold $\Sigma_h \times \mathbf{R}$. To carry this out we must choose a metric on Σ_h. Let $\langle\, ,\, \rangle_s$ be a one parameter family of metrics on Σ such that

$$\langle dh(z)\zeta_0, dh(z)\zeta_1\rangle_s = \langle \zeta_0, \zeta_1\rangle_{s+1}.$$

Then the associated Hodge-$*$-operators $*_s : \Omega^j(\mathfrak{g}_P) \to \Omega^{2-j}(\mathfrak{g}_P)$ satisfy

$$*_{s+1} \circ f^* = f^* \circ *_s. \tag{20}$$

This defines a metric on P_f: whenever $\xi(s), \eta(s) \in \Omega^j(\mathfrak{g}_P)$ with $\xi(s+1) = f^*\xi(s)$ and $\eta(s+1) = f^*\eta(s)$ then the function

$$\langle \xi(s), \eta(s)\rangle_s = \int_\Sigma \langle \xi(s) \wedge *_s\eta(s)\rangle$$

is of period 1 in s. The inner product of ξ and η is defined as the integral of this function over the unit interval

$$\langle \xi, \eta\rangle = \int_0^1 \int_\Sigma \langle \xi(s) \wedge *_s\eta(s)\rangle \, ds.$$

Now recall that the tangent space to $\mathcal{M}(P)$ at A is the quotient $H^1_A(\Sigma) = \ker d_A / \mathrm{im} d_A$. The metrics on Σ determine a family of complex structures

$$J_s = *_s : H^1_A(\Sigma) \to H^1_A(\Sigma)$$

and condition (20) implies that these satisfy (5) with $\phi = \phi_f$. In order for the space of harmonic forms to be invariant under $*_s$ we must use the L^2-adjoint $d_A^* = - *_s d_A *_s$ with respect to the s-metric.

Any smooth function $u : \mathbf{R}^2 \to \mathcal{M}(P)$ lifts to a smooth function $A : \mathbf{R}^2 \to \mathcal{A}_{\mathrm{flat}}(P)$. For any such map the partial derivatives $\partial A/\partial s$ and $\partial A/\partial t$ lie in the kernel of d_A but will in general not be harmonic. To apply a complex structure we must first project these derivatives into the space of harmonic forms corresponding to this complex structure. These projections can be described as $\partial A/\partial s - d_A\Phi$ and $\partial A/\partial t - d_A\Psi$ where $\Phi, \Psi \in \Omega^0(\mathfrak{g}_P)$ are uniquely determined by the requirement

$$d_A *_s \left(\frac{\partial A}{\partial s} - d_A\Phi\right) = 0, \qquad d_A *_s \left(\frac{\partial A}{\partial t} - d_A\Psi\right) = 0.$$

Thus our function $u = [A] : \mathbf{R}^2 \to \mathcal{M}(P)$ satisfies the nonlinear Cauchy-Riemann equations (6) if and only if there exist functions $\Phi, \Psi : \mathbf{R}^2 \to \Omega^0(\mathfrak{g}_P)$ such that

$$\frac{\partial A}{\partial t} - d_A\Psi + *_s \left(\frac{\partial A}{\partial s} - d_A\Phi\right) = 0. \tag{21}$$

Moreover, the periodicity condition (7) with $\phi = \phi_f$ is equivalent to

$$A(s+1,t) = f^*A(s,t), \quad \Phi(s+1,t) = \Phi(s,t) \circ f, \quad \Psi(s+1,t) = \Phi(s,t) \circ f. \tag{22}$$

The limit condition (8) takes the form

$$\lim_{t\to\pm\infty} A(s,t) = A^{\pm}(s), \quad \lim_{t\to\pm\infty} \Phi(s,t) = \Phi^{\pm}(s), \quad \lim_{t\to\pm\infty} \Psi(s,t) = 0 \tag{23}$$

where $A^{\pm}(s) \in \mathcal{A}_{\text{flat}}(P)$ and $\dot{A}^{\pm} = d_{A^{\pm}}\Phi^{\pm}$. This means that $A^{\pm} + \Phi^{\pm}\,ds$ are flat connections on P_f. Strictly speaking, the periodicity conditions (22) need only be satisfied up to even gauge equivalence. However, any such triple A, Φ, Ψ can be transformed so as to obtain (22). (See the proof of Proposition 4.1.)

If two solutions of (21) and (22) are gauge equivalent by a family of even gauge transformations $g(s,t) \in \mathcal{G}_0(P)$ then

$$g(s+1,t) = g(s,t) \circ f.$$

This means that g defines a gauge transformation on the bundle $P_f \times \mathbf{R}$ over the 4-manifold $\Sigma_h \times \mathbf{R}$. Moreover, the action of g on the triple A, Φ, Ψ corresponds to the interpretation of this triple as a connection $A + \Phi\,ds + \Psi\,dt$ on $P_f \times \mathbf{R}$. The curvature of this connection is the 2-form

$$\begin{aligned} F_{A+\Phi\,ds+\Psi\,dt} &= F_A - \left(\frac{\partial A}{\partial s} - d_A\Phi\right) \wedge ds - \left(\frac{\partial A}{\partial t} - d_A\Psi\right) \wedge dt \\ &\quad + \left(\frac{\partial \Psi}{\partial s} - \frac{\partial \Phi}{\partial t} + [\Phi, \Psi]\right) ds \wedge dt. \end{aligned}$$

Hence the connection $A + \Phi\,ds + \Psi\,dt$ is self-dual if and only if

$$\begin{aligned} \frac{\partial A}{\partial t} - d_A\Psi + *_s\left(\frac{\partial A}{\partial s} - d_A\Phi\right) &= 0, \\ \frac{\partial \Phi}{\partial t} - \frac{\partial \Psi}{\partial s} - [\Phi, \Psi] + *_sF_A &= 0. \end{aligned} \tag{24}$$

Note that the first equation in (24) agrees with (21) whereas the second equation replaces the condition on $A(s,t)$ to be flat.

Now the holomorphic curves described by equation (21) with $F_A = 0$ can be viewed as a limit case of the instantons described by equation (24). Following Atiyah [1] we stretch the mapping cylinder Σ_h so that the period converges to infinity. Formally this means that equation (22) is replaced by

$$A(s+1/\varepsilon,t) = f^*A(s,t), \quad \Phi(s+1/\varepsilon,t) = \Phi(s,t) \circ f, \quad \Psi(s+1/\varepsilon,t) = \Phi(s,t) \circ f.$$

and in (24) $*_s$ is replaced by $*_{\varepsilon s}$. Now rescale A, Φ, and Ψ:

$$A^\varepsilon(s,t) = A(s/\varepsilon, t/\varepsilon), \quad \Phi^\varepsilon(s,t) = 1/\varepsilon\, \Phi(s/\varepsilon, t/\varepsilon), \quad \Psi^\varepsilon(s,t) = 1/\varepsilon\, \Psi(s/\varepsilon, t/\varepsilon).$$

The triple $A^\varepsilon, \Phi^\varepsilon, \Psi^\varepsilon$ then satisfies the periodicity condition (22). Moreover, equation (24) becomes

$$\begin{aligned} \frac{\partial A^\varepsilon}{\partial t} - d_{A^\varepsilon}\Psi^\varepsilon + *_s\left(\frac{\partial A^\varepsilon}{\partial s} - d_{A^\varepsilon}\Phi^\varepsilon\right) &= 0, \\ \frac{\partial \Phi^\varepsilon}{\partial t} - \frac{\partial \Psi^\varepsilon}{\partial s} - [\Phi^\varepsilon, \Psi^\varepsilon] + \frac{1}{\varepsilon^2} *_s F_{A^\varepsilon} &= 0. \end{aligned} \tag{25}$$

This is equivalent to conformally rescaling the metric on Σ by the factor ε^2.

It follows from an implicit function theorem that near every solution of (21), (22) and (23) there is a solution of (25), (22) and (23) provided that $\varepsilon > 0$ is sufficiently small. This is a singular perturbation theorem and care must be taken with the dependence of the linearized operators on ε. Conversely, a family of solutions $A^\varepsilon, \Phi^\varepsilon, \Psi^\varepsilon$ of (25), (22) and (23) converges as ε tends to 0 and the curvature of A^ε converges to 0. The key point is an energy estimate. The Yang-Mills action of the rescaled equation is given by

$$\begin{aligned} &\mathcal{YM}_\varepsilon(A^\varepsilon + \Phi^\varepsilon\, ds + \Psi^\varepsilon\, dt) \\ &= \int_{-\infty}^{\infty}\int_0^1 \left(\left\| \frac{\partial A^\varepsilon}{\partial t} - d_{A^\varepsilon}\Psi^\varepsilon \right\|_s^2 + \frac{1}{\varepsilon^2}\|F_{A^\varepsilon}\|_s^2 \right) ds dt \\ &= \mathcal{CS}(A^- + \Phi^-\, ds) - \mathcal{CS}(A^+ + \Phi^+\, ds). \end{aligned} \tag{26}$$

This shows that the L^2-norm of F_{A^ε} on $\Sigma_h \times \mathbf{R}$ converges to 0 as ε tends to 0. Now Uhlenbeck's compactness theorem requires an L^∞-estimate of the form

$$\sup_{\varepsilon > 0}\left(\frac{1}{\varepsilon^2}\|F_{A^\varepsilon}\|_{L^\infty(\Sigma_h\times\mathbf{R})} + \|\partial A^\varepsilon/\partial t - d_{A^\varepsilon}\Psi^\varepsilon\|_{L^\infty(\Sigma_h\times\mathbf{R})}\right) < \infty. \tag{27}$$

The proof of this estimate involves a bubbling argument. Roughly speaking, the estimate (27) may be violated in arbitrarily small neighborhoods of finitely many points and in this case either instantons on S^4 or instantons on $\Sigma \times \mathbf{C}$ or holomorphic spheres in $\mathcal{M}(P)$ will split off. But this can be avoided in the case which is relevant for the construction of the Floer homology groups namely when the relative Morse index is 1:

$$\mu(A^- + \Phi^-\, ds) - \mu(A^+ + \Phi^+\, ds) = 1.$$

Now there are two such relative Morse indices; one in the Chern-Simons theory given by the spectral flow of the operator family $D_{a(t)}$ (section 2) and one in symplectic Floer homology given by the Maslov index (section 3). We must prove that both relative Morse indices agree. We shall address this problem as well as singular perturbation and compactness in a separate paper.

7 Perturbations

The methods we have discussed so far require the assumption that all flat connections on the bundle $Q = P_f$ respectively all fixed points of the symplectomorphism ϕ_f on $\mathcal{M}(P)$ are nondegenerate. The purpose of this section is to show why this assumption is redundant. In particular we wish to apply Theorem 1.1 to $f = \mathrm{id}$ in which case all flat connections are degenerate. Nevertheless we obtain the following

Theorem 7.1 *The instanton homology of the bundle $P \times S^1$ over $\Sigma \times S^1$ is naturally isomorphic to the homology of the moduli space $\mathcal{M}_F(P)$*

$$HF_k^{\mathrm{inst}}(\Sigma \times S^1, P \times S^1) = \bigoplus_{j=k (\mathrm{mod}\ 4)} H_j(\mathcal{M}(P); \mathbf{Z}_2).$$

To construct the symplectic Floer homology groups in the degenerate case we must perturb the nonlinear Cauchy-Riemann equations (6) by a Hamiltonian term as in the proof of Theorem 3.3. Likewise we must consider perturbations of the Chern-Simons functional on the space of connections on P_f as in [11] and [27] to construct the instanton homology groups of section 2.

We shall construct a smooth family of Hamiltonian functions $H_s : \mathcal{A}(P) \to \mathbf{R}$ which satisfy

$$H_s(g^*A) = H_s(A) = H_{s+1}(f^*A) \tag{28}$$

for $A \in \mathcal{A}(P)$ and $g \in \mathcal{G}_0(P)$. The differential of H_s can be represented by a smooth map $X_s : \mathcal{A}(P) \to \Omega^1(\mathfrak{g}_P)$ such that

$$dH_s(A)\alpha = \int_\Sigma \langle X_s(A) \wedge \alpha \rangle.$$

In other words $X_s : \mathcal{A}(P) \to \Omega^1(\mathfrak{g}_P)$ is the Hamiltonian vector field on $\mathcal{A}(P)$ corresponding to the Hamiltonian function H_s. Since H_s is invariant under $\mathcal{G}_0(P)$ the vector fields X_s satisfy

$$X_s(g^*A) = g^{-1}X_s(A)g, \qquad d_A X_s(A) = 0, \qquad X_{s+1}(f^*A) = f^*X_s(A) \tag{29}$$

for $g \in \mathcal{G}_0(P)$ and $A \in \mathcal{A}(P)$. The vector fields X_s that arise from the holonomy will be smooth with respect to the $W^{k,p}$-norm for all k and p and hence give rise to a Hamiltonian flow $\psi_s : \mathcal{A}(P) \to \mathcal{A}(P)$ defined by

$$\frac{d}{ds}\psi_s = X_s \circ \psi_s, \qquad \psi_0 = \mathrm{id}.$$

The diffeomorphisms ψ_s preserve the symplectic structure and are equivariant under the action of $\mathcal{G}_0(P)$

$$\psi_s(g^*A) = g^*\psi_s(A), \qquad F_{\psi_s(A)} = F_A.$$

Moreover,

$$\psi_{s+1} \circ \psi_1{}^{-1}(f^*A) = f^*\psi_s(A)$$

for $A \in \mathcal{A}(P)$. The restriction of this identity to the Marsden-Weinstein quotient $\mathcal{M}(P)$ can be written in the form

$$\psi_{s+1} \circ \phi_{f,H} = \phi_f \circ \psi_s, \qquad \phi_{f,H} = \psi_1{}^{-1} \circ \phi_f.$$

The symplectomorphism $\phi_{f,H} : \mathcal{M}(P) \to \mathcal{M}(P)$ is related to ϕ_f by a Hamiltonian isotopy.

To construct the Hamiltonian functions H_s and the Hamiltonian vector fields X_s we first recall some basic facts about the holonomy of a connection on P. For any loop $\gamma(\theta + 1) = \gamma(\theta) \in P$ the holonomy determines a map $\rho = \rho_\gamma : \mathcal{A}(P) \to \mathrm{SU}(2)$ defined by $\rho(A) = g(1)$ where $g(\theta) \in \mathrm{SU}(2)$ is the unique solution of the ordinary differential equation

$$\dot{g} + A(\dot{\gamma})g = 0, \qquad g(0) = 1.$$

This equation is meaningful since the Lie algebra of SO(3) agrees with the Lie algebra of SU(2). The differential of ρ can be expressed in terms of g:

$$\rho(A)^{-1} d\rho(A)\alpha = -\int_0^1 g^{-1}\alpha(\dot{\gamma})g \, d\theta$$

for $\alpha \in \Omega^1(\mathfrak{g}_P)$.

Now choose $2k$ embeddings $\gamma_j : \mathbf{R}/\mathbf{Z} \times \mathbf{R} \to P$ of the annulus such that the projections $\pi \circ \gamma_j$ are orientation preserving, generate the fundamental group of Σ, and satisfy $\gamma_j(0, \lambda) = p_\lambda$ for every j. Denote by $\rho_\lambda : \mathcal{A}(P) \to \mathrm{SU}(2)^{2k}$ the holonomy along the loops $\theta \mapsto \gamma_j(\theta, \lambda)$ for $\lambda \in \mathbf{R}$. Now choose a smooth family of functions $h_s : \mathrm{SU}(2)^{2k} \to \mathbf{R}$ which are invariant under conjugacy and vanishes for s near 0 and 1. Let $\beta : \mathbf{R} \to \mathbf{R}$ be a smooth cutoff function supported in $[-1, 1]$ with mean value 1 and define $H_s : \mathcal{A}(P) \to \mathbf{R}$ by

$$H_s(A) = \int_{-1}^1 \beta(\lambda) h_s(\rho_\lambda(A)) \, d\lambda$$

for $A \in \mathcal{A}(P)$ and $0 \le s \le 1$.

The partial derivative of h_s with respect to U_j can be represented by a function $\eta_j : [0, 1] \times \mathrm{SU}(2)^{2k} \to \mathfrak{su}(2)$ such that

$$\frac{\partial h_s}{\partial U_j}(U)U_j\xi = \langle \eta_j(s, U), \xi \rangle.$$

Define $X_s : \mathcal{A}(P) \to \Omega^1(\mathfrak{g}_P)$ by

$$X_s(A) = \sum_{j=1}^{2k} X_j(s, A)$$

for $0 \leq s \leq 1$ where $X_j(s, A) \in \Omega^1(\mathfrak{g}_P)$ is supported in $\gamma_j(S^1 \times [-1, 1])$ and

$$\gamma_j^* X_j(s, A) = \beta g_j \eta_j(s, \rho_\lambda(A)) g_j{}^{-1} d\lambda.$$

Here $\theta \mapsto g_j(\theta, \lambda)$ is the holonomy of A along the loop $\theta \mapsto \gamma_j(\theta, \lambda)$. The vector field X_s is related to the Hamiltonian H_s as above for $0 \leq s \leq 1$. Both can be extended to $s \in \mathbf{R}$ by (28) and (29).

Remark 7.2 Any smooth Hamiltonian function $H_s : \mathcal{M}(P) \to \mathbf{R}$ can be represented in the form $H_s = h_s \circ \rho_\lambda : \mathcal{A}_{\text{flat}}(P) \to \mathbf{R}$ where $h_s : \mathrm{SU}(2)^{2k} \to \mathbf{R}$ is invariant under conjugacy. The functions $h_s : \mathrm{SU}(2)^{2k} \to \mathbf{R}$ can be chosen such that $\phi_{f,H}$ has only nondegenerate fixed points. We do not assume here that H_s is invariant under the action of Γ_f. This would require a transversality theorem which takes account of the action of a finite group.

The perturbed Chern-Simons functional $\mathcal{CS}_H : \mathcal{A}(P_f) \to \mathbf{R}$ is defined by

$$\mathcal{CS}_H(a) = \mathcal{CS}(a) - \int_0^1 H_s(A(s))\, ds.$$

for $a = A + \Phi\, ds \in \mathcal{A}(P_f)$. Its differential is given by

$$d\mathcal{CS}_H(A+\Phi\, ds)(\alpha+\phi\, ds) = \int_0^1 \int_\Sigma \Big(\langle (\dot{A} - d_A\Phi - X_s(A)) \wedge \alpha\rangle + \langle F_A \wedge \phi\rangle\Big)\, ds.$$

Hence a connection $A + \Phi\, ds$ on P_f is a critical point of $\mathcal{CS}_H$ if and only if $A(s)$ is flat for every s and

$$\dot{A} - d_A\Phi - X_s(A) = 0. \tag{30}$$

Now the restriction of H_s to the moduli space $\mathcal{M}(P)$ of flat connections is in the class of perturbations for symplectic Floer homology considered in the proof of Theorem 3.3. In other words the restriction of $\mathcal{CS}_H$ to the space of paths $\Omega_{\phi_f} = \mathcal{A}_\Sigma(P_f)/\mathcal{G}_\Sigma(P_f)$ is the perturbed symplectic action functional and this restriction has the same critical points as $\mathcal{CS}_H$. They are in one-to-one correspondence with the fixed points of the symplectomorphism $\phi_{f,H}$. The class of perturbations discussed here is large enough in order to obtain nondegenerate critical points.

A Proof of Lemma 2.3

Let $Q \to M$ be a principal $SO(3)$-bundle over a compact 3-manifold. Here we denote the unit interval by $I = [0, 1]$.

Lemma A.1 *For every integer $k \in \mathbf{Z}$ there exists a gauge transformation $g \in \mathcal{G}(Q)$ with* $\deg(g) = 2k$ *and* $\eta(g) = 0$.

Proof: The condition $\eta(g) = 0$ means that $g : Q \to G$ lifts to a map $\tilde{g} : Q \to SU(2)$. Let $\iota : \Sigma \to M$ be an embedding of an oriented Riemann surface. Then ι extends to an embedding of a tubular neighborhood $\iota : \Sigma \times I \to M$. Cut out a disc $D \subset \Sigma$ and define $\tilde{g} = 1$ outside $\iota(D \times I)$. Now trivialize Q over $\iota(D \times I)$ and choose $\tilde{g} \circ \iota : D \times I \to SU(2)$ of degree k with $\tilde{g} \circ \iota(\partial(D \times I)) = 1$. □

Lemma A.2 *Let $\iota : \Sigma \to M$ be an embedding of an oriented Riemann surface with $w_2(\iota^* Q) = j \in \{0, 1\}$. Then there exists a gauge transformation $g : Q \to SO(3)$ of degree j whose parity*

$$\eta(g) = \eta_\Sigma : \pi_1(M) \to \mathbf{Z}_2$$

is given by the intersection number of a loop with Σ.

Proof: Extend ι to an embedding of a tubular neighborhood $\iota : \Sigma \times I \to M$.

If $w_2(\iota^* Q) = 0$ define $g \circ \iota(z, t) = g_0(t)$ for $z \in \Sigma$ and $t \in I$ where $g_0(t)$ is a nontrivial loop in $SO(3)$ with $g_0(0) = g_0(1) = 1$. Extend by $g(p) = 1$ outside $\iota(\Sigma \times I)$.

If $w_2(\iota^* Q) = 1$ define $g(p) = 1$ for $p \notin \iota(\Sigma \times I)$ as above. Moreover decompose Σ as

$$\Sigma = \Sigma_1 \cup \Sigma_2$$

where $\Sigma_1 = D \subset \Sigma$ is a disc and $\Sigma_2 = \mathrm{cl}(\Sigma \setminus D)$. Choose sections $\sigma_j : \Sigma_j \times I \to Q|_{\iota(\Sigma_j \times I)}$ for $j = 1, 2$ such that

$$\sigma_2(e^{2\pi i\theta}) = \sigma_1(e^{2\pi i\theta}) g_0(\theta).$$

Here $g_0(\theta)$ is the loop in $SO(3)$ covered by

$$\gamma(\theta) = \begin{pmatrix} e^{\pi i\theta} & 0 \\ 0 & e^{-\pi i\theta} \end{pmatrix} \in \mathrm{SU}(2).$$

Define g on $\iota(\Sigma \times I)$ by

$$g \circ \sigma_j(z, t) = g_0(t), \qquad z \in \Sigma_j, \quad t \in I.$$

This is consistent with the patching condition since $g_0(\theta) g_0(t) g_0(\theta)^{-1} = g_0(t)$.

We prove that $h = g \cdot g$ lifts to $\tilde{h} : Q \to SU(2)$ and that $\tilde{h}$ is of degree 1. To see this note that

$$\tilde{h} \circ \sigma_j(z, t) = \gamma(2t), \qquad z \in \Sigma_j, \quad t \in I,$$

and $\tilde{h}(p) = 1$ for $p \notin \iota(\Sigma \times I)$. Now choose a continuous function

$$\beta(r, t) = \begin{pmatrix} a + ib & c \\ -c & a - ib \end{pmatrix} \in \mathrm{SU}(2), \qquad 0 \le t \le 1 \le r \le 2.$$

such that $\beta(1,t) = \gamma(2t)$, $\beta(r,0) = \beta(r,1) = \beta(2,t) = \mathbb{1}$, $a^2+b^2+c^2 = 1$ and $c \geq 0$. (Contract the equator over a hemisphere.) Moreover, assume that $\sigma_1 : D \times I \to Q$ extends to the disc D_2 of radius 2 with the overlap condition $\sigma_2(re^{2\pi it}) = \sigma_1(re^{2\pi it})g_0(t)$ for $1 \leq r \leq 2$. Also assume, up to homotopy, that $\tilde{h} \circ \sigma_2(z,t) = 1$ for $z \notin D_2$ and

$$\tilde{h} \circ \sigma_2(z,t) = \beta(|z|,t), \qquad 1 \leq |z| \leq 2, \quad t \in I.$$

By the overlap condition this implies

$$\tilde{h} \circ \sigma_1(z,t) = \gamma(\theta)\beta(r,t)\gamma(\theta)^{-1} = \begin{pmatrix} a+ib & ce^{2\pi i\theta} \\ -ce^{-2\pi i\theta} & a-ib \end{pmatrix}$$

for $1 \leq |z| \leq 2$ and $t \in I$. Hence it follows from the definition of β that $\tilde{h} \circ \sigma_1(z,t) = 1$ whenever $|z| = 2$ or $t = 0, 1$. Thus $\tilde{h} \circ \sigma_1$ defined a map from $D_2 \times I/\partial(D_2 \times I) \to \mathrm{SU}(2)$. Since $c \geq 0$, this map is of degree 1. □

Lemma A.3 *If $g \in \mathcal{G}(Q)$ with* $\deg(g) = 0$ *and* $\eta(g) = 0$ *then g is homotopic to* 1.

Proof: If $\eta(g) = 0$ then g lifts to $SU(2) = S^3$. Hence the statement follows from the Hopf degree theorem. □

Proof of Lemma 2.3: Statement (1) follows immediately from Lemma A.3.

We prove statement (2). By (H1) every cohomology class $\eta : \pi_1(M) \to \mathbb{Z}_2$ can be represented by finitely many embedded oriented Riemann surfaces $\iota_j : \Sigma_j \to M$. The associated gauge transformations $g_j : Q \to G$ constructed in Lemma A.2 all satisfy (1). This together with Lemma A.1 implies that for every $k \in \mathbb{Z}$ and every $\eta \in H^1(M; \mathbb{Z}_2)$ with $k \equiv w_2(Q) \cdot \eta \pmod 2$ there exists a gauge transformation $g \in \mathcal{G}(Q)$ such that $\deg(g) = k$ and $\eta(g) = \eta$.

Conversely, we must prove that every gauge transformation $g \in \mathcal{G}(Q)$ satisfies (1). By Lemma A.2 we may assume that $\eta(g) = 0$. This implies that g lifts to a map $\tilde{g} : Q \to SU(2)$ and hence $\deg(g)$ is even.

We prove statement (3). By (H2) there exists an embedding $\iota : \Sigma \to M$ of an oriented Riemann surface such that $w_2(\iota^* Q) = 1$. So it follows from Lemma A.2 that there is a gauge transformation of degree 1. □

References

[1] M.F. Atiyah, New invariants of three and four dimensional manifolds, *Proc. Symp. Pure Math.* **48** (1988).

[2] M.F. Atiyah and R. Bott, The Yang Mills equations over Riemann surfaces, *Phil. Trans. R. Soc. Lond. A* **308** (1982), 523–615.

[3] M.F. Atiyah, V.K. Patodi and I.M. Singer, Spectral asymmetry and Riemannian geometry III, *Math. Proc. Camb. Phil. Soc.* **79** (1976), 71–99.

[4] S.E. Cappell, R. Lee and E.Y. Miller, Self-adjoint elliptic operators and manifold decomposition I: general techniques and applications to Casson's invariant, Preprint, 1990.

[5] C.C. Conley and E. Zehnder, Morse-type index theory for flows and periodic solutions of Hamiltonian equations, *Commun. Pure Appl. Math.* **37** (1984), 207–253.

[6] G.D. Daskalopoulos and K.K. Uhlenbeck, An application of transversality to the topology of the moduli space of stable bundles, Preprint, 1990.

[7] S. Dostoglou and D.A. Salamon, Cauchy-Riemann operators, self-duality, and the spectral flow, Preprint, 1992.

[8] S. Donaldson, M. Furuta and D. Kotschick, *Floer homology groups in Yang-Mills theory*, in preparation.

[9] A. Floer, Morse theory for Lagrangian intersections, *J. Diff. Geom.* **28** (1988), 513–547.

[10] A. Floer, A relative Morse index for the symplectic action, *Commun. Pure Appl. Math.* **41** (1988), 393–407.

[11] A. Floer, An instanton invariant for 3-manifolds, *Commun. Math. Phys.* **118** (1988), 215–240.

[12] A. Floer, Symplectic fixed points and holomorphic spheres, *Commun. Math. Phys.* **120** (1989), 575–611.

[13] A. Floer, Instanton homology and Dehn surgery, Preprint 1991.

[14] A. Floer, Instanton homology for knots, Preprint 1991.

[15] A. Floer and H. Hofer, Coherent orientations for periodic orbit problems in symplectic geometry, Preprint, Ruhr-Universität Bochum, 1990.

[16] M. Gromov, Pseudoholomorphic curves in symplectic manifolds, *Invent. Math.* **82** (1985), 307–347.

[17] H. Hofer and D.A. Salamon, Floer homology and Novikov rings, Preprint, 1991.

[18] J. Jones, J. Rawnsley and D. Salamon, Instanton homology and Donaldson polynomials, Preprint, 1991.

[19] T. Kato, *Perturbation Theory for Linear Operators* Springer-Verlag, 1976.

[20] D. McDuff, Elliptic methods in symplectic geometry, *Bull. A.M.AS.* **23** (1990), 311–358.

[21] P.E. Newsteadt, Topological properties of some spaces of stable bundles, *Topology* **6** (1967), 241–262.

[22] T.R. Ramadas, I.M. Singer and J. Weitsman, Some comments on Chern-Simons gauge theory *Commun. Math. Phys.* **126** (1989), 421–431.

[23] J. Sacks and K. Uhlenbeck, The existence of minimal immersions of 2-spheres, *Annals of Math.* **113** (1981), 1–24.

[24] D. Salamon, Morse theory, the Conley index and Floer homology, *Bull. L.M.S.* **22** (1990), 113–140.

[25] D. Salamon and E. Zehnder, Floer homology, the Maslov index and periodic orbits of Hamiltonian equations, *'Analysis et cetera'*, edited by P.H. Rabinowitz and E. Zehnder, Academic Press, 1990, pp. 573–600.

[26] S. Smale, An infinite dimensional version of Sard's theorem, *Am. J. Math.* **87** (1973), 213–221.

[27] C.H. Taubes, Casson's invariant and gauge theory, *J. Diff. Geom.* **31** (1990), 547–599.

[28] K. Uhlenbeck, Connections with L^p bounds on curvature *Commun. Math. Phys.* **83** (1982), 31–42.

[29] C. Viterbo, Intersections de sous-variétés Lagrangiennes, fonctionelles d'action et indice des systèmes Hamiltoniens, *Bull. Soc. Math. France* **115** (1987), 361–390.

[30] E. Witten, Supersymmetry and Morse theory, *J. Diff. Geom.* **17** (1982), 661–692.

[31] J.G. Wolfson, Gromov's compactness of pseudoholomorphic curves and symplectic geometry, *J. Diff. Geom.* **23** (1988), 383–405.

[32] T. Yoshida, Floer homology and splittings of manifolds, *Ann. of Math.* **134** (1991), 277–323.

An Energy–Capacity Inequality for the Symplectic Holonomy of Hypersurfaces Flat at Infinity

Y. Eliashberg*and H. Hofer†
Department of Mathematics*
Stanford University
California, 94305, USA
and
Institut für Mathematik†
Ruhr-Universität Bochum
4630 Bochum, Germany

March 12, 1993

Contents

1 Introduction

The main purpose of this paper is the energy–capacity inequality for a special class of so–called flat at infinity hypersurfaces. The need for such an inequality, which generalizes some results from [7] and [12], appeared in our study of

*supported by NSF Grant DMS - 900 6179
†supported partially by DFG–SFB 237 and DAAD–Procope

boundaries of symplectic manifolds (see [2]). The Corollary 1.4 of Theorem 1.2 plays the crucial role in [4].

Consider $\mathbf{C}^{n+1} = \mathbf{C}^n \oplus \mathbf{C}$ as a real vector space. We denote by $(*,*)$ the standard–$\mathbf{C}$–valued Hermitian inner product and by $\omega = -\operatorname{Im}(*,*)$ the associated symplectic form.

A flat at infinity hypersurface (abbreviated **FIH** in the following) Σ in $\mathbf{C}^{n+1}$ is a hypersurface Σ diffeomorphic to $\mathbf{R}^{2n+1}$ such that there exists a compact subset K of $\mathbf{C}^{n+1}$ with

$$\Sigma \setminus K = (\mathbf{C}^n \oplus \mathbf{R}) \setminus K \ .$$

We denote by $\mathcal{D}$ the group of compactly supported symplectic diffeomorphisms in $\mathbf{C}^{n+1}$. We define the energy $e(\Sigma)$ of an asymptotically flat hypersurface as follows

$$e(\Sigma) = \inf\left\{2a(c^+ - c^-) \mid a, c^+ \text{ and } c^- \text{ satisfy (1)}\right\} \ ,$$

where

$$\begin{aligned}&\text{There exists } \Psi \in \mathcal{D} \text{ such that}\\ &\Psi(\Sigma) \subset (\mathbf{C}^n \oplus \mathbf{R}) \cup (\mathbf{C}^n \oplus (-a, a) \oplus i(c^-, c^+)) \ .\end{aligned} \tag{1}$$

The energy is a very rough quantity measuring the C^0–size of Σ compared with $\mathbf{C}^n \oplus \mathbf{R}$. Surprisingly, $e(\Sigma)$ also measures some C^1–properties of Σ. The goal of this paper will be to show this.

We define a line bundle $\mathcal{L}_\Sigma \longrightarrow \Sigma$ over Σ, where the fibre over $z \in \Sigma$ consists of all $\xi \in T_z\Sigma$ such that

$$\omega(\xi, \eta) = 0 \quad \text{for all} \quad \eta \in T_z\Sigma \ . \tag{2}$$

The one–dimensional distribution $\mathcal{L}_\Sigma$ defines a foliation on Σ. The leaf through some point $z \in \Sigma$ will be denoted by $L_\Sigma(z)$.

In order to introduce the holonomy associated to Σ, pick a number $a > 0$ (which is supposed to be large enough) and define

$$H_a^\pm = \left\{z \in \mathbf{C}^{n+1} \mid z_{n+1} \in (\pm a) + i\mathbf{R}\right\} \ . \tag{3}$$

For all large $a > 0$ we readily see that the leaves of the foliation are transversal to H_a^- and H_a^+. For $z \in \Sigma \cap H_a^- = (\mathbf{C}^n \oplus \mathbf{R}) \cap H_a^-$ we consider $L_\Sigma(a) \cap H_a^+$. We denote the collection of all points in $H_a^- \cap \Sigma$ with $L_\Sigma(z) \cap H_a^+ \neq \emptyset$ by U_Σ. If $a > 0$ is large enough the latter intersection will consist of precisely one point denoted by $\varphi_\Sigma(z)$. With the obvious identification we have $U_\Sigma \subset \mathbf{C}^n$ and consider φ_Σ as a map from U_Σ into $\mathbf{C}^n$. One easily verifies that U_Σ is open and dense and φ_Σ is the identity outside of a compact set. We call φ_Σ the holonomy of Σ and note that φ_Σ is a symplectic map.

Next let us recall the definition of a symplectic capacity as given in [9] (see [2, 3, 6, 10, 11] for more information). We adapt the definition to our special situation. Denote by $\mathcal{H}$ the collection of all smooth Hamiltonians $H : \mathbf{C}^n \longrightarrow [0, +\infty)$ satisfying:

$$\begin{array}{l}\text{There exists a number } m(H) > 0 \text{ such that } 0 \leq H \leq m(H).\\ \text{Moreover } m(H) - H \text{ has compact support and there exists}\\ \text{an open nonempty set } U \text{ with } H|U \equiv 0 \ .\end{array} \tag{4}$$

Given $H \in \mathcal{H}$ we consider the Hamiltonian system

$$\dot{x} = X_H(x) \tag{5}$$

and study its T–periodic solutions with $0 < T \leq 1$. We call H admissible and write $H \in \mathcal{H}_{ad}$ provided every T–periodic solution of (5) with $T \in (0, 1]$ is constant. It is an easy exercise to show that $\mathcal{H}_{ad}$ is nonempty. Given a nonempty domain $W \subset \mathbf{C}^n$ we denote by $\mathcal{H}_{ad}(W)$ the subset of $\mathcal{H}_{ad}$ consisting of all $H \in \mathcal{H}_{ad}$ with $\text{supp}(m(H) - H) \subset W$. Again $\mathcal{H}_{ad}(W) \neq \emptyset$. Finally, we define the following [9]:

$$c(W) = \sup\{m(H) \mid H \in \mathcal{H}_{ad}(W)\} \ . \tag{6}$$

It has been proved in [9] that

Proposition 1.1 *The number* $c(W) \in (0, +\infty]$ *satisfies*

(i) $c(B^{2n}(1)) = c(B^2(1) \times \mathbf{C}^{n-1}) = \pi$.

(ii) *If* $\Psi : W \hookrightarrow V$ *is an embedding satisfying* $\Psi^*\omega = \alpha\omega$ *for some* $\alpha \neq 0$, *then* $c(V) \geq |\alpha| \, c(W)$.

$c(W)$ is called the (symplectic) capacity of W. For more information see [2, 3, 7, 6, 9, 12]. Our main result is the following:

Theorem 1.2 *Let* Σ *be an asymptotically flat hypersurface in* $\mathbf{C}^{n+1}$ *with holonomy* $\varphi_\Sigma : U_\Sigma \longrightarrow \mathbf{C}^n$ *and energy* $e(\Sigma)$. *Then for every open domain* $\Omega \subset \mathbf{C}^n$ *with* $\overline{\Omega} \subset U_\Sigma$ *and capacity* $c(\Omega)$ *satisfying*

$$c(\Omega) > e(\Sigma) \ , \tag{7}$$

we have

$$\varphi_\Sigma(\overline{\Omega}) \cap \overline{\Omega} \neq \emptyset \ . \tag{8}$$

Our first corollary is

Corollary 1.3 *For every asymptotically flat hypersurface* Σ *we have the inequality*

$$\begin{array}{ll}\inf\{\alpha > 0 & | \quad \text{For all domains } \Omega \subset \mathbf{C}^n \\ & \quad\ \ \text{with } \overline{\Omega} \subset U_\Sigma \text{ and } c(\Omega) \geq \alpha \\ & \quad\ \ \text{we have } \varphi_\Sigma(\overline{\Omega}) \cap \overline{\Omega} \neq \emptyset\} \\ \leq e(\Sigma) \ . & \end{array}$$

We call the above inequality the *energy–capacity inequality*. Further, Theorem 1.2 implies

Corollary 1.4 *Suppose (Σ_k) is a sequence of asymptotically flat hypersurfaces such that $e(\Sigma_k) \longrightarrow 0$. Assume there exists an open dense set $U \subset \mathbf{C}^n$ such that $U_{\Sigma_k} \supset U$ for every $k \in \mathbf{N}$, and φ_{Σ_k} converges in the compact open topology to some continuous map $\varphi : U \longrightarrow \mathbf{C}^n$. Then $\varphi = \mathrm{Id}_U$.*

The main result of this paper will be important in [4], where we study the question to which extent the symplectic topology of the interior of a domain determines the symplectic properties of the boundary.

The present paper, in particular Corollary 1.4 is strongly related to results in [7] and [12], where an affirmative answer is given to a conjecture of the first author.

The present set up, however, is much more general.

2 Functional Analysis of the Action Integral

We define an unbounded selfadjoint operator L with domain $D(L)$ in $L^2(0,2;\mathbf{C}^{n+1})$ by

$$D(L) = \{ \begin{array}{l} z \in H^{1,2}(0,2;\mathbf{C}^{n+1}) \mid z_k(0) = z_k(2) \text{ for} \\ k = 1,\dots,n \text{ and } \mathrm{Re} z_{n+1}(0) = \mathrm{Re} z_{n+1}(2) = 0\} \end{array} \tag{9}$$

and

$$Lz = -i\dot{z} := -i\frac{d}{dt}z \; . \tag{10}$$

The splitting $L^2(0,2;\mathbf{C}^{n+1}) = L^2(0,2;\mathbf{C}^n) \oplus L^2(0,2;\mathbf{C})$ gives a splitting $L = L_1 \oplus L_2$ with

$$\begin{aligned} D(L_1) &= \{z \in H^{1,2}(0,2;\mathbf{C}^n) \mid z(0) = z(2)\} \\ D(L_2) &= \{z \in H^{1,2}(0,2;\mathbf{C}) \mid \mathrm{Re} z(0) = \mathrm{Re} z(1) = 0\} \; . \end{aligned}$$

We have for the spectrum of L_1 and L_2

$$\begin{aligned} \sigma(L_1) &= \pi\mathbf{Z} \\ \sigma(L_2) &= \frac{\pi}{2}\mathbf{Z} \; , \end{aligned}$$

where every element of $\sigma(L_j)$ is an eigenvalue of multiplicity $2n$ if $j = 1$ and multiplicity 1 if $j = 2$. If we write $z = (x, y)$ according to our splitting we have a Fourier expansion

$$\begin{array}{ll} x = \sum_{k\in\mathbf{Z}} x^{(k)} e^{\pi ikt} \quad , & x^{(k)} \in \mathbf{C}^n \\ y = \sum_{k\in\mathbf{Z}} y^{(k)} i e^{\frac{\pi}{2}ikt} \quad , & y^{(k)} \in \mathbf{R} \; , \end{array} \tag{11}$$

with the summability condition

$$\begin{aligned} \sum \left|x^{(k)}\right|^2 &< \infty \\ \sum \left|y^{(k)}\right|^2 &< \infty \end{aligned} \tag{12}$$

for $z \in L^2(0,2;\mathbf{C}^{n+1})$. We denote by $E = E_1 \oplus E_2$ the domain of $|L|^{\frac{1}{2}} = |L_1|^{\frac{1}{2}} \oplus |L_2|^{\frac{1}{2}}$. On E we define a Hilbertspace norm by

$$\begin{aligned} \|z\|^2 &= 2\pi\left(\sum |k| \left|x^{(k)}\right|^2\right) + \pi\left(\sum |k| \left|y^{(k)}\right|^2\right) \\ &\quad +2\left(\left|x^{(0)}\right|^2 + \left|y^{(0)}\right|^2\right) . \end{aligned} \tag{13}$$

The norm $\|\ \ \|$ is induced by an inner product $(*,*)$.E also admits a splitting according to the parts of $\sigma(L)$ contained in $(-\infty,0), \{0\}$, and $(0,+\infty)$. We write this as

$$E = E^- \oplus E^0 \oplus E^+ .$$

Note that for example the elements in E^- are Fourier series with summation over $k < 0$.

We define a smooth map $\Phi : E \longrightarrow \mathbf{R}$ by

$$\Phi(z) = \frac{1}{2}\|z^-\|^2 - \frac{1}{2}\|z^+\|^2 + a\int_0^2 \mathrm{Im}(z_{n+1}(t))dt ,$$

where $a > 0$ is some fixed number. We note that Φ does not have any critical points. Our proof of the main result will rely on a careful study of suitable compact perturbations of Φ for a suitable choice of the constant $a > 0$. The underlying nonlinear Functional analysis goes back to [1, 5] and our approach is close to [4, 7, 8, 9].

We begin by defining a suitable group $\mathcal{B}$ of homeomorphism $h : E \longrightarrow E$. A homeomorphism h belongs to $\mathcal{B}$ provided there exist continuous maps $\gamma^{\pm} : E \longrightarrow \mathbf{R}$ and $k : E \longrightarrow E$ such that γ^+, γ^- and k map bounded sets into compact sets. Further, there exists a constant $R = R(h) > 0$ such that $\gamma^{\pm}(z) = 0, k(z) = 0$ for $z \in E^-$ with $\|z\| \geq R$. Moreover, h can be written in the form

$$\begin{aligned} h(z) &= e^{\gamma^-(z)}z^- + z^0 + e^{\gamma^+(z)}z^+ - k(z) \\ z &= z^- + z^0 + z^+ \in E^- \oplus E^0 \oplus E^+ . \end{aligned} \tag{14}$$

The group $\mathcal{B}$ is actually very large and contains all necessary deformations for a variational theory of a suitable global pertubation of Φ. We note that $\mathcal{B}$ differs somewhat from the group introduced in [2]. We need

Lemma 2.1 *For every $h \in \mathcal{B}$ we have*

$$h(E^-) \cap (E^0 \oplus E^+) \neq \emptyset .$$

Proof: We have to find $z \in E^-$ with $P^- h(z) = 0$, where $P^- : E \longrightarrow E^-$ is the orthogonal projection. This equation is equivalent to

$$z = e^{-\gamma^-(z)} P^- k(z) =: T(z)$$
$$z \in E^- \ .$$

Since $\overline{T(E^-)}$ is compact, by our assumptions on $h \in \mathcal{B}$, we find a fixed point by Schauder's fixed point theorem. □

We say a functional $\Psi \in C^\infty(E, \mathbf{R})$ satisfies the weak Palais–Smale condition (WPS) on level $d \in \mathbf{R}$ provided the following holds

$$\begin{array}{l} \text{If } (z_k) \subset E \text{ with } \Psi(z_k) \to d \text{ and } \Psi'(z_k) \to 0 \\ \text{then } d \text{ is a critical level of } \Psi. \end{array} \tag{15}$$

Let $b : E \longrightarrow \mathbf{R}$ be a smooth map and assume $b' : E \longrightarrow E$ maps bounded sets into compact sets. We define Φ_b by

$$\Phi_b = \Phi - b \ .$$

We assume that

$$\Phi_b(z) \longrightarrow +\infty \quad \text{for} \quad \|z\| \longrightarrow +\infty, z \in E^-$$

and

$$\operatorname{supp} \Phi_b(E^0 \oplus E^+) < +\infty \ .$$

We define a number d by

$$d = \sup_{h \in \mathcal{B}} \inf_{z \in E^-} \Phi_b(h(z)) \ . \tag{16}$$

Our main existence result for critical points is

Proposition 2.2 *Assume b is as described above. Then $d \in \mathbf{R}$ and if $(WPS)_d$ holds d is a critical level.*

Proof: By our assumption on b we have

$$\sup \Phi_b(E^+ \oplus E^0) < +\infty \ , \ \inf \Phi_b(E^-) > -\infty \ .$$

By Lemma 2.1 we know that

$$h(E^-) \cap (E^0 \oplus E^+) \neq \emptyset \ .$$

Hence

$$\inf \Phi_b(h(E^-)) \leq \sup \Phi_b(E^0 \oplus E^+) < +\infty \ .$$

This proves

$$-\infty < \inf \Phi_b(E^-) \le d \le \sup \Phi_b(E^0 \oplus E^+) < +\infty \ .$$

Thus we know $d \in \mathbf{R}$. Arguing indirectly let us assume that d is not a critical level. In view of $(WPS)_d$ we find $\epsilon > 0$ such that

$$\|\Psi'(z)\| \ \ge \epsilon \quad \text{if} \quad \Psi(z) \in [d-\epsilon, d+\epsilon] \ , \tag{17}$$

where $\Psi = \Phi - b$. We find a smooth map $\beta : E \longrightarrow [0,1]$ satisfying

$$\begin{array}{lll} \beta(z) = 0 & \text{if} & \Psi(z) \notin [d-2\epsilon, d+2\epsilon] \\ \beta(z) = 1 & \text{if} & \Psi(z) \in [d-\epsilon, d+\epsilon] \ . \end{array} \tag{18}$$

Finally define a globally Lipschitz continuous map $\delta : \mathbf{R} \longrightarrow [0,1]$ by

$$\delta(s) = \left[\begin{array}{lll} 1 & \text{if} & s \le 1 \\ \frac{1}{s} & \text{if} & s > 1 \ . \end{array} \right. \tag{19}$$

Using Ψ', β and δ we define a bounded locally Lipschitz continuous vector field by

$$G(z) = \beta(z)\delta(\ \|\Psi'(z)\|\ )\,\Psi'(z) \ . \tag{20}$$

By our assumption on b we find $R > 0$ such that $\Psi(z) \ge d + 2\epsilon$ for $z \in E^-$ and $\|z\| \ \ge R$. Hence $G(z) = 0$ for $z \in E^-$ with $\|z\| \ \ge R$. In view of the estimate (17) we find a number $T > 0$, such that for $h : E \longrightarrow E$ being the time–T–map of the flow associated to the ordinary differential equation

$$\dot{z} = G(z) \ , \tag{21}$$

we have

$$\Phi_b(h(z)) \ge d + \epsilon \text{ provided } \Phi_b(z) \ge d - \epsilon \ . \tag{22}$$

Using the variation of constant formula we deduce the following representation for h

$$h(z) = e^{\gamma^-(z)} z^- + z^0 + e^{\gamma^+(z)} z^+ - k(z) \ , \tag{23}$$

see [2]. Hence $h \in \mathcal{B}$. By the definition of d we find a $h_0 \in \mathcal{B}$ such that

$$\Psi(h_0(E^-)) \subset [d-\epsilon, +\infty) \ . \tag{24}$$

Combining (22) and (23) gives

$$\Psi(h \circ h_0(E^-)) \subset [d+\epsilon, +\infty) \ . \tag{25}$$

Hence

$$d \ge \inf \Psi(h \circ h_0(E^-)) \ge d + \epsilon \ .$$

Since $d \in \mathbf{R}$ this gives a contradiction in view of $h \circ h_0 \in \mathcal{B}$. □

The following corollary will be frequently used.

Corollary 2.3 *Let b be as described above and assume Φ_b has a unique critical level d_0 and satisfies $(WPS)_d$ for every $d \in \mathbf{R}$. Then we have the identity*

$$d_0 = \sup_{h \in \mathcal{B}} \inf_{z \in E^-} \Phi_b(h(z)) \ . \tag{26}$$

Proof: In view of Proposition 2.2 the righthand side of (26) has to be a critical level d. By uniqueness $d = d_0$. □

3 A Weak PS–Condition for a Class of Functionals

Let $a > 0$ in the definition of Φ be fixed. Consider a smooth Hamiltonian $K : \mathbf{C}^{n+1} \longrightarrow [0, +\infty)$ having the following properties:

There exists a constant $R_1 < 0$ such that $K(z) = 0$ for $\mathrm{Im}(z_{n+1}) \leq R_1$. (27)

Moreover:

There exists a constant $R_2 > 0$ such that $K(z) = \varphi(\mathrm{Im} z_{n+1})$ for $\mathrm{Im}(z_{n+1}) \geq R_2$ or $\sum_{k=1}^n |z_k|^2 \geq R_2^2$, where $\varphi : \mathbf{R} \to [0, +\infty)$ is a suitable smooth map satisfying $\varphi'(s) = \text{const} > 2a$ for all large $s > 0$. (28)

Next we choose a $H \in \mathcal{H}_{ad}$ and define $b : E \longrightarrow \mathbf{R}$ by

$$b(z) = \int_0^1 H(z_1, \ldots, z_n) dt + \int_1^2 K(z + \alpha) dt \ , \tag{29}$$

where

$$\alpha(t) = (0, \ldots, 0, -a + ta) \ . \tag{30}$$

Under our assumption on H and K the map $b : E \longrightarrow \mathbf{R}$ is smooth and extends to a C^1–map on $L^2(0, 2; \mathbf{C}^{n+1})$.

Denoting the L^2–gradient by ∇b we have the following relation between b' and ∇b

$$b'(z) = \left(|L| + P^0 \right)^{-1} \nabla b(z) \, , \; z \in E \ . \tag{31}$$

In view of (27) and (28) we see that $\nabla b(L^2)$ is L^2–bounded. This implies that

$$b'(E) \text{ is bounded in } D(L) \subset E, \tag{32}$$

where $D(L)$ is equipped with the graph norm.

Lemma 3.1 *Suppose (27) and (28) hold. Then $\Psi = \Phi - b$ satisfies*

$$\begin{aligned} &\Psi(z) \to +\infty \text{ for } z \in E^- \text{ with } \|z\| \to +\infty \\ &\sup \Psi(E^0 \oplus E^+) < +\infty \ . \end{aligned} \tag{33}$$

Proof: For $z \in E^-$ we estimate

$$\begin{aligned}\Psi(z) &= \frac{1}{2}\|z^-\|^2 - \frac{1}{2}\|z^+\|^2 + a\int_0^2 \operatorname{Im}(z_{n+1}(t))dt - b(z)\\ &\geq \frac{1}{2}\|z^-\|^2 - \text{const}\,(1 + \|z^-\|)\ ,\end{aligned}$$

for a suitable constant *const* not depending on $z \in E^-$. Here we used that (27), (28) imply that K has linear growth. Further we find a constant such that

$$2a\operatorname{Im}(z_{n+1}) - K(z + \alpha(t)) \leq \text{const}\ ,$$

for all $z \in \mathbf{C}^{n+1}$ and $t \in [0,2]$. For $z \in E^0 \oplus E^+$ we estimate

$$\begin{aligned}\Psi(z) &= -\tfrac{1}{2}\|z^+\|^2 + a\textstyle\int_0^2 \operatorname{Im}(z_{n+1}(t))dt\\ &\quad - \textstyle\int_0^1 H(z_1(t),\ldots,z_n(t))dt - \int_1^2 K(z+\alpha)dt\\ &\leq -\tfrac{1}{2}\|z^+\|^2 + a\textstyle\int_0^2 \operatorname{Im}(z_{n+1}(t))dt\\ &\quad -2a\textstyle\int_1^2 \operatorname{Im}(z_{n+1}(t))dt + \text{const}\\ &\leq -\tfrac{1}{2}\|z^+\|^2 + a\textstyle\int_0^2 \operatorname{Im}(P^0 z_{n+1})dt\\ &\quad -2a\textstyle\int_1^2 \operatorname{Im}(P^0 z_{n+1})dt\\ &\quad + \text{const}(1 + \|z^+\|)\\ &= -\tfrac{1}{2}\|z^+\|^2 + \text{const}(1 + \|z^+\|)\ .\end{aligned} \tag{34}$$

□

The next step is to show that $\Psi = \Phi - b$ satisfies $(WPS)_d$ for every $d \in \mathbf{R}$ provided (27) and (28) hold.

Proposition 3.2 *Let $(z(j)) \subset E$ be a sequence such that*

$$\Psi'(z(j)) \longrightarrow 0 \text{ and } \Psi(z(j)) \longrightarrow d \in \mathbf{R}. \tag{35}$$

Then d is a critical level of Ψ.

Proof: By (32) we know that $b'(E)$ is bounded in $D(L)$. Hence $(P^- + P^+)b'(E)$ is bouded in $D(L)$. Since $\Psi'(z(j)) \longrightarrow 0$ in E we see that $(P^- + P^+)z(j)$ can be written as

$$(P^- + P^+)z(j) = u(j) + v(j)$$

with $(v(j))$ bounded in $D(L)$ and $u(j) \longrightarrow 0$ in E. By the compact embedding $D(L) \hookrightarrow E$ we may assume without loss of generality (after passing to a subsequence)

$$\begin{aligned}v(j) &\rightharpoonup v \in D(L)\ , v(j) \to v \text{ in } E\\ u(j) &\to 0 \text{ in } E.\end{aligned}$$

Hence

$$(P^- + P^+)(z(j)) \longrightarrow v \text{ in } E$$
$$\text{with } v \in D(L).$$

If $(P^0 u(j))$ has a bounded subsequence we are done since $\dim P^0 E < \infty$. In that case our discussion implies that $(z(j))$ has a subsequence converging in E to some z. Taking the limits in (35) implies then the desired result. Hence we may assume $\|P^0 z(j)\| \longrightarrow +\infty$. Assume first $(P^0 z(j))$ has a subsequence such that the $(n+1)$–component tends in norm to $+\infty$. So, after passing to a subsequence let us assume $|(P^0 z(j))_{n+1}| \longrightarrow +\infty$. Using (27) and (28) we see immediately that

$$\left| a \int_0^2 \mathrm{Im}(z(j)_{n+1}) dt - b(z(j)) \right| \longrightarrow +\infty$$

as $j \longrightarrow +\infty$. Since $(\frac{1}{2}\, \|z(j)^-\|^2 - \frac{1}{2}\, \|z(j)^+\|^2\,)$ is bounded we infer

$$|\Psi(z(j))| \longrightarrow +\infty \ ,$$

contradicting our assumption $\Psi(z(j)) \longrightarrow d \in \mathbf{R}$. Hence we may assume that the $(n+1)$–component of $(P^0 z(j))$ is bounded. Let us write $(P^0 z(j)_r)$ for the first n components. Hence we know so far

$$\begin{array}{rcl} \|P^0 z(j)_{n+1}\| & \leq & \text{const} \\ \|P^0 z(j)_r\| & \longrightarrow & +\infty \end{array}$$

and may assume after taking a sequence

$$P^0 z(j)_{n+1} \longrightarrow i\tau \ ,$$

where $\tau \in \mathbf{R}$. Hence, using (28)

$$K'(z(j)) \longrightarrow (0, \ldots, 0, i\varphi'(\tau))$$

in $L^2(1,2;\mathbf{C}^{n+1})$ and

$$H'(z(j)_1, \ldots, z(j)_n) \longrightarrow 0$$

in $L^2(0,1;\mathbf{C}^n)$. This implies

$$b'(z(j)) \longrightarrow \left(|L| \, + P^0 \right)^{-1} (\chi(0, \ldots, 0, i\varphi'(\tau)))$$

in $D(L)$, where χ is the characteristic function of the interval $[1,2]$. Moreover

$$\Phi'(z(j)) = z(j)^- - z(j)^+ + \left(|L| \, + P^0 \right)^{-1} (0, \ldots, 0, ia) \ .$$

Since $\Psi'(z(j)) \longrightarrow 0$ we conclude

$$\begin{aligned} & v^- - v^+ + \left(|L| + P^0\right)^{-1}(0,\ldots,0,ia) \\ = & \left(|L| + P^0\right)^{-1}(\chi(0,\ldots,0,i\varphi'(\tau))) \ . \end{aligned}$$

Applying $|L| + P^0$ gives

$$\begin{aligned} & -Lv + (0,\ldots,ia) = \chi(0,\ldots,0,i\varphi'(\tau)) \\ & v \in D(L) \ . \end{aligned}$$

This precisely means

$$\begin{aligned} & i\left(\frac{dv}{dt} + \frac{d\alpha}{dt}\right) = \chi(0,\ldots,0,i\varphi'(\tau)) \\ & v \in D(L) \ . \end{aligned}$$

Let us put $u(t) = v(t) + \alpha(t)$. Then

$$\begin{aligned} \dot{u} &= \chi(t)(0,\ldots,0,\varphi'(\tau)) \text{ on } (0,2) \\ u_k(0) &= u_k(2) \text{ for } k = 1,\ldots,n \\ \operatorname{Re}(u_{n+1})(0) &= -a \,, \operatorname{Re}(u_{n+1}(2)) = a \ . \end{aligned}$$

This implies $u_1 = \ldots = u_n =$ const and $u_{n+1}(t) = -a + i\tau$ for $t \in [0,1], u_{n+1}(t) = -a + 2(t-1)a + i\tau$ for $t \in [1,2]$. Consequently $\varphi'(\tau) = 2a$.

It is clear by (27) and (28) that

$$d = \lim \Psi(z(j)) = 2a\tau - \varphi(\tau) - m(H) \ .$$

It remains to show that d is a critical level. Let $\hat{u}(t) = (0,\ldots,0,u_{n+1}(t))$, where u_{n+1} is as just defined. Let $z(t)$ be given by $z(t) = (0,\ldots,u_{n+1}(t) + a - ta) = \hat{u}(t) - \alpha(t)$. Finally consider

$$\gamma e_1 + z(t)$$

for $\gamma \geq 0$ a real number. For γ large we have

$$\begin{aligned} K'(\gamma e_1 + z(t)) &= (0,\ldots,0,i\varphi'(\tau)) \\ H'((\gamma e_1 + z(t))_\tau) &= 0 \ . \end{aligned}$$

Hence, for γ large the map $\gamma e_1 + z(t)$ which belongs to $D(L)$ is a critical point for Ψ with critical level

$$\Psi(\gamma e_1 + z) = 2a\tau - \varphi(\tau) - m(H) \ .$$

This completes the proof of Proposition (3.2). □

4 Some Estimates for Max–Min–Levels

Let $l \in \mathbf{R}$ and assume $\varphi : \mathbf{R} \longrightarrow [0, +\infty)$ is a smooth map satisfying

$$\begin{aligned} \varphi(s) &= 0 \text{ for } s \leq l \\ \varphi'(s) &> 0 \text{ for } s > l \\ \varphi'(s) &= \text{const} > 2a \text{ for } s \text{ large} \\ \varphi'(s_0) &= 2a \text{ has a unique solution.} \end{aligned} \tag{36}$$

Here "a" is again the number occurring in the definition of Φ. We denote by θ_l the collection of all φ's satisfying (36). We define $b : E \longrightarrow \mathbf{R}$ by

$$b(z) = \int_1^2 \varphi(\mathrm{Im} z_{n+1}(t))dt \ . \tag{37}$$

Then in view of 3.1 and 3.2 the functional Φ_b satisfies $(WPS)_d$ for every $d \in \mathbf{R}$ and

$$d = \sup_{h \in \mathcal{B}} \inf_{z \in E^-} \Phi_b(h(z))$$

is a real number and a critical level. A straight forward calculation shows that Φ_b has precisely one critical level d which satisfies $\Phi_b(z) = d$ with z of the form

$$z(t) = (c_1, \ldots, c_n, z_{n+1}(t)) \ ,$$

where $c_1, \ldots, c_n \in \mathbf{C}^n$ are any constants and

$$\begin{aligned} z_{n+1}(t) &= i\tau - ta \text{ for } t \in [0,1] \\ z_{n+1}(t) &= i\tau + ta - 2a \text{ for } t \in [1,2] \ , \end{aligned}$$

with $\varphi'(\tau) = 2a$ (which determines τ uniquely). We compute easily that

$$d = \Phi_b(z) = 2a\tau - \varphi(\tau) \ .$$

We have

Proposition 4.1 *For fixed l consider for $\varphi \in \theta_l$ the corresponding* min max *value $d(\varphi)$. Then*

$$\inf_{\varphi \in \theta_l} d(\varphi) = 2al \ .$$

Proof: By the previous discussion Φ_b with $b = b(\varphi)$ has a unique critical level

$$d(\varphi) = 2a\tau - \varphi(\tau) \ ,$$

where τ is determined by the equation

$$\varphi'(\tau) = 2a \ .$$

Given $\delta > 0$ and $\varphi \in \theta_l$ we find $\tilde{\varphi} \in \theta_l$ with

$$\tilde{\varphi} \geq \varphi \tag{38}$$

and

$$\tilde{\varphi}'(\tau) = 2a \Rightarrow \tilde{\varphi}(\tau) \leq \delta \ . \tag{39}$$

By the definition of $d(\varphi)$ it follows immediately that

$$d(\tilde{\varphi}) \leq d(\varphi) \ . \tag{40}$$

Moreover, (39) implies in view of $\tau > l$

$$\begin{aligned} d(\tilde{\varphi}) &= 2a\tau - \tilde{\varphi}(\tau) \\ &\geq 2al - \delta \ . \end{aligned} \tag{41}$$

Since δ was arbitrarily choosen we find in view of (40) and (41)

$$\inf_{\varphi \in \theta_l} d(\varphi) \geq 2al \ .$$

Since for given $\delta > 0$, we find $\tilde{\varphi} \in \theta_l$ with $\tilde{\varphi}'(\tau) = 2a$ for $\tau < l + \delta$, we deduce

$$\begin{aligned} d(\tilde{\varphi}) &\leq 2a(l+\delta) - \tilde{\varphi}(\tau) \\ &\leq 2al + 2a\delta \ . \end{aligned}$$

This implies the desired equality

$$\inf_{\varphi \in \theta_l} d(\varphi) = 2al$$

with

$$d(\varphi) = \sup_{h \in \mathcal{B}} \inf_{z \in E^-} \left[\Phi(h(z)) - \int_1^2 \varphi(\operatorname{Im} h(z)_{n+1}) dt \right] \ .$$

$\square$

Next assume $H \in \mathcal{H}_{ad}$. For $\tau \in [0,1]$ and $\varphi \in \theta_l$ we define b_τ by

$$\begin{aligned} b_\tau(z) &:= b_{\tau,\varphi,H}(z) \\ &= \tau \int_0^1 H(z_1, \ldots, z_n) dt + \int_1^2 \varphi(\operatorname{Im} z_{n+1}(t)) dt \ . \end{aligned}$$

Finally we put $\Phi_\tau := \Phi - b_\tau$. We have in view of Proposition 4.1 the estimate

$$\begin{aligned} & -m(H) + \Phi(z) - \textstyle\int_1^2 \varphi(\operatorname{Im} z_{n+1}) dt \\ \leq \ & \Phi(z) - \textstyle\int_0^1 H(z_1, \ldots, z_n) dt - \int_1^2 \varphi(\operatorname{Im} z_{n+1}) dt \\ \leq \ & \Phi(z) - \textstyle\int_1^2 \varphi(\operatorname{Im} z_{n+1}) dt \ . \end{aligned} \tag{42}$$

Moreover for all $z \in E$:

$$|\Phi_{\tau_2}(z) - \Phi_{\tau_1}(z)| \ \leq m(H) \, |\tau_2 - \tau_1| \ . \tag{43}$$

We define $d_\tau := d_{\tau,\varphi,H}$ by

$$d_\tau = \sup_{h\in\mathcal{B}} \inf_{z\in E^-} \Phi_\tau(h(z)) \ . \tag{44}$$

In view of (43) the map $\tau \longrightarrow d_\tau$ is continuous and non increasing. The crucial point is

Lemma 4.2 *$\tau \longrightarrow d_\tau$ is a constant map. In particular we have*

$$d(\varphi) = \sup_{h\in\mathcal{B}} \inf_{z\in h(E^-)} \left[\Phi(z) - \int_0^1 H(z_1,\ldots,z_n)dt - \int_1^2 \varphi(\operatorname{Im} z_{n+1})dt\right]$$

for every $H \in \mathcal{H}_{ad}$ and $\varphi \in \theta_l$.

Proof: H vanishes on some nonempty set U. Let $\delta \in U$. We define H_0 by

$$H_0(z) = \frac{\pi}{3}\left|(z_1,\ldots,z_n) - \delta\right|^2 \ . \tag{45}$$

For $\tau > 0$ small we have

$$H_0 \geq \tau H \ . \tag{46}$$

We define $\tilde{\Phi} : E \longrightarrow \mathbf{R}$ by

$$\tilde{\Phi}(z) = \Phi(z) - \int_0^1 H_0(z(t))dt - \int_1^2 \varphi(\operatorname{Im} z_{n+1})dt \ .$$

One estimates easily in view of (45) (recall $\sigma(L_1)$)

$$\begin{aligned} &\tilde{\Phi}(z) \longrightarrow +\infty \text{ for } \|z\| \rightarrow \infty \, , \ z \in E^- \\ &\sup \tilde{\Phi}(E^+ \oplus E^0) < \infty \ . \end{aligned} \tag{47}$$

Also $\tilde{\Phi}$ satisfies $(WPS)_d$ for every $d \in \mathbf{R}$. (In fact it satisfies even the usual Palais–Smale condition.) The fact that it satisfies (PS) follows as in Proposition 3.2 up to the point that the form (45) of H_0 even implies that $(\ \|P^0 z(j)\|\ )$ is bounded on a Palais–Smale sequence

$$\tilde{\Phi}(z(j)) \longrightarrow d, \tilde{\Phi}'(z(j)) \longrightarrow 0 \ .$$

(The argument is similar as in [8]. Further $\tilde{\Phi}$ has precisely one critical point z satisfying

$$\tilde{\Phi}(z) = d(\varphi) \ .$$

Hence in view of (46) for $\tau \geq 0$ small

$$\begin{aligned} d(\varphi) &= \sup_{h\in\mathcal{B}} \inf_{z\in E^-} \tilde{\Phi}(h(z)) \\ &\leq \sup_{h\in\mathcal{B}} \inf_{z\in E^-} \Phi_\tau(h(z)) \\ &\leq d(\varphi) \ . \end{aligned}$$

Consequently $\tau \longrightarrow d_\tau$ is non increasing, continuous and constant for τ close to 0. The critical levels of Φ_τ for $\tau \in [0,1]$ are given by the set

$$\Gamma_\tau = \{d(\varphi) - \tau H(m) \mid dH(m) = 0\} \ .$$

By Sard's theorem Γ_τ has measure zero. Moreover, it is evidently compact. Summing up, for some $\tau_0 > 0$ small

$$\begin{aligned} &d_\tau = d(\varphi) \text{ for } 0 \le \tau \le \tau_0 \\ &d_\tau \in \Gamma_\tau \text{ for } \tau \in [0,1] \ . \end{aligned} \tag{48}$$

The second part of (48) and the properties of Γ_τ imply

$$d_\tau = d(\varphi) - \tau H(m) \ , \ \tau \in [0,1] \tag{49}$$

for some $m \in \mathbf{C}^n$. The first part of (48) implies $H(m) = 0$. This completes the proof of 4.2. □

Summing up the previous discussion we have the following crucial "a priori equality":

Proposition 4.3 *For any $H \in \mathcal{H}_{ad}$ we have the equality*

$$2al = \inf_{\varphi \in \theta_l} \sup_{h \in B} \inf_{z \in h(E^-)} \left[\Phi(z) - \int_0^1 H(z_1,\ldots,z_n)dt - \int_1^2 \varphi(\operatorname{Im} z_{n+1})dt\right] \ .$$

5 Proof of the Main Result

In this section we prove Theorem 1.2. Let $\Omega \subset \mathbf{C}^n$ with $\overline{\Omega} \subset U_\Sigma$ and assume $e(\Sigma) < c(\Omega)$. If Ω is unbounded we have $\varphi_\Sigma(\Omega) \cap \Omega \neq \emptyset$ since $\varphi_\Sigma = \mathrm{Id}$ outside of a compact subset K of $\mathbf{C}^n$. So we may assume without loss of generality that Ω is bounded. Arguing indirectly let us assume $\varphi_\Sigma(\overline{\Omega}) \cap \overline{\Omega} = \emptyset$.$\mathbf{C}^{n+1} \setminus \Sigma$ has two connected components C^+ and C^-, where C^+ is distinguished by the fact that it contains $(0,\ldots,0,-a+i)$ for all large $a \in \mathbf{R}$. Fix $\delta > 0$. We find $a > 0$ and $c^- < 0, c^+ > 0$ and $\Psi \in \mathcal{D}$ such that

$$\begin{aligned} &\Psi(\Sigma) \subset (\mathbf{C}^n \oplus \mathbf{R}) \cup (\mathbf{C}^n \oplus (-a,a) \oplus i(c_-,c_+)) \\ &2a(c^+ - c^-) < e(\Sigma) + \delta \ . \end{aligned} \tag{50}$$

We note that $\Psi(\Sigma)$ has the same holonomy. So we may assume without loss of generality that $\Psi(\Sigma)$ is Σ. That is

$$\begin{aligned} &\Sigma \subset (\mathbf{C}^n \oplus \mathbf{R}) \cup (\mathbf{C}^n \oplus (-a,a) \oplus i(c_-,c_+)) \\ &2a(c^+ - c^-) < e(\Sigma) + \delta \ . \end{aligned} \tag{51}$$

We foliate an open neighbourhood in $C^+ \cup \Sigma$ by hyperplanes $\Sigma_\epsilon, 0 \le \epsilon \le \epsilon_0$ with $\Sigma_0 = \Sigma$ and

$$\Sigma_\epsilon \subset (\mathbf{C}^n \oplus \mathbf{R} \oplus i\epsilon) \cup (\mathbf{C}^n \oplus (-a,a) \oplus i(c_-,c_+)) \ . \tag{52}$$

Here $\epsilon_0 > 0$ is supposed to be sufficiently small. With the obvious identification every Σ_ϵ defines a holonomy

$$\varphi_\epsilon : U_\epsilon \longrightarrow \mathbf{C}^n . \tag{53}$$

If $\epsilon_0 > 0$ is small enough we have $\overline{\Omega} \subset U_\epsilon$ for every $\epsilon \in [0, \epsilon_0]$ and $\varphi_\epsilon(\overline{\Omega}) \cap \overline{\Omega} = \emptyset$, since the same is true for $\varphi_0 = \varphi_\Sigma$. Pick a $H \in \mathcal{H}_{ad}(\Omega)$ with

$$m(H) \geq c(\Omega) - \delta . \tag{54}$$

Pick a $\varphi_+ \in \theta_{c^+}$ with

$$d(\varphi_+) \leq 2ac_+ + \delta . \tag{55}$$

Choose a smooth map $\gamma_1 : \mathbf{R} \longrightarrow [0, +\infty)$ satisfying (assuming $\epsilon_0 < c_+$)

$$\begin{array}{rcll} \gamma_1(s) & = & 0 & \text{for } s \leq 0 \\ \gamma_1'(s) & > & 0 & \text{for } s \in (0, \epsilon_0) \\ \gamma_1(s) & = & c_1 & \text{for } s \geq \epsilon_0 \end{array} \tag{56}$$

with c_1 a constant bigger than $2a(c_+ - c_- + 2)$. Next choose a smooth map $\gamma_2 : \mathbf{R} \longrightarrow [c_1, +\infty)$ satisfying

$$\begin{array}{l} \gamma_2(s) = c_1 \text{ for } s \leq c_+ \\ \gamma_2'(s) > 0 \text{ for } s > c_+ \\ \gamma_2'(s) = \text{const} > 2a \text{ for } s \text{ large} \\ \gamma_2'(s) = 2a \text{ has a unique solution} \\ \text{which satisfies } s_0 \leq c_+ + 1 \\ \gamma_2(s) > \varphi_+(s) \text{ for } s \geq c_+ \ . \end{array} \tag{57}$$

Now we define a smooth Hamiltonian $K : \mathbf{C}^{n+1} \longrightarrow [0, +\infty)$ by

$$\begin{array}{rcl} K(z) & = & 0 \text{ for } z \in C^- \\ K(z) & = & \gamma_1(\epsilon) \text{ for } z \in \Sigma_\epsilon,\ 0 \leq \epsilon \leq \epsilon_0 \\ K(z) & = & c_1 \text{ for } z \in C^+ \setminus \cup\Sigma_\epsilon),\ \operatorname{Im} z_{n+1} \leq c_+ \\ K(z) & = & \gamma_2(\operatorname{Im} z_{n+1}) \text{ for } \operatorname{Im} z_{n+1} > c_+ \ . \end{array} \tag{58}$$

By construction we have

$$K(z) \geq \varphi_+(\operatorname{Im} z_{n+1}) \ , \ z \in \mathbf{C}^{n+1} \ . \tag{59}$$

Finally take a $\varphi_- \in \theta_{c_-}$ with

$$\varphi_-(\operatorname{Im} z_{n+1}) \geq K(z) \ , \ z \in \mathbf{C}^{n+1} \ . \tag{60}$$

Defining $\varphi : \mathbf{R} \longrightarrow [0, +\infty)$ by

$$\begin{array}{rcll} \varphi(s) & = & 0 & \text{for } s \leq 0 \\ \varphi(s) & = & \gamma_1(s) & \text{for } 0 \leq s \leq \epsilon_0 \\ \varphi(s) & = & c_1 & \text{for } \epsilon_0 \leq s \leq c_+ \\ \varphi(s) & = & \gamma_2(s) & \text{for } s \geq c_+ \end{array} \tag{61}$$

we see that K satisfies (27) and (28)of section III with the φ defined above for suitable constants R_1 and $R_2 > 0$. We define now a smooth functional $\tilde{\Phi} : E \longrightarrow \mathbf{R}$ by (with $\alpha(t) = (0, \ldots, 0, -a + ta)$)

$$\tilde{\Phi}(z) = \Phi(z) - \int_0^1 H(z_1, \ldots, z_n)dt - \int_1^2 K(z + \alpha)dt \ , \tag{62}$$

and comparison functionals Φ_+ and Φ_- by

$$\begin{aligned} \Phi_+(z) &= \Phi(z) - \textstyle\int_1^2 \varphi_+(\operatorname{Im} z_{n+1})dt \\ \Phi_-(z) &= \Phi(z) - \textstyle\int_1^2 \varphi_-(\operatorname{Im} z_{n+1})dt \ . \end{aligned} \tag{63}$$

The max–min values of Φ_+ and Φ_- are $d(\varphi_+)$ and $d(\varphi_-)$ respectively. In view of Proposition 4.3 these values remain the same if we substract from Φ_+ and Φ_- the functional $z \longrightarrow \int_2^1 H(z_1, \ldots, z_n)dt$ for $H \in \mathcal{H}_{ad}(\Omega)$. Next observe that for $z \in E$

$$\begin{aligned} & \textstyle\int_0^1 H(z_1, \ldots, z_n)dt - \int_1^2 \varphi_+(\operatorname{Im} z_{n+1})dt \\ = \ & \textstyle\int_0^1 H(z_1, \ldots, z_n)dt - \int_1^2 \varphi_+(\operatorname{Im}(z_{n+1} + (-a + ta)))dt \\ \geq \ & \textstyle\int_0^1 H(z_1, \ldots, z_n)dt - \int_1^2 K(z(t) + \alpha(t))dt \\ \geq \ & \textstyle\int_0^1 H(z_1, \ldots, z_n)dt - \int_1^2 \varphi_-(\operatorname{Im}(z_{n+1}))dt \ . \end{aligned} \tag{64}$$

As a consequence

$$d(\varphi_-) \leq \sup_{h \in \mathcal{B}} \inf_{z \in E^-} \tilde{\Phi}(h(z)) \leq d(\varphi_+) \ . \tag{65}$$

Consequently in view of previous results

$$2ac_- \leq \sup_{h \in \mathcal{B}} \inf_{z \in E^-} \tilde{\Phi}(h(z)) \leq 2ac_+ + \delta \ . \tag{66}$$

This result is true for any $H \in \mathcal{H}_{ad}(\Omega)$. Now let (54) hold and let $\tilde{\Phi}_\tau$ be the functional obtained from $\tilde{\Phi}$ by replacing H by τH for $\tau \in [0, 1]$. Since

$$\left|\tilde{\Phi}_{\tau_1}(z) - \tilde{\Phi}_{\tau_2}(z)\right| \leq m(H) \left|\tau_1 - \tau_2\right|$$

we see that

$$\tilde{d}_\tau := \sup_{h \in \mathcal{B}} \inf_{z \in E^-} \tilde{\Phi}_\tau(h(z)) \tag{67}$$

is a continuous, non increasing function of $\tau \in [0, 1]$, with range in the interval $[2ac_-, 2ac_+ + \delta]$. Since $\tilde{\Phi}_\tau$ satisfies $(WPS)_d$ for every $d \in \mathbf{R}$ in view of the results in section III we find for every $\tau \in [0, 1]$ a critical point z^τ of $\tilde{\Phi}_\tau$. Defining $u(t) = \alpha(t) + z^\tau(t)$ we see that u solves

$$\begin{aligned} i\tfrac{d}{dt}u &= (H'(u_1, \ldots, u_n), 0) \text{ on } [0, 1] \\ i\tfrac{d}{dt}u &= K'(u) \text{ on } [1, 2] \\ u_k(0) &= u_k(2) \text{ for } k = 1, \ldots, n \\ \operatorname{Re} u_{n+1}(0) &= -a \ , \ \operatorname{Re} u_{n+1}(2) = a \ . \end{aligned} \tag{68}$$

Assume first for some $t \in [1,2]$ we have $\operatorname{Im} z_{n+1} \geq c_+$. Then we must have

$$u(t) = (*, \ldots, *, -a, +2a(t-1) + \tau i)$$

for $t \in [1,2]$ with $*$ being constants. Here τ satisfies (and is uniquely determined by) $\gamma_2'(\tau) = 2a$. We calculate using the properties of K

$$\begin{aligned} \tilde{\Phi}_\tau(z^\tau) &\leq 2a(c_+ + 1) - c_1 \\ &\leq 2a(c_+ + 1) - 2a(c_+ - c_- + 2) \\ &= 2ac_- - 2a \\ &< 2ac_- . \end{aligned}$$

This contradicts (66) if $\tilde{\Phi}_\tau(z^\tau) \equiv \sup_{h \in \mathcal{B}} \inf_{z \in E^-} \tilde{\Phi}_\tau(h(z))$.

Using (68) again we see that u cannot be constant on $[1,2]$. The possibility left is that $u(t) \in \Sigma_\epsilon$ for some $\epsilon \in (0, \epsilon_0)$. We note that

$$\varphi_\epsilon(u_1(1), \ldots, u_n(1)) = (u_1(2), \ldots, u_n(2)) . \tag{69}$$

By the properties of $H \in \mathcal{H}_{ad}(\Omega)$ we see that in case $(u_1(1), \ldots, u_n(1)) \in \overline{\Omega}$ we must have $(u_1(0), \ldots, u_n(0)) \in \overline{\Omega}$ implying in view of the boundary condition on u that $\varphi_\epsilon(u_1(1), \ldots, u_n(1)) \in \overline{\Omega}$. This, however, contradicts our assumption. So we must have $(u_1(1), \ldots, u_n(1)) \notin \overline{\Omega}$. Consequently u is constant on $[0,1]$. Hence u solves

$$\begin{aligned} & i\tfrac{d}{dt}u = K'(u) \text{ on } [1,2] \\ & u_k(1) = u_k(2) \text{ for } k = 1, \ldots, n \\ & \operatorname{Re} u_{n+1}(1) = -a, \ \operatorname{Re} u_{n+1}(2) = a \\ & (u_1(1), \ldots, u_n(1)) \notin \overline{\Omega} . \end{aligned} \tag{70}$$

Hence z^τ is a critical point of the smooth functional $\hat{\Phi}$ defined by

$$\hat{\Phi}(z) = \Phi(z) - \int_1^2 K(z)dt .$$

The gradient of $\hat{\Phi}$ at every point is a nonlinear Fredholm operator. Given any critical point z we can construct a finite dimensional manifold $M_z \subset E$ containing all critical points of $\hat{\Phi}$ close to z and the critical points of $\hat{\Phi} \mid M_z$ coincide. Using the *Morse–Sard–theorem* it follows that the critical levels of $\hat{\Phi}$ are a closed set of measure zero. Denote it by Γ. Clearly

$$\tilde{\Phi}_\tau(z_\tau) \in -\tau m(H) + \Gamma .$$

Since $\tau \longrightarrow \tilde{d}_\tau = \tilde{\Phi}_\tau(z^\tau)$ is continuous and non increasing we find

$$\tilde{d}_\tau = -\tau m(H) + \tilde{d}_0 . \tag{71}$$

In view of (66)

$$2ac_- \le -\tau m(H) + \tilde{d}_0 \le 2ac_+ + \delta \tag{72}$$

for every $\tau \in [0,1]$. Hence $\tilde{d}_0 \le 2ac_+ + \delta$ and $2ac_- \le -m(H) + \tilde{d}_0$. This implies

$$m(H) \le 2a(c_+ - c_-) + \delta \ . \tag{73}$$

Since by (54) $m(H) \ge c(\Omega) - \delta$ we deduce

$$\begin{aligned} c(\Omega) &\le 2a(c_+ - c_-) + \delta + \delta \\ &\le e(\Sigma) + 3\delta \ . \end{aligned}$$

Since δ was arbitrarily chosen

$$c(\Omega) \le e(\Sigma) \ .$$

This, however, contradicts our assumption $e(\Sigma) < c(\Omega)$. This proves Theorem 1.2. □

References

[1] Benci, V. and Rabinowitz, P., Critical point theory for indefinite functionals, *Invent. Math.*, **52**, (197)9, 241–273.

[2] Ekeland, I. and Hofer, H., Symplectic topology and Hamiltonian dynamics, *Math. Zeit.*, **200**, (1990), 355–378.

[3] Ekeland, I. and Hofer, H., Symplectic topology and Hamiltonian dynamics II, *Math. Zeit.*, **203**, (1990), 553–567.

[4] Eliashberg, Y. and Hofer, H., Towards the definition of symplectic boundary, to appear GAFA.

[5] Hofer, H., On strongly indefinite functionals with applications, *TAMS*, **275**, (1983) 185–214.

[6] Hofer, H., Symplectic capacities, in *Geometry of low–dimensional manifolds, Vol. 2*, edited by S. K. Donaldson and C. B. Thomas, pp. 15–34, Cambridge University Press, 1990.

[7] Hofer, H., On the topological properties of symplectic maps, *Proceedings of the Royal Society of Edinburgh*, **115 A**, (1990), 25–38.

[8] Hofer, H. and Zehnder, E., Periodic solutions on hypersurfaces and a result by C. Viterbo, *Inv. Math.*, **90**, (1987), 1–7

[9] Hofer, H. and Zehnder, E., A new capacity for symplectic manifolds, in *Analysis et cetera*, edited by P.H. Rabinowitz and E. Zehnder, 405–428, Academic press, 1990.

[10] Sikorav, J.-C., Systèmes Hamiltoniens et Topologie Symplectique, preprint, Dipartimento di Matematica dell' Università di Pisa, 1990.

[11] Viterbo, C., Capacités symplectiques et applications, *Astéristique*, Séminaire Bourbaki, 695, 1989.

[12] Viterbo, C., Symplectic topology as the geometry of generating functions, preprint, January, 1991.

Caustics D_k at points of interface between two media

M. Kazarian

The object of our investigation is singularities of caustics at points of interface between two media. When a beam of light passes through the interface, it is refracted according to a well-known law: the ratio of sines of inclination angles of incoming and outgoing beams is equal to the ratio of speeds of propagation of waves in the media (fig. 1). The bigger the inclination angle of the incoming beam the bigger the one of the outgoing beam, and for some angle we have an effect of complete reflection: if the inclination angle is larger then the beam is reflected. If the caustic approaches the interface at a point where there is no complete reflection it is only refracted preserving all its singularities. Quite a different situation occurs at points of complete reflection. At this points a caustic arises even if there was no caustic in the first medium (fig. 2). Let us call the caustic new if there was no caustic in the first medium.

Theorem 1. Suppose that the initial wave front and interface are in general position. Then the analytic continuation of the new caustic at a point of complete reflection together with the interface form a singularity which is diffeomorphic to the caustic D_k of the Lagrangian mapping given by the generating family

$$F(x,y,z) = x^2y + y^{k-1} + q_{k-1}y^{k-2} + \ldots + q_2y + q_1x$$

together with the plane $q_1 = 0$ for some $k \geq 3$ (for $k = 3$ it is better to write A_3 instead of D_3). □

Corollary. In three-dimensional space points of interface where refracted beams touch the interface form a line of complete reflection.

At a generic point of the line of complete reflection analytic continuation of the new caustic together with the interface form a singularity diffeomorphic to the cuspidal edge together with the tangent plane passing through the edge. Physically either one of the branches of the caustic is visible (fig. 2a) or the caustic is invisible at all (fig. 2b).

At those points of the line of complete reflection, at which the reflected beam touches the line itself the analytic continuation of the new caustic together with the interface form a singularity diffeomorphic to the caustic

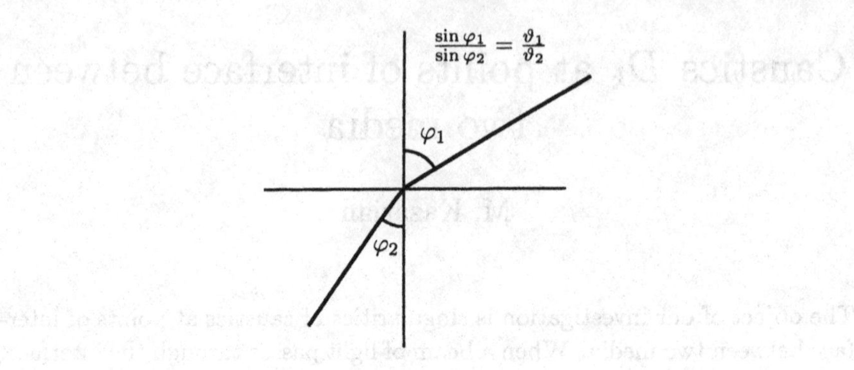

Figure 1:

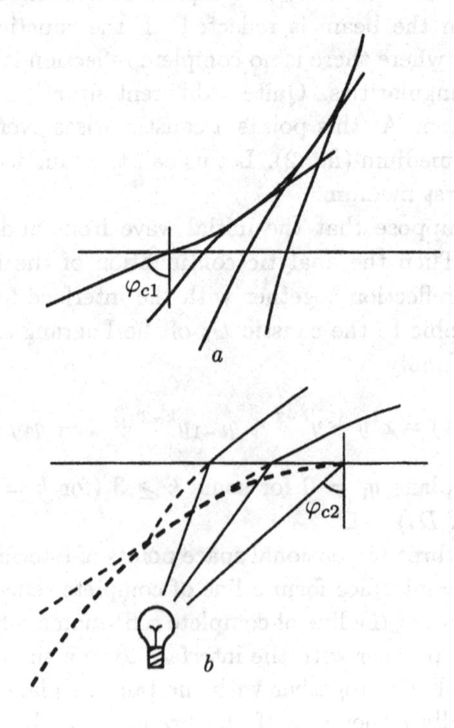

Figure 2:

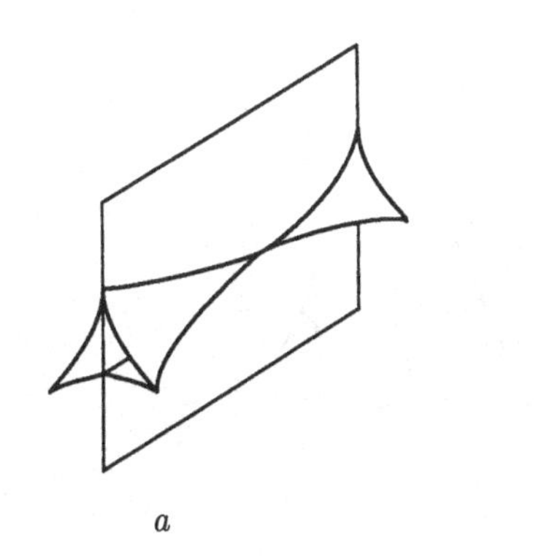

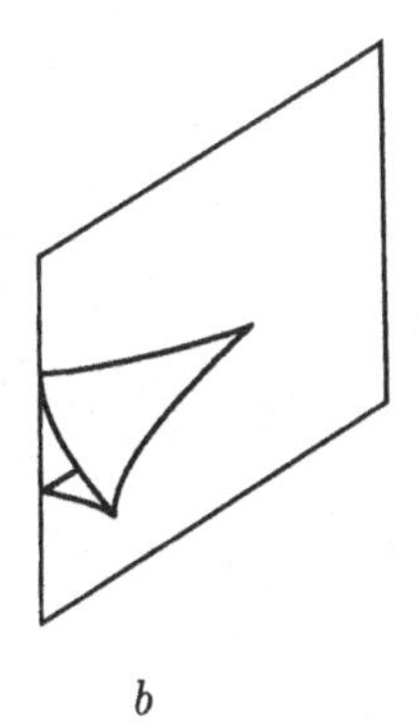

Figure 3:

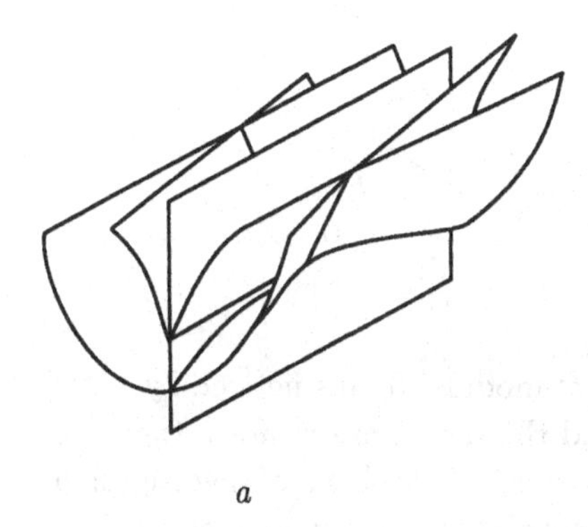

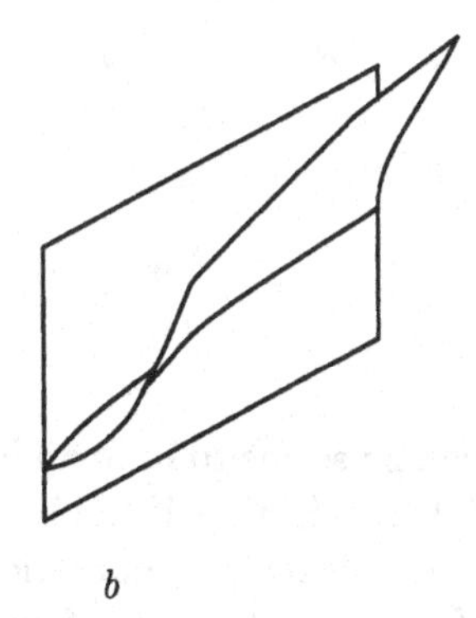

Figure 4:

D_4 together with the tangent plane passing through one of the cuspidal edges (fig. 3a,4a). Physically visible parts of caustics are represented on fig. 3b,4b. □

In three-dimensional space one more singularity may occur. Namely, suppose that there is a caustic in the first medium. Its smooth part intersects the interface transversally along some line. Along this line the inclination angle may change and at some points the caustic itself may have complete reflection. In such points the refracted caustic resembles a "symmetric butterfly", that is a caustic with the generating family

$$x^6 + q_3x^4 + q_2x^2 + q_1x$$

(fig. 5, physically visible parts are represented on fig. 5b,c. They resemble correspondent parts of the caustic D_4 but are not diffeomorphic to them). In [4] such a caustic arises as a stable caustic under the condition of symmetry. In our case a symmetry is not supposed and the differential type of this

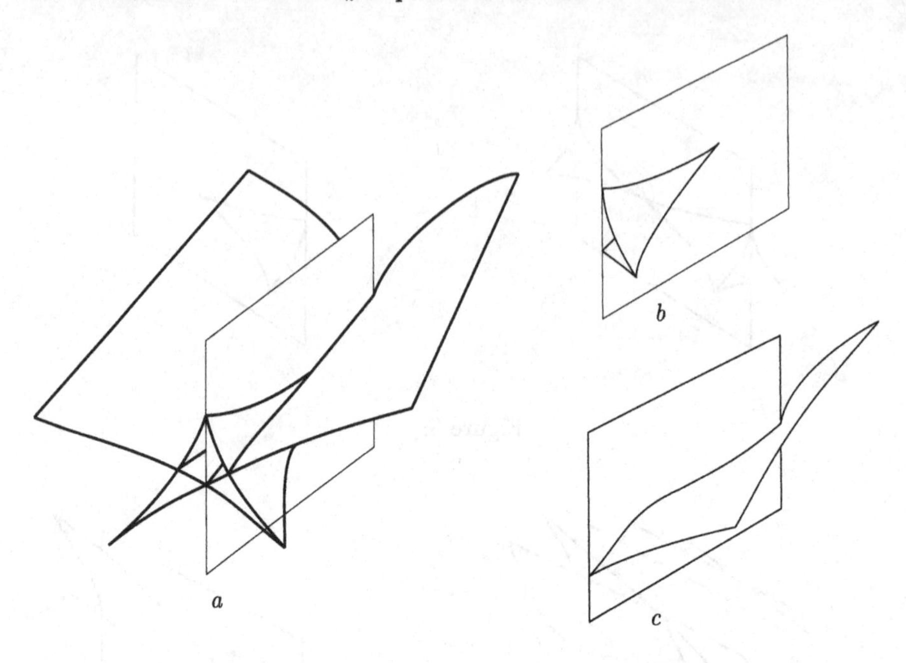

Figure 5:

caustic seems to have an infinite number of moduli. It was noticed by V.M. Zakalyukin that if the interface is flat and the media are uniform then the caustic is really symmetric. Thus we can regard the problem of investigation of caustics at points of an interface as a problem of broken symmetry.

In Section 1 we formulate the problem mathematically as a problem for Lagrangian mappings.

In Section 2 we discuss the genericity condition and calculate the normal form for initial data of the propagation of waves in the second medium.

In Section 3 we solve an auxiliary problem of classification of Lagrangian mappings with given boundary conditions. The results of this classification are used in the proof of Theorem 1.

The author is grateful to V.M. Zakalyukin for stating the problem and constant attention to the work. The author also expresses his thanks to V.I. Arnold for valuable observations.

1 Lagrangian manifolds at points of refraction

As we are interested in local questions only, in what follows we mean by manifolds and their submanifolds germs of the corresponding manifolds. We consider a smooth manifold M which is divided into two parts M_1 and M_2

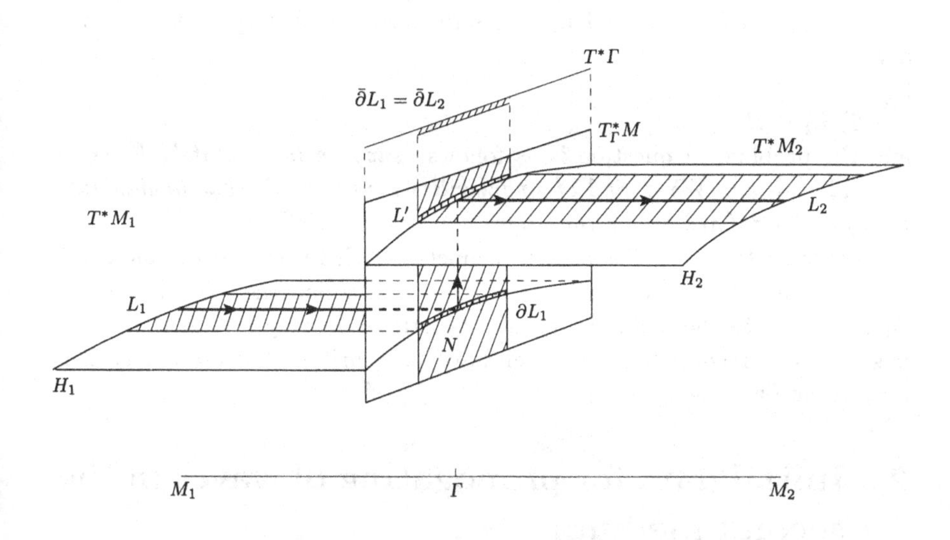

Figure 6:

by a smooth hypersurface Γ. Two Hamiltonians $h_i : T^*M_i \to \mathbb{R}, i = 1,2$ give rise to a propagation of waves in the media M_i. We assume h_i to be convex on impulses and have smooth continuations over neighbourhoods of M_i in T^*M. Let $H_i = \{h_i = 1\}$ be the level surfaces of the Hamiltonians.

A disturbance propagates in medium M_1 along the characteristics of H_1. When it comes to the boundary $T^*_\Gamma M$ it changes by a jump in order to come to the surface H_2 of the second Hamiltonian. This jump is done along the characteristics of $T^*_\Gamma M$ which are covectors that are orthogonal to the tangent space $T_*\Gamma$ of the interface. Then the disturbance propagates along the characteristics of H_2.

A wave state in M_1 corresponds to a Lagrangian submanifold $L_1 \subset T^*M_1$. The construction described above gives the following. The boundary $\partial L_1 = L_1 \cap T^*_\Gamma M$ is an isotropic submanifold of $T^*_\Gamma M$. The union of all the characteristics of $T^*_\Gamma M$ passing through ∂L_1 is a Lagrangian submanifold N. Then $N \cap H_2 = L'$ is an isotropic submanifold of H_2 which forms the initial data for the propagation of waves in M_2, and the union of all the characteristics of H_2 passing through L' is a Lagrangian submanifold L_2 which corresponds to the wave state in M_2 (fig. 6).

This construction will perhaps be clearer if we recall the notion of Lagrangian boundary. The manifold of characteristics of $T^*_\Gamma M$ is naturally identified with $T^*\Gamma$. The projection of the boundary $\partial L = L \cap T^*_\Gamma M$ of a

Lagrangian submanifold $L \subset T^*M$ is a Lagrangian submanifold $\bar{\partial} L \subset T^*\Gamma$ which is called a Lagrangian boundary of L. Now given the manifold L_1, the manifold L_2 is defined uniquely by the following conditions:

(1) $\bar{\partial} L_1 = \bar{\partial} L_2$ (L_1 and L_2 define the same wave state on the boundary),

(*)

(2) $L_2 \subset H_2$,

and the problem in question is as follows: *suppose the manifold L_1 is in general position. What kind of singularities may have the Lagrangian submanifold L_2 defined by the conditions (*) as a caustic?*

Although these conditions are symmetric with respect to the indices 1 and 2 (if we send the disturbance in inverse direction then L_2 determines L_1 according to these conditions) the caustic of L_2 may not be in general position with respect to Γ and even have singularities that are not typical for generic Lagrangian mappings.

2 Initial data for propogating of waves in the second medium

We use standard notation $q_1, \ldots, q_n$ for coordinates on M and $p_1, \ldots, p_n$ for impulses. Let Γ be given by the equation $q_1 = 0$.

A wave state in the whole space defines a wave state on an interface. From genericity conditions we can suppose that the projection of the boundary $\bar{\partial} L_1 = \bar{\partial} L_2 \subset T^*\Gamma$ onto Γ has singularities which are typical for generic Lagrangian mappings. We will consider two cases:

A. The projection has no singularities, and $\bar{\partial} L_1$ is reduced by a Lagrangian equivalence (i.e. a symplectomorphism of $T^*\Gamma$ preserving the foliation) to the normal form $p_2 = \ldots = p_n = 0$.

B. The projection has a fold as a singularity and L is reduced to the normal form $q_2 = p_2^2, p_3 = \ldots = p_n = 0$.

Proposition 2. In three-dimensional space on an interface at a point of complete reflection in the case of general position a Lagrangian mapping has no singularity other than A and B above. □

Proof. For a two-dimensional interface possible singularities are foldings along some line and cusps at distinct points. But by a small perturbation of the interface we can obtain that a complete reflection does not occur at a cusp. □

Let us see now to which normal form one can reduce $L' = \partial L_2$. Consider a projection of $H_2 \cap T^*_\Gamma M$ along the characteristics of $T^*_\Gamma M$. Let H' be the set of critical values of this projection. The following lemma follows from the convexity condition.

Lemma 3. H' is a smooth hypersurface of $T^*\Gamma$ which has a folding over Γ. □

We can call H' the *surface of complete reflection* because complete reflections occur just at the points corresponding to $H' \cap \bar{\partial} L_1$.

Proposition 4(a) If there is no complete reflection one can reduce L' by a Lagrangian equivalence of T^*M to the form:

$aA : p_1 = \ldots = p_n = q_1 = 0;$

$aB : q_2 = p_2, p_1 = p_3 = \ldots = p_n = q_1 = 0.$

(b) In the case of complete reflection one can reduce L' to the form:

$bA : q_2 = p_1^2, p_2 = \ldots = p_n = q_1 = 0;$

$bB : p_1^2 = p_2 + q_3, q_2 = p_2^2, p_3 = \ldots = p_n = q_1 = 0.$ □

We represent $T_\Gamma^* M$ as $T^*\Gamma \times \mathbf{R}$ with coordinates $p_2, \ldots, p_n, q_2, \ldots, q_n$ on $T^*\Gamma$ and p_1 on $\mathbf{R}$.

Lemma 5. The restriction of a Lagrangian equivalence of T^*M preserving Γ to $T_\Gamma^* M$ can be reduced to the following transformations:

(1) A Lagrangian equivalence of $T^*\Gamma$ (at coordinates $p_2, \ldots, p_n, q_2, \ldots, q_n$).

(2) Transformations of the form

$$p_1 \mapsto \tilde{p}_1 = b_1(q_2, \ldots, q_n) p_1 + \sum_2^n b_k(q_2, \ldots, q_n) p_k + c(q_2, \ldots, q_n), b_1(0) \neq 0$$

(i.e. transformations which are linear in p). □

To prove this lemma one has to express a Lagrangian equivalence of T^*M as a power series in q_1.

Proof of Proposition 4. If there is no complete reflection then L' is projected diffeomorphically onto $\bar{\partial} L_1$ along the p_1-axis. Consequently p_1 is a function of the point of the point of $\bar{\partial} L_1$. If we choose coordinates $q_2, \ldots, q_n$ on $\bar{\partial} L_1$ in the case aA and coordinates $p_2, q_3, \ldots, q_n$ in the case aB we get

$$\begin{aligned} A : p_1 &= f(q_2, \ldots, q_n); \\ B : p_1 &= f(p_2, q_3, \ldots, q_n) = f_1(p_2^2, q_3, \ldots, q_n) + p_2 f_2(p_2^2, q_3, \ldots, q_n) \\ &\qquad = f_1(q_2, \ldots, q_n) + p_2 f_2(q_2, \ldots, q_n). \end{aligned}$$

And we see that L' is reduced to the required form $p_1 = 0$ by means of a transformation of the form (2) in Lemma 5.

In the case of complete reflection the projection $L' \to \bar{\partial} L_1$ is 2–1 mapping. Critical values of this projection form a smooth submanifold $K^e \bar{\partial} L_1$. In the case bA one can reduce K to the form $q_2 = 0$ by means of a diffeomorphism of Γ. Then L' is given by the equation

$$bA : (f_1(q_2, \ldots, q_n) p_1 + f_2(q_2, \ldots, q_n))^2 = q_2, f_1(0) \neq 0,$$

which is reduced to the required form $p_1^2 = q_2$ by means of transformations of the form (2) in Lemma 5.

In the case bB we have on $\bar{\partial}L_1$ a surface K of critical points of the projection onto Γ (the inverse image of the caustic) $p_2 = 0$ and a field of kernels of this projection $\partial/\partial p_2$. Suppose that at a generic point of K this surface is transversal to this field. In this case K can be reduced by means of a diffeomorphism of Γ to the form $p_2 + q_3 = 0$. Hence L is given by the equation

$$bB : (p_1 + f_1(p_2, q_3, \ldots, q_n))^2 = (p_2 + q_3) f_2(p_2, q_3, \ldots, q_n), \quad f_2(0) \neq 0.$$

The right-hand side is reduced by a transformation of the form (1) in Lemma 5 to the form $p_2 + q_3$, and the left hand side is reduced to the form p_1^2 by a transformation of the form (2) in Lemma 5 and we get the equation for L' in the required form $p_1^2 = p_2 + q_3$.

3 Lagrangian mappings with fixed boundary conditions

Proposition 4 leads us to the following auxiliary problem of the classification of Lagrangian mappings with fixed boundary conditions. Consider a diagram:

$$\begin{array}{ccccc} L' & \hookrightarrow & T^*_\Gamma M & \to & \Gamma \\ \cap & & \cap & & \cap \\ L & \hookrightarrow & T^*M & \to & M, \end{array}$$

where Γ, M, L' are fixed smooth manifolds, Γ being given by the equation $q_1 = 0$, L' being taken from Proposition 4 and L being a Lagrangian submanifold containing L' as a submanifold.

Theorem 6. A generic Lagrangian submanifold L in the class of Lagrangian submanifolds containing L' may be reduced by means of a Lagrangian equivalence of T^*M preserving Γ to a form with one of the following generating families (the equation for Γ is $q_1 = 0$) :

$aA : B_k : F(x, q) = x^{k+1} + q_k x^k + \ldots + q_2 x^2 + q_1 x;$

aB : A generic such germ can be reduced to the form

$$F(x, y, q) = x^3 + y^2 + q_2 x + q_1 y;$$

A set of such germs of codimension 1 can be reduced to the form

$$F(x, y, q) = x^3 + xy^2 + q_3 y^2 + q_2 x + q_1 y.$$

All the other germs have higher codimension;

$$bA : D_k : F(x, q) = x^2 y + y^{k-1} + q_{k-1} y^{k-2} + \ldots + q_2 y + q_1 x.$$

Proof. aA Because

$$T_0L \subset (T_0L')^{\perp} = \mathrm{Span}\langle \partial/\partial p_1, \partial/\partial q_1, \ldots, \partial/\partial q_n \rangle,$$

we can choose either $(q_1, \ldots, q_n)$ or $(p_1, q_2, \ldots, q_n)$ as coordinates on L. In the first case there is no singularity of the projection of L onto M, and L can be brought to the normal form $p_1 = \ldots = p_n = 0$. In the second case there exists a generating function $S(p_1, q_2, \ldots, q_n)$ such that L is given by the equations

$$q_1 = -\frac{\partial S}{\partial p_1}, \quad p_2 = \frac{\partial S}{\partial q_2}, \quad \ldots, \quad p_n = \frac{\partial S}{\partial q_n}.$$

The condition $L' \subset L$ gives rise to the following: all the partial derivatives of S are known at the hypersurface $p_1 = 0$ of the space of parameters (namely, they are equal to zero). This implies that S is of the form

$$S = p_1^2 \cdot \varphi(p_1, q_2, \ldots, q_n),$$

where φ is a germ of some function. Hence L is given by the generating family

$$F(x, q) = S(x, q_2, \ldots, q_n) + xq_1 = x^2\varphi(x, q_2, \ldots, q_n) + xq_1.$$

Consequently for a generic φ the function $f(x) = F(x, 0)$ is of the type A_k for some k, and F is a versal deformation of f, which is R_+-equivalent to the family

$$F(x, q) = x^{k+1} + q_{k-1}x^{k-1} + \ldots + q_2x^2 + q_1x.$$

Because we can only make transformations of the q-variables preserving $q_1 = 0$, indeed we may only reduce the family to the form

$$F(x, q) = x^{k+1} + q_kx^k + \ldots + q_2x^2 + q_1x.$$

Analogous calculations in the cases aB and bA show that the generating family may be chosen in the form

$aB : F(x, y, q) = x^3 + xq_2 + yq_1 + y^2\varphi(x, y, q_3, \ldots, q_n);$

$bA : F(x, y, q) = x^2y + xq_2 + yq_1 + y^2\varphi(x, y, q_3, \ldots, q_n).$

Standard classification theorems of critical point theory show that these families are versal deformations of the corresponding functions and R_+-equivalent to the families given in the theorem. Moreover, a reduction to the normal form may be made by means of changes of the variables q preserving $q_1 = 0$. □

The case bB which is not considered in Theorem 6 is much more complicated. In this case (and in the more general cases) the generating family is not a versal deformation. Calculations show that the generating family in this case is of the form

$$F(x,q) = \tfrac{1}{6}x^6 + \tfrac{1}{4}x^4 q_3 + \tfrac{1}{2}x^2 q_2 + (x^4 + q_3 x^2 + q_2)^2 \varphi(x, q_2, \ldots, q_n) + x q_1.$$

We are interested in the problem, to which form this family may be brought by means of $\mathbf{R}_+$-transformations. Even if we are not interested in the Lagrangian sub-manifold itself but only in its caustic, the problem is to which simplest form a germ of the family

$$\frac{\partial F}{\partial x} = (x^4 + x^2 q_3 + q_2)[x + (8x^3 + 4q_3 x)\varphi + (x^4 + q_3 x^2 + q_2)\varphi'] + q_1$$

may be brought by means of a V-equivalence.

If we could bring this family to the same form with $\varphi = 0$, we would have obtained a "symmetric butterfly" (fig. 5, see [4]) but we cannot guarantee this.

Proof of Theorem 1. According to condition (*) the manifold L_2 is known to be contained in the level surface H_2 of the Hamiltonian h_2. We can obtain any small perturbation of L_2 by means of a corresponding perturbation of H_2. So we may assume L_2 to be in general position and make use of Theorem 6. However, the condition of convexity of the Hamiltonian implies a prohibition of some singularities from Theorem 6. Namely, if there is no complete intersection then the characteristics of H_2 are transversal to the boundary $T^*_\Gamma M$. This means that L_2 is transversal to $T^*_\Gamma M$ as well. From the list of Theorem 6 this condition is satisfied only by the simplest singularities: $F = x^2 + xq_1$ (the caustic is absent) in the case aA and $F = x^3 + y^2 + q_2 x + q_1 y$ (the caustic is smooth and transversal to the boundary) in the case aB.

At the points of complete reflection L_2 may not be (and indeed is not) transversal to the boundary, and of all the singularities described in Theorem 6, only the case Ba may occur, which completes the proof. □

References

[1] Arnold V.I., Critical points of functions and classification of functics, Uspechi math. nauk, v.29, 1974, 2, 11–49 (in Russian).

[2] Arnold V.I,. Gusein-Zade, S.M. and Varchenko, A.N., Singularities of differentiable maps, I, Birkhauses, Boston, 1985.

[3] Janeczko, S. and Kowalczyk, A., Classification of generic 3-dimentional Lagrangian singularities with $(Z_2)^1$-symmetries, preprint (1988).

[4] Janeczko, S. and Roberts, R.M., Classification of symmetric caustics I: Symplectic equivalence, in "Singularity theory and Applications, Warwick 1989, Part II" edited by R.M. Robert and I.N. Stewart, LNM 1463, Springer-Verlag, Heidelberg, Berlin, 1991, pp. 193–219.

[5] Nye, J.F., The catastrophe optics of liquid drop lenses, Proc. R. Soc. London A403 (1986), 27–50.

[6] Zakalyukin, V.M., On Lagrangian and Legendrean singularities, Funct. anal. i pril., 1976, 10, 1, 26–36.

Examples of Singular Reduction

Eugene Lerman Richard Montgomery Reyer Sjamaar

Mathematisch Instituut der Rijksuniversiteit te Utrecht *

PO Box 80.010, 3508 TA Utrecht, The Netherlands

January 1991

Introduction

The construction of the reduced space for a symplectic manifold with symmetry, as formalized by Marsden and Weinstein [13], has proved to be very useful in many areas of mathematics ranging from classical mechanics to algebraic geometry. In the ideal situation, which requires the value of the moment map to be weakly regular, the reduced space is again a symplectic manifold. A lot of work has been done in the last ten years in the hope of finding a 'correct' reduction procedure in the case of singular values. For example, Arms, Gotay and Jennings describe several approaches to reduction in [4]. At some point it has also been observed by workers in the field that in all examples the level set of a moment map modulo the appropriate group action is a union of symplectic manifolds. Recently Otto has proved that something similar does indeed hold, namely that such a quotient is a union of symplectic *orbifolds* [16]. Independently two of us, R. Sjamaar and E. Lerman, have proved a stronger result [21]. We proved that in the case of proper actions the reduced space, which we simply took to be the level set modulo the action, is a stratified symplectic space. Thereby we obtained a global description of the possible dynamics, a procedure for lifting the dynamics to the original space and a local characterization of the singularities of the reduced space. (The precise definitions will be given below.) The goal of this paper is twofold. First of all, we would like to present a number of examples that illustrate the general theory. Secondly, in computing the examples we have noticed that many familiar methods for computing reduced spaces work nicely in the singular situations. For instance, in the case of a lifted action on a cotangent bundle the reduced space at the zero level is the 'cotangent bundle' of the

*Current addresses: Univ. of Pennsylvania; Univ. of California at Santa Cruz; Mass. Inst. of Technology

orbit space. And in some cases the reduced space can be identified with the closure of a coadjoint orbit.

1 A Simple Example

Consider the standard action of the circle group $SO(2)$ on $\mathbf{R}^2$, and lift this action to $T^*\mathbf{R}^2 \simeq \mathbf{R}^2 \times \mathbf{R}^2$. In coordinates,

$$\begin{pmatrix} q^1 \\ q^2 \\ p_1 \\ p_2 \end{pmatrix} \mapsto \left(\begin{array}{cc|cc} \cos\theta & -\sin\theta & \multicolumn{2}{c}{\mathbf{0}} \\ \sin\theta & \cos\theta & & \\ \hline \multicolumn{2}{c|}{\mathbf{0}} & \cos\theta & -\sin\theta \\ & & \sin\theta & \cos\theta \end{array}\right) \begin{pmatrix} q^1 \\ q^2 \\ p_1 \\ p_2 \end{pmatrix},$$

and the canonical symplectic form is $\omega = dq^1 \wedge dp_1 + dq^2 \wedge dp_2$. The corresponding momentum map J is the angular momentum $J(q,p) = q^1 p_2 - q^2 p_1$. Zero is a singular value of J. Let us compute the reduced space at zero, $(T^*\mathbf{R}^2)_0$, which we will take to be the quotient $J^{-1}(0)/SO(2)$. The zero level set $J^{-1}(0)$ is a union of a point, 0, and of a hypersurface

$$Z = \{q^1 p_2 - q^2 p_1 = 0 : (q^1, q^2, p_1, p_2) \neq 0\}.$$

The hypersurface is a $SO(2)$-invariant coisotropic submanifold of $T^*\mathbf{R}^2$. The group $SO(2)$ acts freely on Z and the null directions of the restriction of the symplectic form ω to Z are precisely the orbital directions (just as in the regular case). Consequently the quotient $C_1 = Z/SO(2)$ is a symplectic manifold. The other piece of the zero level set, the origin 0, is fixed by the action of $SO(2)$ and we may consider the quotient $C_0 = \{0\}/SO(2)$ as a zero-dimensional symplectic manifold. Thus the reduced space $(T^*\mathbf{R}^2)_0$ is a disjoint union of two symplectic manifolds,

$$(T^*\mathbf{R}^2)_0 = C_0 \coprod C_1. \tag{1}$$

Let us give a more concrete description of the reduced space. We claim that C_1 is $\mathbf{R}^2 \backslash \{0\}$ with the standard symplectic structure and that the reduced space as a whole is diffeomorphic to the orbifold $\mathbf{R}^2/\mathbf{Z}_2$, where the action of $\mathbf{Z}_2$ is generated by the reflection $(x^1, x^2) \to (-x^1, -x^2)$.

1.1 Digression: Smooth Structures on Reduced Spaces

Let us explain what is meant by $(T^*\mathbf{R}^2)_0$ being diffeomorphic to $\mathbf{R}^2/\mathbf{Z}_2$. In general, let (M, ω) be a Hamiltonian G-space with corresponding moment map $\Phi : M \to \mathfrak{g}^*$ and let us assume that G acts properly on M. (In all the examples that follow the group G is going to be compact and for compact groups the properness of the action is automatic.) For us the reduced space at zero, M_0, is the topological space formed by dividing the zero level set $\Phi^{-1}(0)$ by the group action, i.e.,

$$M_0 = \Phi^{-1}(0)/G.$$

(We will see later that M_0 has a lot of structure, not just a topology.) As we have just seen, $\Phi^{-1}(0)$ need not be a manifold and the action of G on the zero level set need not be free. Thus there is no reason for the reduced space so defined to be a manifold (or even an orbifold). However, as Arms et al. have observed [3], it makes sense to single out a certain subset of the set of continous functions on M_0 as follows. Call a function $f : M_0 \to \mathbf{R}$ *smooth* if there exists a smooth G-invariant function $\bar{f}$ on M whose restriction to the zero level set $\Phi^{-1}(0)$ equals the pullback of f to $\Phi^{-1}(0)$ by the orbit map $\pi : \Phi^{-1}(0) \to \Phi^{-1}(0)/G = M_0$, i.e.,

$$\bar{f}|_{\Phi^{-1}(0)} = \pi^* f.$$

Let us denote the set of smooth functions by $C^\infty(M_0)$. A map $F : M_0 \to N$, where N is a manifold (or an orbifold, or another reduced space), is *smooth* if for any function $\phi \in C^\infty(N)$ the pullback $F^*\phi$ is a smooth function on M_0, $\phi \circ F \in C^\infty(M_0)$. It is now clear what we mean by two singular spaces being diffeomorphic.

1.1. REMARK. If G is a discrete group acting symplectically on a manifold (M, ω), it makes sense to define the corresponding moment map to be the zero map, since the Lie algebra of G is trivial. The reduced space is then a symplectic orbifold M/G. (See [18] or [15] for the definition of an orbifold.) For example, the action of $\mathbf{Z}_2$ on $\mathbf{R}^2$ described above preserves the standard symplectic form $dx^1 \wedge dx^2$ and the reduced space is the symplectic orbifold $\mathbf{R}^2/\mathbf{Z}_2$ with ring of smooth functions isomorphic to the collection of the smooth *even* functions on $\mathbf{R}^2$.

1.2 The Reduced Space $(T^*\mathbf{R}^2)_0$ as an Orbifold

Let us now go back to our example. Consider the 2-plane

$$\Lambda = \{ (q^1, q^2, p_1, p_2) \in T^*\mathbf{R}^2 : q^2 = 0,\ p_2 = 0 \}.$$

This plane is symplectic, it is completely contained in the zero level set of the moment map J and the $SO(2)$-orbit of any point $(q, p) \in J^{-1}(0)$ intersects Λ in exactly two points. Indeed, a point (q, p) lies in the zero level set if and only if q and p are collinear as vectors in $\mathbf{R}^2$. Consequently, $J^{-1}(0)/SO(2)$ is homeomorphic to $\Lambda/\mathbf{Z}_2$.

What about the two smooth structures? Clearly any $SO(2)$-invariant function on $T^*\mathbf{R}^2$ restricts to a $\mathbf{Z}_2$-invariant function on Λ. So the map

$\Lambda/\mathbf{Z}_2 \to J^{-1}(0)/SO(2)$ is smooth. To show that this map is a diffeomorphism it suffices to prove that any (smooth) $\mathbf{Z}_2$-invariant function on Λ extends to a (smooth) $SO(2)$-invariant function on $T^*\mathbf{R}^2$. By Schwarz's theorem [20] any smooth $\mathbf{Z}_2$-invariant function on Λ is a smooth function of the invariants $(q^1)^2$, ${p_1}^2$ and $q^1 p_1$ (these functions are a set of generators of the $\mathbf{Z}_2$-invariant polynomials on Λ). Now $(q^1)^2$ is the restriction to Λ of the $SO(2)$-invariant $(q^1)^2 + (q^2)^2$. Similarly,

$$\begin{aligned} {p_1}^2 &= \left.({p_1}^2 + {p_2}^2)\right|_\Lambda \\ q^1 p_1 &= \left.(q^1 p_1 + q^2 p_2)\right|_\Lambda . \end{aligned}$$

Consequently the map $J^{-1}(0)/SO(2) \to \Lambda/\mathbf{Z}_2$ is smooth as well and, therefore, the two reduced spaces are diffeomorphic.

Note that the $\mathbf{Z}_2$-invariant functions on Λ form a Poisson subalgebra of $C^\infty(\Lambda)$. So the smooth functions on the reduced space $(T^*\mathbf{R}^2)_0$ form a Poisson algebra. This is an example of the fact proved by Arms et al. (loc. cit.) that the set of smooth functions on a reduced space M_0 has a well-defined Poisson bracket induced by the bracket on the original manifold M.

The Poisson bracket of $C^\infty((T^*\mathbf{R}^2)_0)$ is compatible with the symplectic structure of the pieces C_1 and C_0 of the reduced space (see (1)) in the following sense. A pair of functions f and g in $C^\infty((T^*\mathbf{R}^2)_0)$ restrict to a pair of smooth functions on the symplectic manifold C_1. The symplectic structure of C_1 defines a Poisson bracket $\{\cdot,\cdot\}_{C_1}$. It is easy to check that this new bracket coincides with the bracket induced by the Poisson structure on $C^\infty((T^*\mathbf{R}^2)_0)$, i.e.,

$$\{f|_{C_1}, g|_{C_1}\}_{C_1} = \left.\{f,g\}_{(T^*\mathbf{R}^2)_0}\right|_{C_1} .$$

Similarly, one can show that

$$\left.\{f,g\}_{(T^*\mathbf{R}^2)_0}\right|_{C_0} = 0,$$

which is consistent with viewing C_0 as a zero-dimensional symplectic manifold. We thus see that the Poisson bracket of $C^\infty((T^*\mathbf{R}^2)_0)$ and the decomposition (1) of the reduced space into symplectic manifolds are intimately related.

1.3 Reduction via Invariants

Let us present a different calculation of the reduced space $(T^*\mathbf{R}^2)_0$. The calculation uses invariant theory, an approach advocated by R. Cushman. We will realize the reduced space as a subspace of $\mathbf{R}^4$ cut out by the equations

$$\begin{cases} {x_1}^2 &= {x_2}^2 + {x_3}^2 \\ x_4 &= 0 \\ x_1 &\geq 0 \end{cases} \tag{2}$$

In words, this reduced space is diffeomorphic to the top half, with vertex included, of the standard cone in $\mathbf{R}^3$. Consider a change of variables

$$\begin{cases} u_1 &= \frac{1}{2}(q^2 - p_1) \\ u_2 &= \frac{1}{2}(q^1 - p_2) \\ u_3 &= \frac{1}{2}(q^1 + p_2) \\ u_4 &= \frac{1}{2}(q^2 + p_1) \end{cases}$$

and set

$$\begin{cases} z_1 &= u_1 + iv_1 \\ z_2 &= u_2 + iv_2. \end{cases}$$

We have thus identified $T^*\mathbf{R}^2$ with $\mathbf{C}^2$. In these complex coordinates the symplectic form is given by $\omega = i\,(dz_1 \wedge d\bar{z}_1 + dz_2 \wedge d\bar{z}_2)$, the action of the circle group $SO(2) \simeq U(1)$ by

$$e^{i\theta} \cdot (z_1, z_2) = (e^{-i\theta} z_1, e^{i\theta} z_2)$$

and the moment map by $J(z_1, z_2) = |z_2|^2 - |z_1|^2$. It is easy to see that the set of (real) invariant polynomials is generated by four polynomials:

$$\begin{aligned} \sigma_1 &= |z_2|^2 + |z_1|^2, \\ \sigma_2 &= z_1 z_2 + \overline{z_1 z_2}, \\ \sigma_3 &= i(z_1 z_2 - \overline{z_1 z_2}), \\ \sigma_4 &= |z_2|^2 - |z_1|^2. \end{aligned}$$

The map $\sigma = (\sigma_1, \sigma_2, \sigma_3, \sigma_4) : \mathbf{C}^2 \to \mathbf{R}^4$ pushes down to an injective map $\bar{\sigma} : \mathbf{C}^2/SO(2) \to \mathbf{R}^4$. The invariants satisfy the relations

$$\begin{cases} \sigma_1^2 - \sigma_4^2 &= \sigma_2^2 + \sigma_3^2 \\ \sigma_1 &\geq 0. \end{cases} \tag{3}$$

Consequently the image of $\bar{\sigma}$ is a subset of $\mathbf{R}^4$ cut out by the equations

$$\begin{cases} {x_1}^2 &= {x_2}^2 + {x_3}^2 \\ x_1 &\geq 0 \end{cases}$$

Therefore the reduced space $(T^*\mathbf{R}^2)_0 := \{\sigma_4 = 0\}/SO(2)$ embeds in $\mathbf{R}^4$ as the subset cut out by (2) as claimed. If we ignore the fourth coordinate, we see that the reduced space is simply a round cone in $\mathbf{R}^3$. Since the invariants

$\sigma_1, \ldots, \sigma_4$ are quadratic, their linear span in $C^\infty(\mathbf{C}^2)$ forms a four-dimensional Lie algebra under the standard Poisson bracket. Alternatively, it is enough to note that

$$\begin{aligned} \{\sigma_i, \sigma_4\} &= 0 \quad \text{for } i = 1, \ldots, 4 \\ \{\sigma_1, \sigma_2\} &= 2\sigma_3 \\ \{\sigma_1, \sigma_3\} &= -2\sigma_2 \\ \{\sigma_2, \sigma_3\} &= 2\sigma_1. \end{aligned}$$

Therefore, the correspondence

$$\sigma_4 \leftrightarrow \begin{pmatrix} 1 & 0 \\ 0 & 1 \end{pmatrix} \qquad \sigma_1 \leftrightarrow \begin{pmatrix} 0 & 1 \\ 1 & 0 \end{pmatrix}$$

$$\sigma_2 \leftrightarrow \begin{pmatrix} -1 & 0 \\ 0 & 1 \end{pmatrix} \qquad \sigma_3 \leftrightarrow \begin{pmatrix} 0 & -1 \\ 1 & 0 \end{pmatrix}$$

establishes an isomorphism between the Lie algebra spanned by the generators of the invariants and $\mathfrak{gl}(2, \mathbf{R})$. The image cut out by (2) is nothing more than half of the nilpotent cone, the closure of the connected component of the principal nilpotent orbit in $\mathfrak{gl}(2, \mathbf{R})$.

More intrinsically this can be seen as follows. The moment map for the action of $Sp(T^*\mathbf{R}^2, \omega) \simeq Sp(2, \mathbf{R})$ on $T^*\mathbf{R}^2 \simeq \mathbf{R}^4$ identifies $\mathfrak{sp}(2, \mathbf{R})$ with the Poisson algebra of quadratic polynomials. The polynomials that commute with σ_4 then get identified with $\mathfrak{u}(1,1)$, which is isomorphic to $\mathfrak{gl}(2, \mathbf{R})$. We will come back to this point in Section 5, Remark 5.4.

2 A Summary of the General Theory

The goal of this section is to introduce the notion of a stratified symplectic space, to explain how this notion arises naturally in reduction and to describe some properties of reduced spaces.

2.1 Stratifications

The main idea of a stratification is that of a partition of a nice topological space into a disjoint union of manifolds. Thus a manifold is trivially a stratified space. A more interesting example of a stratified space is that of a *cone* on a manifold: given a manifold M the open cone $\mathring{C}M$ on M is the product $M \times [0, \infty)$ modulo the relation $(x, 0) \sim (y, 0)$ for all $x, y \in M$. That is, $\mathring{C}M$ is $M \times [0, \infty)$ with the boundary collapsed to a point, the vertex $*$ of the cone. The cone $\mathring{C}M$ is a disjoint union of two manifolds: $M \times (0, \infty)$ and the vertex $*$. Similarly one can consider the cone $\mathring{C}(\mathring{C}M)$ on the cone $\mathring{C}M$,

$$\mathring{C}(\mathring{C}M) = \left(\mathring{C}M \times [0,\infty)\right) / \sim .$$

The space $\mathring{C}(\mathring{C}M)$ is a union of three manifolds:

the vertex $*$ of $\mathring{C}(\mathring{C}M)$;
the open half line $\{*\} \times (0,\infty)$ through the vertex of $\mathring{C}M$;
the manifold $(M \times (0,\infty)) \times (0,\infty)$.

In general we will see that locally a stratified space is a cone on a cone on a cone .... Let us now make this precise.

2.1. DEFINITION. A *decomposed space* is a Hausdorff paracompact topological space X equipped with a locally finite partition $X = \coprod_{i \in \mathcal{I}} S_i$ into locally closed subsets S_i called *pieces*, each of which is a manifold.

We shall only consider decompositions each of whose pieces has the structure of a *smooth* manifold. A given space may be decomposed in a number of different ways.

2.2. EXAMPLE. Consider the subset of $\mathbf{R}^2$

$$Y = \{ (x^1, x^2) \in \mathbf{R}^2 : x^2 = 0 \} \cup \{ (x^1, x^2) \in \mathbf{R}^2 : x^1 \geq 0, x^2 \geq 0 \}.$$

The space Y can be broken up into a union of manifolds as

$$Y = \{x^2 = 0\} \cup \{x^1 > 0, x^2 > 0\} \cup \{x^1 = 0, x^2 > 0\} \tag{4}$$

or as

$$\begin{aligned} Y &= \{x^1 > 0, x^2 > 0\} \cup \{x^1 = 0, x^2 > 0\} \cup \{(0,0)\} \\ &\quad \cup \{x^1 < 0, x^2 = 0\} \cup \{x^1 > 0, x^2 = 0\}. \end{aligned} \tag{5}$$

2.3. EXAMPLE. A triangulated space is a decomposed space, if we declare the strata to be the (combinatorial) interiors of the simplexes.

2.4. EXAMPLE. If $X = \coprod_{i \in \mathcal{I}} S_i$ is a decomposed space, the cone $\mathring{C}X$ has a natural decomposition

$$\mathring{C}X = \{*\} \cup \coprod_{i \in \mathcal{I}} S_i \times (0,\infty).$$

2.5. EXAMPLE. The product of two decomposed spaces $X = \coprod S_i$ and $Y = \coprod P_j$ is a decomposed space

$$X \times Y = \coprod_{i,j} S_i \times P_j.$$

Define the *dimension* of a decomposed space X to be $\dim X = \sup_{i \in \mathcal{I}} \dim S_i$. We shall only consider finite-dimensional spaces. A stratification is a particular kind of decomposition. Its definition is recursive on the dimension of a decomposed space.

2.6. DEFINITION (cf. [7]). A decomposed space $X = \{S_i\}_{i \in \mathcal{I}}$ is called a *stratified space* if the pieces of X, called *strata*, satisfy the following local condition:

Given a point x in a piece S there exist an open neighbourhood U of x in X, an open ball B around x in S, a compact stratified space L, called the *link* of x, and a homeomorphism $\varphi : B \times \mathring{C}L \to U$ that preserves the decomposition, i.e., maps pieces onto pieces.

2.7. REMARK. We say that a decomposed space X satisfies the *condition of the frontier* if the closure of each piece is a union of connected components of pieces of X. It follows easily from Definition 2.6 that stratified spaces satisfy the condition of the frontier.

2.8. EXAMPLE. The decomposition (5) satisfies the frontier condition while (4) does not. So decomposition (4) is not a stratification. We leave it to the reader to check that decomposition (5) is a stratification.

2.9. EXAMPLE. A triangulated space is stratified by the interiors of its simplexes. The proof is an elementary exercise in PL-topology.

We are now in a position to define a stratified symplectic space.

2.10. DEFINITION. A *stratified symplectic space* is a stratified space X together with a distinguished subalgebra $C^\infty(X)$ (a *smooth structure*) of the algebra of continuous functions on X such that:

(i) each stratum S is a symplectic manifold;
(ii) $C^\infty(X)$ is a Poisson algebra;
(iii) the embeddings $S \hookrightarrow X$ are Poisson.

Condition (iii) means that given two functions $f, g \in C^\infty(X)$ their restrictions, $f|_S$ and $g|_S$, to a stratum S are smooth functions on S and their Poisson bracket at the points of S coincides with the Poisson brackets of the restrictions defined by the symplectic structure on S: $\{f,g\}\big|_S = \{f|_S, g|_S\}_S$.

2.11. THEOREM (cf. [21]). *Let (M,ω) be a Hamiltonian G-space with moment map $J : M \to \mathfrak{g}^*$ and suppose that the action of the Lie group G is proper. Then given an orbit $\mathcal{O} \in \mathfrak{g}^*$ the reduced space $M_\mathcal{O} := J^{-1}(\mathcal{O})/G$ is a stratified symplectic space.*

2.12. THEOREM (loc. cit.). *Assume that the level set $J^{-1}(\mathcal{O})$ is connected. Then the reduced space $M_{\mathcal{O}}$ has a unique open stratum. It is connected and dense.*

2.13. REMARK. We note two important cases when the level set is connected. First, if M is a symplectic vector space and G acts linearly on M then the zero level set is conical and so is connected. Secondly, F. Kirwan has proved [11] that if the moment map J is proper (for example if M is compact) and M is connected then the zero level set $J^{-1}(0)$ is connected. It follows then from the shifting trick, Proposition 2.16 below, that the level set $J^{-1}(\mathcal{O})$ is connected for any compact orbit $\mathcal{O}$.

The symplectic structure on the dense open stratum determines the Poisson structure on the whole reduced space and, therefore, the symplectic structures on all the lower-dimensional strata by condition (iii) of Definition 2.10. We will refer to the dense open stratum as the *top* stratum. Condition (i) also has some interesting consequences. Suppose that the top stratum is two-dimensional as in Section 1. Then all the other strata are zero-dimensional, i.e., they are isolated points. There is a temptation in view of Theorem 2.12 to discard all the lower-dimensional strata. We will see in the next section that giving in to such a temptation leads to a loss of interesting information.

2.2 Hamiltonian Mechanics on a Stratified Symplectic Space

Just as we defined in Section 1.1 a diffeomorphism between two reduced spaces, one can define an isomorphism between two stratified symplectic spaces.

2.14. DEFINITION. Let X and Y be two stratified symplectic spaces. A map $\phi : X \to Y$ is an *isomorphism* if ϕ is a homeomorphism and the pullback map $\phi^* : C^\infty(Y) \to C^\infty(X)$, $f \mapsto f \circ \phi$ is an isomorphism of Poisson algebras.

Note that we do not explicitly require that ϕ be strata-preserving. The reason for this is that the stratification of a stratified symplectic space X is completely determined by the Poisson algebra structure on the space of smooth functions on X, as we shall see shortly.

2.15. EXAMPLE (the 'shifting trick'). Let M be a Hamiltonian G-space with momentum map $J : M \to \mathfrak{g}^*$ and let $\mathcal{O}$ be any coadjoint orbit of G. Consider the symplectic manifold $M \times \mathcal{O}^-$, the symplectic product of M with the coadjoint orbit $\mathcal{O}$, endowed with the opposite of the Kirillov symplectic form. The diagonal action of G on $M \times \mathcal{O}^-$ is Hamiltonian with momentum map $J_{\mathcal{O}}$ given by $J_{\mathcal{O}}(m, \nu) = J(m) - \nu$. It is easy to check that the cartesian projection

$\Pi : M \times \mathcal{O}^- \to M$ restricts to an equivariant bijection $J_{\mathcal{O}}^{-1}(0) \cong J^{-1}(\mathcal{O})$. As a result, Π descends to a bijection between reduced spaces,

$$\tilde{\Pi} : (M \times \mathcal{O}^-)_0 \xrightarrow{\sim} M_{\mathcal{O}}.$$

2.16. PROPOSITION. *Assume that the orbit $\mathcal{O}$ is a closed subset of $\mathfrak{g}^*$. Then the map $\tilde{\Pi}$ is an isomorphism of stratified symplectic spaces.*

See [5] for a proof.

2.17. DEFINITION. A *flow* $\{\phi_t\}$ on a stratified symplectic space X is a one-parameter family of isomorphisms $\phi_t : X \to X$, $t \in \mathbf{R}$, such that $\phi_{t+s} = \phi_t \circ \phi_s$ for all t and s.

2.18. DEFINITION. Let h be a smooth function on a stratified symplectic space X, $h \in C^\infty(X)$. A *Hamiltonian flow* of h is a flow $\{\phi_t\}$ having the property that for any function $f \in C^\infty(X)$

$$\frac{d}{dt}(f \circ \phi_t) = \{f, h\} \circ \phi_t. \tag{6}$$

This is Heisenberg's form of Hamilton's equations. Since the space X is not necessarily a manifold, (6) cannot be reduced to a system of ordinary differential equations. For this reason the existence and uniqueness of the Hamiltonian flow is not immediately obvious. If X is a reduced space, the Hamiltonian flow does indeed exist and is unique [21]. Moreover, the following lemma holds.

2.19. LEMMA (cf. [21]). *Let $M_{\mathcal{O}}$ be the reduced space of a Hamiltonian G-space M at a coadjoint orbit $\mathcal{O}$ of G. The Hamiltonian flow of a smooth function $h \in C^\infty(M_{\mathcal{O}})$ preserves the stratification. The restriction of the flow of h to a stratum S equals the Hamiltonian flow of the restriction $h|_S$.*

The connected components of the strata are the symplectic leaves of $M_{\mathcal{O}}$, i.e., given any pair of points p, q in a connected component of a stratum of $M_{\mathcal{O}}$, there exists a piecewise smooth path joining p to q, consisting of a finite number of Hamiltonian trajectories of smooth functions on $M_{\mathcal{O}}$. Thus the Poisson structure of $C^\infty(M_{\mathcal{O}})$ determines the stratification of $M_{\mathcal{O}}$.

2.20. REMARK. It follows that a zero-dimensional stratum of the reduced space $M_{\mathcal{O}}$ is automatically a fixed point of *any* Hamiltonian flow. Thus the zero-dimensional strata of $M_{\mathcal{O}}$ determine relative equilibria in the original space M.

2.3 Orbit Types

We now explain where the stratification of a reduced space comes from and how it can be computed. Let G be a Lie group acting properly on a manifold M. (For example if G is compact then its action is automatically proper.) For a subgroup H of G denote by $M_{(H)}$ the set of all points whose stabilizer is conjugate to H,

$$M_{(H)} = \{ m \in M : G_m \text{ is conjugate to } H \}.$$

By virtue of the slice theorem for proper actions (see e.g. Palais [17]), the set $M_{(H)}$ is a smooth submanifold of M, called the manifold of *orbit type* (H). Thus we have a decomposition $M = \coprod_{H<G} M_{(H)}$ of M into a disjoint union of manifolds. Theorem 2.11 can now be restated as follows.

2.21. THEOREM. *Let (M,ω) be a Hamiltonian G-space with moment map $J : M \to \mathfrak{g}^*$ and let $\mathcal{O}$ be a coadjoint orbit of G. Assume that the action of G on $J^{-1}(\mathcal{O})$ is proper. Then the intersection of the preimage of the orbit $J^{-1}(\mathcal{O})$ with a manifold of the form $M_{(H)}$, $H < G$, is a manifold. The orbit space*

$$(M_{\mathcal{O}})_{(H)} := (J^{-1}(\mathcal{O}) \cap M_{(H)})/G$$

is also a manifold. There exists a unique symplectic form $\omega_{(H)}$ on $(M_{\mathcal{O}})_{(H)}$ such that the pullback of $\omega_{(H)}$ by the orbit map $J^{-1}(\mathcal{O}) \cap M_{(H)} \to (M_{\mathcal{O}})_{(H)}$ coincides with the restriction to $J^{-1}(\mathcal{O}) \cap M_{(H)}$ of the symplectic form ω. Finally, the decomposition of $M_{\mathcal{O}} := J^{-1}(\mathcal{O})/G$, the reduced space of M at the orbit $\mathcal{O}$, given by

$$M_{\mathcal{O}} = \coprod_{H<G} (M_{\mathcal{O}})_{(H)}$$

is a symplectic stratification of $M_{\mathcal{O}}$.

It is a curious fact that each stratum $(M_{\mathcal{O}})_{(H)}$ may also obtained by a regular Marsden-Weinstein reduction. To keep the discussion simple let us assume that $\mathcal{O}$ is the zero orbit. (This is no loss of generality by virtue of the shifting trick.) For a subgroup H of G define

$$M_H = \{ m \in M : G_m \text{ is exactly } H \}$$

It is well-known that M_H is a symplectic submanifold of M. The action of G does not preserve the manifold M_H. However, the smaller group $L = N_G(H)/H$ does act on M_H, where $N_G(H)$ denotes the normalizer of H in G. Moreover, the action of L is Hamiltonian and the corresponding moment map $J_L : M_H \to \mathfrak{l}^*$ is essentially the restriction of the moment map $J : M \to \mathfrak{g}^*$ to M_H.

2.22. THEOREM (cf. [21]). *Zero is a regular value of the moment map* J_L. *The Marsden-Weinstein reduced space* $(J_L)^{-1}(0)/L$ *is symplectically isomorphic to the stratum* $(M_0)_{(H)}$.

This theorem provides us with a simple recipe for lifting integral curves of a reduced Hamiltonian flow on the reduced space M_0 to the level set $J^{-1}(0)$. Namely, let $\bar{h}$ be an invariant smooth function on the manifold M, and let h be the smooth function on the reduced space induced by $\bar{h}$. Let $\bar{\Phi}_t$ and Φ_t, denote the Hamiltonian flows of $\bar{h}$ and h, respectively. If $\gamma(t)$ is an integral curve of the function h, then it lies inside some stratum $(M_0)_{(H)}$, and the classical recipe for lifting a reduced flow (see e.g. [1]) can be used to lift $\gamma(t)$ to an integral curve of the Hamiltonian $\bar{h}$, lying in the manifold M_H.

2.4 The Closure of a Coadjoint Orbit as a Stratified Symplectic Space

The object of this section is to show that for a large class of Lie groups the closure of every coadjoint orbit is a stratified symplectic space. In Section 4 we shall see that in some cases a reduced space of a Hamiltonian space can be identified with the closure of a coadjoint orbit of a different group.

2.23. THEOREM. *Let* H *be a reductive Lie group and let* $\mathcal{O} \subset \mathfrak{h}^*$ *be a coadjoint orbit of* H. *Then the closure* $\bar{\mathcal{O}}$ *of* $\mathcal{O}$ *is a stratified symplectic space. The strata are the* H*-orbits in* $\bar{\mathcal{O}}$.

PROOF. We take the space $C^\infty(\bar{\mathcal{O}})$ of smooth functions on $\bar{\mathcal{O}}$ to be the space of *Whitney smooth* functions. Recall that a continuous map $f : \bar{\mathcal{O}} \to \mathbf{R}$ is called Whitney smooth if and only if there exists a function $F \in C^\infty(\mathfrak{h}^*)$ such that $F|_{\bar{\mathcal{O}}} = f$.

It is easy to see that $C^\infty(\bar{\mathcal{O}})$ is naturally a Poisson algebra. Indeed, since the coadjoint orbits are the symplectic leaves of the Poisson structure on $\mathfrak{h}^*$, for all $F, G \in C^\infty(\mathfrak{h}^*)$ and $x \in \bar{\mathcal{O}}$ the bracket $\{F, G\}(x)$ depends only on the restrictions of F and G to the coadjoint orbit of x, which is contained in $\bar{\mathcal{O}}$. Thus the bracket $\{\cdot,\cdot\}_{\bar{\mathcal{O}}}$ given by

$$\{F|_{\bar{\mathcal{O}}}, G|_{\bar{\mathcal{O}}}\}_{\bar{\mathcal{O}}}(x) := \{F, G\}(x)$$

is well-defined. The partition of $\bar{\mathcal{O}}$ into coadjoint orbits is a decomposition. The local finiteness follows from the assumption that H is reductive. The proof of the fact that $\bar{\mathcal{O}}$ is a *stratified* space requires some machinery.

2.24. DEFINITION. Let X be a subspace of $\mathbf{R}^n$. A decomposition of X is called a *Whitney stratification* if the pieces of X are smooth submanifolds of $\mathbf{R}^n$ and if for each pair of pieces P, Q with $P \leq Q$ the following condition of Whitney holds:

WHITNEY'S CONDITION B. Let p be an arbitrary point in P and let $\{p_i\}$ and $\{q_i\}$ be sequences in P, resp. Q, both converging to p. Assume that the lines l_i joining p_i and q_i converge (in the projective space $\mathbf{R}P^{n-1}$) to a line l, and that the tangent planes $T_{q_i}Q$ converge (in the Grassmannian of $(\dim Q)$-planes in $\mathbf{R}^n$) to a plane τ. Then l is contained in τ.

It follows from Mather's theory of control data (see [14]) that a Whitney stratified subset of Euclidean space is a stratified space in the sense of our Definition 2.6. An outline of the argument can be found in [8, page 40]. So it suffices to show that $\bar{\mathcal{O}}$ is a Whitney stratified space.

Since H is reductive, the coadjoint representation $Ad^* : H \to Gl(\mathfrak{h}^*)$ is algebraic, i.e., the image $Ad^*(H)$ is an algebraic subgroup of $Gl(\mathfrak{h}^*)$ and the coadjoint action of $Ad^*(H)$ on $\mathfrak{h}^*$ is algebraic (see e.g. [23] for a proof). Now a coadjoint orbit $Ad^*(H)\cdot q$, $q \in \mathfrak{h}^*$, is semialgebraic by the Seidenberg-Tarski theorem, since it is the image of $Ad^*(H)$ under the algebraic map 'evaluation at q', which sends $a \in Ad^*(H)$ to $a \cdot q$. Let $\mathcal{O}_1$ and $\mathcal{O}_2$ be two orbits in $\bar{\mathcal{O}}$ with $\mathcal{O}_1$ contained in the closure of $\mathcal{O}_2$. The two orbits are smooth and semialgebraic. Therefore a theorem of Wall [24, p. 337] applies. In this case the theorem says that Whitney's condition B for the pair $(\mathcal{O}_1, \mathcal{O}_2)$ holds at all points of $\mathcal{O}_1$ except possibly for the points in a semialgebraic subvariety of dimension strictly less than the dimension of $\mathcal{O}_1$. In particular condition B holds at *some* point of $\mathcal{O}_1$. But the pair $(\mathcal{O}_1, \mathcal{O}_2)$ is H-homogeneous, so condition B holds everywhere. This proves that $\bar{\mathcal{O}}$ is a Whitney stratified space. □

3 Reduction of Cotangent Bundles

3.1 The Cotangent Bundle of a Quotient Variety

We have seen in Section 1 that the singular reduced space $(T^*\mathbf{R}^2)_0$ is a symplectic orbifold. There are a few other interesting examples of singular reduced spaces coming from reduction of cotangent bundles which turn out to be orbifolds. In order to understand what makes these examples work it will be helpful to consider lifted actions on cotangent bundles in general. (We caution the reader that not every reduced space is an orbifold; see [5] for a counterexample.) Let G be a Lie group acting smoothly and properly on a smooth manifold X. Let x be a point in X and H the stabilizer of x in G. Since the action is proper H is compact. Therefore there exists an H-equivariant splitting of the tangent space to X at x:

$$T_xX = T_x(G \cdot x) \oplus V$$

where V is some subspace of the tangent space. Let B be a small H-invariant ball in V centered at the origin. The slice theorem asserts that a neighbourhood of the orbit $G \cdot x$ in X is G-equivariantly diffeomorphic to the associated bundle $G \times_H B$.

If the action of G is free, it follows from the slice theorem that X is a principal G-bundle over the orbit space $Q = X/G$. Lift the action of G to an action on the cotangent bundle. It is well-known (see e.g. [1]) that in this case the reduced space at the zero level is simply the cotangent bundle of the base, $(T^*X)_0 = T^*Q$. This result has been recently generalized by Emmrich and Römer [6] to the case when the action of G on X is of constant orbit type, that is, there exists a subgroup H of G such that for any $x \in X$ the orbit $G \cdot x$ is diffeomorphic to the homogeneous space G/H. Alternatively, by virtue of the slice theorem, the action of G on X is of constant orbit type if and only if X is a fibre bundle over the orbit space $Q = X/G$ with typical fibre G/H. Emmrich and Römer showed that in this case the reduced space $(T^*X)_0$ is again T^*Q, the cotangent bundle of the orbit space.

Let us now consider the general case of an action of G on X, that is, we make no assumption concerning the structure of the orbits. Lift the action of G to an action on the cotangent bundle T^*X and let $J : T^*X \to \mathfrak{g}^*$ be the corresponding moment map. Recall that for $(x, \eta) \in T_x^*X$ the value of J is defined by

$$\langle \xi, J(x, \eta) \rangle = -\langle \xi_X(x), \eta \rangle, \tag{7}$$

where $\langle \cdot, \cdot \rangle$ on the left hand side of the equation denotes the pairing between the Lie algebra $\mathfrak{g}$ and its dual, and on the right hand side the pairing between the tangent and the cotangent spaces of X at x, while $\xi_X(x)$ is the vector obtained by evaluating at x the vector field defined by the infinitesimal action ξ on X. Let us compute the zero level set of the moment map. It follows from (7) that

$$J^{-1}(0) \cap T_x^*X = \{\, \eta : \langle \xi_X(x), \eta \rangle = 0 \text{ for all } \xi \in \mathfrak{g} \,\}.$$

We have proved:

3.1. Lemma. *Let $J : T^*X \to \mathfrak{g}^*$ be the moment map induced by the lift of the action G on X to an action on T^*X. Then the intersection of the zero level set of the moment map with the fibre of the cotangent space at a point $x \in X$ is $(T_x(G \cdot x))^\circ$, the annihilator of the tangent space to the orbit through x. Consequently,*

$$J^{-1}(0) = \coprod_{x \in X} (T_x(G \cdot x))^\circ. \tag{8}$$

3.2. Remark. It follows from the description (8) of the zero level set that it retracts onto X. In particular, if X is connected then the level set $J^{-1}(0)$ is connected as well.

For a point x in X we call the orbit space $(T_x(G\cdot x))^\circ/G_x$ the *cotangent cone* of X/G at the point $G\cdot x \in X/G$. It is easy to see that this definition does not depend on the choice of the point $x \in G\cdot x$, i.e., if $x' = a\cdot x$ for some $a \in G$, then multiplication by a induces an isomorphism between the orbit spaces $(T_x(G\cdot x))^\circ/G_x$ and $(T_{x'}(G\cdot x'))^\circ/G_{x'}$. Moreover, the quotient

$$J^{-1}(0)/G = \Big(\coprod_{x\in X}(T_x(G\cdot x))^\circ\Big)\Big/G$$

is set-theoretically the disjoint union of all cotangent cones to X/G. Therefore the following definition makes sense.

3.3. DEFINITION. The *cotangent bundle* of an orbit space X/G is the stratified symplectic space $T^*(X/G) := J^{-1}(0)/G$.

3.4. EXAMPLE. Suppose that G is finite. Then $J = 0$, so $T^*(X/G) := J^{-1}(0)/G = T^*(X)/G$.

3.5. REMARK. The cotangent bundle $T^*(X/G)$ is not a locally trivial bundle over the base variety X/G, since the fibres may vary from point to point. Nor is the projection $T^*(X/G) \to X/G$ a stratification-preserving map.

3.6. EXAMPLE. Let $X = \mathbf{R}^2$ and let $G = SO(2)$ act on X in the standard way. Then the quotient X/G is a closed half-line $[0,\infty)$. It consists of two strata: the end-point $\{0\}$ and the open half-line $(0,\infty)$. We saw in Section 1 that the cotangent bundle $T^*(X/G)$ of the half-line is a cone. The fibre $\pi^{-1}(x)$ of the projection $\pi : T^*(X/G) \to X/G$ is a line if $x \in (0,\infty)$, but it is a closed half-line if $x = 0$. So $T^*(X/G)$ is not a locally trivial bundle over X/G. Notice that $\pi^{-1}(0)$ intersects the top stratum of $T^*(X/G)$. So the preimage of the stratum $\{0\}$ is not a union of strata.

It seems unlikely to us that the smooth structure of a cotangent bundle $T^*(X/G)$ depends on the way in which the orbit space X/G is written as a quotient. More precisely, we make the following

3.7. CONJECTURE. *Let G and H be Lie groups and let X, resp. Y, be smooth manifolds on which G, resp. H act properly. Assume that the orbit spaces X/G and Y/H are diffeomorphic in the sense that there exists a homeomorphism $\phi : X/G \to Y/H$ such that the pullback map ϕ^* is an isomorphism from $C^\infty(Y/H) := C^\infty(Y)^H$ to $C^\infty(X/G) := C^\infty(X)^G$. Then the cotangent bundles of X/G and Y/H are isomorphic in the sense of Definition 2.14.*

In his unpublished thesis [19], Schwarz showed that modulo some assumptions $T^*(X/G)$ and $T^*(Y/H)$ are *homeomorphic* if X/G and Y/H are diffeomorphic. In the next sections we prove a version of this result and provide some experimental evidence for Conjecture 3.7.

3.2 Cross-sections

Let X be a smooth manifold and G a Lie group acting on X. Often one can compute the cotangent bundle of the quotient variety X/G by means of a *cross-section* of the G-action, i.e., a pair (Y, H), where Y is an embedded submanifold of X and H a Lie group acting on Y such that every G-orbit in X intersects Y in exactly one H-orbit. If (Y, H) is a cross-section, it is easy to see that the natural map $Y/H \to X/G$ is a homeomorphism. On an additional assumption we show now that the cotangent bundles $T^*(X/G)$ and $T^*(Y/H)$ are also homeomorphic.

3.8. PROPOSITION. *Let X be a Riemannian G-manifold. Assume that (Y, H) is a cross-section of the G-action on X. Assume further that the cross-section is orthogonal in the sense that for all y in Y*

$$T_y(G \cdot y) = \big((T_yY)^\perp \cap (T_y G \cdot y)\big) \oplus \big(T_yY \cap T_y(G \cdot y)\big). \tag{9}$$

Then the inclusion $Y \subset X$ induces a homeomorphism $(T^*Y)_0 \xrightarrow{\cong} (T^*X)_0$.

3.9. REMARK. Suppose the cross-section Y is the set of fixed points for some subgroup K of G. Let H be the 'Weyl group' $N(K)/K$. The statement (9) regarding the orthogonality of the intersections of the G-orbits with Y holds automatically in this case. This follows easily from the proof of the slice theorem.

PROOF. The metric allows us to identify equivariantly tangent and cotangent bundles of X and of Y, giving rise to a symplectic embedding

$$T^*Y \simeq TY \hookrightarrow TX \simeq T^*X.$$

Let $J_X : TX \to \mathfrak{g}^*$ and $J_Y : TY \to \mathfrak{h}^*$ denote the moment maps. Let y be a point in Y. Since the orbit $G \cdot y$ intersects Y in a single H-orbit, (9) implies that

$$T_y(G \cdot y) = \big((T_yY)^\perp \cap (T_y G \cdot y)\big) \oplus (T_y(H \cdot y))$$

is an orthogonal decomposition. Hence $(T_y(G \cdot y))^\perp = V_1 \oplus V_2$, where $V_1 = (T_y H \cdot y)^\perp \cap T_yY$ and $V_2 = (T_y G \cdot y)^\perp \cap (T_yY)^\perp$, so that

$$J_X^{-1}(0) \cap T_y^*X = V_1 \oplus V_2$$

and

$$J_Y^{-1}(0) \cap T_y^*Y = V_1.$$

This gives us an inclusion $J_Y^{-1}(0) \to J_X^{-1}(0)$. Composing with the orbit map $J_X^{-1}(0) \to J_X^{-1}(0)/G = (T^*X)_0$ gives us a map from $J_Y^{-1}(0)$ to $(T^*X)_0$. We claim that this map descends to a map from $J_Y^{-1}(0)/H$ to $(T^*X)_0$. Indeed, suppose (y',η') and (y,η) are two points in $J_Y^{-1}(0)$ and $a\cdot(y',\eta')=(y,\eta)$ for some $a\in H$. By assumption $G\cdot y\cap Y=H\cdot y$, so there is $b\in G$ with $b\cdot y'=y$. It is therefore no loss of generality to assume that $y=y'$. In this case η and η' both lie in V_1 and $a\in H_y$. Locally near y the space Y is H-equivariantly diffeomorphic to the associated bundle $H\times_{H_y}V_1$, so locally

$$Y/H\simeq(H\times_{H_y}V_1)/H=V_1/H_y.$$

Here H_y denotes the stabilizer of y in H. Similarly,

$$X/G\simeq(G\times_{G_y}(V_1\oplus V_2))/G=(V_1\oplus V_2)/G_y,$$

where G_y denotes the stabilizer of y in G. We have assumed that (Y,H) is a cross-section for the G-action, and therefore $X/G\simeq Y/H$. It follows that $\eta,\eta'\in V_1$ lie in the same H_y-orbit if and only if $\eta,\eta'\in V=V_1\oplus V_2$ lie in the same G_y orbit. We conclude that there is $c\in G$ with $c\cdot(y,\eta)=(y',\eta')$, thereby proving the existence of a continuous map

$$\varphi:(T^*Y)_0=J_Y^{-1}(0)/H\to(T^*X)_0.$$

A similar argument shows that φ is bijective and that φ^{-1} is continuous.

□

3.10. REMARK. This proof shows that each G-orbit in $J_X^{-1}(0)$ intersects $J_Y^{-1}(0)$ in a single H-orbit, in other words, that the pair $(J_Y^{-1}(0),H)$ is a cross-section of the G-action on $J_X^{-1}(0)$.

3.3 Row, Row, Row your Boat

Let X be the unit two-sphere in $\mathbf{R}^3$ and let G be the circle acting on X by rotations on the z-axis. The space X is the configuration space of the spherical pendulum and G is its group of symmetries. Now let Y be a great circle through the poles and let H be the group $\mathbf{Z}_2$ acting on Y by reflection in the z-axis. Then the pair (Y,H) is obviously an orthogonal cross-section of the G-action on X. Let J_X be the momentum map of the lifted action of G on T^*X. The lifted action of H on T^*Y has trivial momentum map, since H is finite. By Proposition 3.8, the pair (T^*Y,H) is a cross-section for the G-action on $J_X^{-1}(0)$. The physical meaning of this fact is that a spherical pendulum with zero angular momentum is just a planar pendulum.

Let us describe the orbifold $(T^*Y)/H$ in some detail. We identify the meridian Y with $S^1=\{e^{i\theta}:\theta\in\mathbf{R}\}$ in such a manner that the south pole is mapped to $1\in S^1$. We cover $(T^*Y)/H=(S^1\times\mathbf{R})/\mathbf{Z}_2$ with two orbifold charts. The domain of both charts is the strip $D=(-\pi,\pi)\times\mathbf{R}\subset\mathbf{R}^2$

equipped with the $\mathbf{Z}_2$-action generated by reflection in the origin. The chart maps ψ_1 and ψ_2 are given by:

$$\psi_1 : D \to (S^1 \times \mathbf{R})/\mathbf{Z}_2, \quad (\theta, r) \mapsto [e^{i\theta}, r],$$
$$\psi_2 : D \to (S^1 \times \mathbf{R})/\mathbf{Z}_2, \quad (\theta, r) \mapsto [-e^{i\theta}, r],$$

where $[x, y]$ denotes the equivalence class of $(x, y) \in S^1 \times \mathbf{R}$. It is easy to write down the transition map from one chart to the other. The resulting space has the shape of a 'canoe' with two isolated conical singularities. We encourage the reader to construct this orbifold with paper and glue.

We claim that the natural homeomorphism

$$\phi : (T^*Y)/H = T^*(Y/H) \to T^*(X/G) \tag{10}$$

is an isomorphism of reduced spaces. It obviously suffices to show that $\phi : O_i \to \phi(O_i)$ is an isomorphism, where $O_i = \psi_i(D)$ for $i = 1, 2$. Note that $\phi(\psi_i(D))$ is the space obtained by reducing $T^*(X - \{*\})$ at zero, where $\{*\}$ is either the south or the north pole of the sphere X, depending on whether $i = 1$ or 2. But $X - \{*\}$ is G-equivariantly diffeomorphic to the plane $\mathbf{R}^2$, if we let $G = SO(2)$ act on $\mathbf{R}^2$ in the standard fashion. So the maps $\phi : O_i \to \phi(O_i)$ are, up to changes of coordinates, equal to the map $\mathbf{R}^2/\mathbf{Z}_2 \to (T^*\mathbf{R}^2)_0$ exhibited in Section 1, which is an isomorphism. Therefore, the map (10) is also an isomorphism.

The two isolated singularities of the 'canoe' are relative equilibria of the spherical pendulum. Both are actually absolute equilibria, corresponding to the pendulum pointing straight up or down. For an alternative computation of the 'canoe' using invariant polynomials, see [3].

3.4 Reduction of the Cotangent Bundle of a Symmetric Space

Consider the special orthogonal group $SO(n)$ acting by conjugation on $S^2(\mathbf{R}^n)$ the space of real symmetric $n \times n$-matrices. Let S_n denote the symmetric group on n letters acting on $\mathbf{R}^n$ by permuting the coordinates and hence on $T^*\mathbf{R}^n$ by permuting the coordinates in pairs. Note that $\mathbf{R}^n$ embeds into $S^2(\mathbf{R}^n)$ as the set of diagonal matrices. Since any symmetric matrix is diagonalizable, the pair $(\mathbf{R}^n, S_n)$ is a cross-section of the $SO(n)$-action on $S^2(\mathbf{R}^n)$. Therefore $S^2(\mathbf{R}^n)/SO(n)$ is homeomorphic to $\mathbf{R}^n/S_n$. The vector space $S^2(\mathbf{R}^n)$ has a natural $SO(n)$-invariant inner product:

$$((a_{ij}), (b_{kl})) = \operatorname{trace}((a_{ij})(b_{kl})) = \sum_{ij} a_{ij} b_{ij}.$$

Remark 3.9 implies that the cross-section $(\mathbf{R}^n, S_n)$ is orthogonal. Therefore Proposition 3.8 provides us with a homeomorphism $\phi : (T^*\mathbf{R}^n)_0 \to (T^*S^2(\mathbf{R}^n))_0$. We contend that ϕ is an isomorphism of reduced spaces. Since the group S_n is finite, the zero level set of the S_n-moment map is the whole space $\mathbf{R}^n \times \mathbf{R}^n$, which embeds naturally into $T^*S^2(\mathbf{R}^n) \simeq S^2(\mathbf{R}^n) \times S^2(\mathbf{R}^n)$. In fact $\mathbf{R}^n \times \mathbf{R}^n$ is a subset of the zero level set of the $SO(n)$-moment map $J : T^*S^2(\mathbf{R}^n) \to \mathfrak{so}(n)^*$. Clearly any $SO(n)$-invariant function on $T^*S^2(\mathbf{R}^n)$ restricts to an S_n-invariant function on $\mathbf{R}^n \times \mathbf{R}^n$. This implies that $\phi^*C^\infty((T^*S^2(\mathbf{R}^n))_0)$ is contained in $C^\infty((T^*\mathbf{R}^n)_0) = C^\infty(\mathbf{R}^n \times \mathbf{R}^n)^{S_n}$. To show that ϕ is an isomorphism of reduced spaces we need to prove that $\phi^*C^\infty(T^*S^2(\mathbf{R}^n))$ is equal to $C^\infty((T^*\mathbf{R}^n)_0)$. By the same argument as the one we have used in the example of Section 1, it is enough to show that there is a set $\{\sigma_{ij}\}$ of polynomials that generates the S_n-invariant polynomials on $\mathbf{R}^n \times \mathbf{R}^n$ and has the property that each σ_{ij} is the restriction of an $SO(n)$-invariant polynomial on $S^2(\mathbf{R}^n) \times S^2(\mathbf{R}^n)$. According to Weyl [25], the polynomials

$$\sigma_{kl}(x,y) = \sum_{ij} x_i^k y_j^l, \quad 1 \leq k, l \leq n, \tag{11}$$

generate the S_n-invariant polynomials on $\mathbf{R}^n \times \mathbf{R}^n$. On the other hand, σ_{kl} is the restriction of the $SO(n)$-invariant polynomial $\tau_{kl}(A,B) = \mathrm{trace}(A^k B^l)$, so the polynomials (11) are the required set. We have thus proved that

$$(T^*S^2(\mathbf{R}^n))_0 \simeq (\mathbf{R}^n \times \mathbf{R}^n)/S_n$$

as stratified symplectic spaces.

More generally, let G be a semisimple Lie group over $\mathbf{R}$, K a maximal compact subgroup of G and $\mathfrak{g} = \mathfrak{k} \oplus \mathfrak{p}$ a Cartan decomposition of $\mathfrak{g} = Lie(G)$. Then K acts on $\mathfrak{p}$ by conjugation. Pick a maximal abelian subspace $\mathfrak{a}$ of $\mathfrak{p}$ and let $W = N(\mathfrak{a})/C(\mathfrak{a})$ denote the Weyl group. It is well-known that the restriction map $\mathbf{R}[\mathfrak{p}] \to \mathbf{R}[\mathfrak{a}]$ from polynomials on $\mathfrak{p}$ to polynomials on $\mathfrak{a}$ gives rise to an isomorphism $\mathbf{R}[\mathfrak{p}]^K \to \mathbf{R}[\mathfrak{a}]^W$. The quotient spaces $\mathfrak{a}/W$ (the Weyl chamber) and $\mathfrak{p}/K$ are therefore isomorphic. The computation above verifies Conjecture 3.7 in the special case $G = Sl(n, \mathbf{R})$, showing that we have an isomorphism of cotangent bundles, $(T^*\mathfrak{a})/W = T^*(\mathfrak{a}/W) \cong T^*(\mathfrak{p}/K)$. For arbitrary G, Conjecture 3.7 would follow from (but is not equivalent to):

3.11. CONJECTURE. *The restriction map* $\mathbf{R}[\mathfrak{p} \times \mathfrak{p}]^K \to \mathbf{R}[\mathfrak{a} \times \mathfrak{a}]^W$ *is surjective.*

4 Poisson Embeddings of Reduced Spaces

The goal of this section is to show that in some cases a reduced space of a symplectic representation space can be realized as the closure of a coadjoint orbit in the dual of some Lie algebra (cf. Section 2.4). For the remainder of this section, let K be a compact group acting linearly on a symplectic vector space V and preserving its symplectic form ω. Then the action of K is

Hamiltonian. Let $J : V \to \mathfrak{k}^*$ denotes the corresponding moment map. The ring of invariant polynomials $\mathbf{R}[V]^K$ is finitely generated. We now make the following assumption:

ASSUMPTION Q. The ring of all K-invariant polynomials on V is generated by the homogeneous quadratic K-invariant polynomials.

The space of homogeneous quadratic polynomials, $\mathbf{R}_2[V]$, and the space of invariant polynomials are both closed under the Poisson bracket. It follows that their intersection,

$$\mathfrak{h} := \mathbf{R}_2[V]^K,$$

which is the space of invariant homogeneous quadratic polynomials, is also closed under the Poisson bracket. The algebra $\mathbf{R}_2[V]$ is canonically isomorphic to the Lie algebra $\mathfrak{sp}(V)$ of all infinitesimally symplectic linear transformations: the isomorphism takes a quadratic polynomial to its associated Hamiltonian vector field. The inverse map sends $\xi \in \mathfrak{sp}(V, \omega)$ to the polynomial $1/2\,\omega(\xi v, v)$. Thus we can view $\mathfrak{h}$ as a subalgebra of $\mathfrak{sp}(V)$.

Consider the map $\sigma : V \to \mathfrak{h}^*$ defined by

$$\langle \sigma(v), P \rangle = P(v)$$

where $P \in \mathfrak{h}$ and $\langle \cdot, \cdot \rangle$ denotes the canonical pairing of a vector space with its dual. This is the Hilbert map of classical invariant theory. It is manifestly K-invariant, and so induces a map $\bar{\sigma} : V/K \to \mathfrak{h}^*$. Assumption Q above implies that σ separates K-orbits. Thus $\bar{\sigma}$ is a homeomorphism onto its image $\sigma(V) \subset \mathfrak{h}^*$. Let H be the connected subgroup of $Sp(V)$ whose Lie algebra is $\mathfrak{h}$. Note that the map σ is the momentum map for the H action on V. It is H-equivariant. (Here H acts on $\mathfrak{h}^*$ by the coadjoint action.)

4.1. REMARK. It is perhaps helpful to rephrase the above discussion in coordinates. Let $\sigma_1, \ldots, \sigma_N$ be a basis for the space $\mathfrak{h}$ of invariant homogeneous quadratic polynomials. The Poisson bracket of any two generators is again a homogeneous quadratic K-invariant polynomial (or zero), which demonstrates that $\mathfrak{h}$ is a Lie algebra. This Lie algebra has structure constants $c_{ij}^k \in \mathbf{R}$ defined by

$$\{\sigma_i, \sigma_j\} = \sum_k c_{ij}^k \sigma_k.$$

The map σ, in terms of this choice of coordinates on $\mathfrak{h}^*$ is

$$\sigma : V \to \mathbf{R}^N, \quad v \mapsto (\sigma_1(v), \ldots, \sigma_N(v)),$$

where we have identified $\mathbf{R}^N$ with $\mathfrak{h}^*$, the isomorphism being the one associated to choosing the basis of $\mathfrak{h}^*$ which is dual to the basis σ_i.

4.2. REMARK. Motivated by problems in representation theory, Howe [9] defined a *reductive dual pair* to be a pair of reductive subgroups of $Sp(V)$ that are each other's centralizers. The groups K and H above clearly commute with each other and it is easy to see that H is (the identity component of) the centralizer of K. It is not true in general that K is the centralizer of H, as the example of $K = SU(2)$ acting on $V = \mathbf{C}^2$ clearly indicates. One can get around the problem of K not being the full centralizer of H in $Sp(V)$ by replacing it with $K' :=$ the centralizer of H in $Sp(V)$. However, it is not at all clear why K' and H should be reductive. Also, given a dual pair (K, H) with K compact, it is not clear whether the quadratic polynomials corresponding to $\mathfrak{h} = Lie(H)$ generate $\mathbf{R}[V]^K$.

However, in three interesting physical examples of symplectic representations of K satisfying condition Q, the groups K and H do form a reductive dual pair:

1. the planar N-body problem ($SO(2)$ acting diagonally on $(T^*\mathbf{R}^2)^N$);
2. the d-dimensional N-body problem ($O(d)$ acting diagonally on $(T^*\mathbf{R}^d)^N$), this example is worked out in the next section;
3. $U(p)$ acting on $\mathbf{C}^p \otimes \mathbf{C}^q$.

These examples seem to hint at an interesting connection between reductive dual pairs and condition Q.

Now let $\mathcal{O}$ be a coadjoint orbit of K. Consider the corresponding reduced space $V_{\mathcal{O}} = J^{-1}(\mathcal{O})/K$. We claim that the map $\bar{\sigma} : V_{\mathcal{O}} \to \mathfrak{h}^*$ induced by the H-momentum map σ is a Poisson embedding in the following sense.

4.3. DEFINITION. Let X be a stratified symplectic space and let P be a Poisson manifold. A *proper Poisson embedding* of X into P is a proper injective map $j : X \to P$ such that

i. the pullback by j of every smooth function on P is a smooth function on X;
ii. the pullback map $j^* : C^\infty(P) \to C^\infty(X)$ is surjective;
iii. the pullback map j^* is a morphism of Poisson algebras.

We mention a few obvious consequences of this definition: the image of a proper Poisson embedding $j : X \to P$ is closed; j is a homeomorphism onto its image; the kernel of j^*, which is the set of smooth functions vanishing on the image $j(X)$, is a Poisson ideal inside $C^\infty(P)$; and the set of Whitney smooth functions on $j(X)$ is a Poisson algebra, which is isomorphic to $C^\infty(X)$. Therefore $j(X)$ is a stratified symplectic space (stratified by the images of the strata in X) and $j : X \to j(X)$ is an isomorphism of stratified symplectic spaces.

4.4. THEOREM. *Suppose that assumption Q holds. Let H be the closed connected Lie subgroup of $Sp(V)$ described above, and $\sigma : V \to \mathfrak{h}^*$ its associated momentum map. Let $\mathcal{O}$ be an arbitrary coadjoint orbit of K. Then the following statements hold.*

1. *The map $\bar{\sigma} : V_{\mathcal{O}} \to \mathfrak{h}^*$ is a proper Poisson embedding of the K-reduced space $V_{\mathcal{O}}$ (where the bracket on $\mathfrak{h}^*$ is the usual Lie-Poisson bracket);*
2. *Each connected component of a symplectic stratum of $V_{\mathcal{O}}$ is mapped symplectomorphically by $\bar{\sigma}$ onto a coadjoint orbit of H contained in $\bar{\sigma}(V_{\mathcal{O}})$;*
3. *The image $\bar{\sigma}(V_{\mathcal{O}})$ of the Poisson embedding is the closure of a single coadjoint orbit of H.*

PROOF. 1. We check the conditions of Definition 4.3. The square of the distance to the origin in V is a K-invariant polynomial function. From this it follows easily that the Hilbert map σ is proper. Hence the map $\bar{\sigma} : V_{\mathcal{O}} \to \mathfrak{h}^*$ is proper. It is injective because the Hilbert map separates the K-orbits. It is not hard to see from the definition of smooth functions on $V_{\mathcal{O}}$ that $\bar{\sigma}$ pulls back smooth functions to smooth functions. That the pullback map $\bar{\sigma}^* : C^\infty(\mathfrak{h}^*) \to C^\infty(V_{\mathcal{O}})$ is surjective is an easy consequence of Schwarz's theorem [20]. It is a homomorphism of Poisson algebras, because the Hilbert map σ, being the H-momentum map, is a Poisson map.
2. The connected components of the symplectic strata are the symplectic leaves of the reduced space $V_{\mathcal{O}}$, i.e., they are swept out by the Hamiltonian flows of smooth functions (see Lemma 2.19). Since the Poisson algebras $C^\infty(V_{\mathcal{O}})$ and $C^\infty(j(V_{\mathcal{O}}))$ are isomorphic, the embedding j maps leaves onto leaves. But the leaves of $\mathfrak{h}^*$ are simply the coadjoint H-orbits. (Here we use that H is connected.)
3. Theorem 4.6 below states that the level set $J^{-1}(\mathcal{O})$ is connected. It follows now from Theorem 2.12 that the reduced space $V_{\mathcal{O}}$ has a connected open dense stratum S_{top}; so the set $\bar{\sigma}(V_{\mathcal{O}})$ has to be the closure of $\bar{\sigma}(S_{top})$, which is a single coadjoint orbit by statement 2 of this theorem.

□

4.5. REMARK. Denote the stratified symplectic space $\bar{\sigma}(V_{\mathcal{O}})$ by $X_{\mathcal{O}}$. If the group H is semisimple then we use a Killing form to identify $\mathfrak{h}$ with $\mathfrak{h}^*$ in an H-equivariant way. If $\mathcal{O}$ is the zero orbit, then the image $X_0 = \sigma(J^{-1}(0))$ described in Theorem 4.4 is neccessarily the closure of a *nilpotent* orbit. This is because $J(0) = 0$ and $\sigma(0) = 0$ so that $0 \in \sigma(J^{-1}(0))$. And the only orbits whose closure contains 0 are the nilpotent ones.

More generally $X_{\mathcal{O}}$ contains a single semisimple orbit orbit Q and any other orbit P contained in $X_{\mathcal{O}}$ fibres over Q. The fibration $\pi_P : P \to Q$ is simply the projection of $\eta \in P$ onto its semisimple part, $\pi_P(\eta) = \eta_{ss}$. The fibre of π_P is the orbit of the nilpotent part η_n of η under the action of the stabilizer group $H_{\eta_{ss}}$ of η_{ss}. Note that η_n is nilpotent in $Lie(H_{\eta_{ss}})$. It follows that one can view the map $\pi : X_{\mathcal{O}} \to Q$ as a fibre bundle with typical fibre being the closure of a nilpotent orbit in some smaller reductive group. These facts about the structure of orbits of a semisimple group are well-known and we refer the reader to [23] for proofs and further references. It was shown in [12] that if a (co)adjoint orbit P fibres over an semisimple orbit Q then the fibration is symplectic. Thus the map $\pi : X_{\mathcal{O}} \to Q$ can be viewed as a fibration of stratified symplectic spaces.

To conclude this section, we prove the connectivity statement used in the proof of Theorem 4.4. This result does not use assumption Q.

4.6. THEOREM. *Let K be a compact group acting linearly on a symplectic vector space V and preserving its symplectic form ω. Let $J : V \to \mathfrak{k}^*$ denotes the corresponding moment map. Then for any coadjoint orbit $\mathcal{O}$ of K the set $J^{-1}(\mathcal{O})$ is connected.*

PROOF. Without loss of generality we may assume that V is $\mathbf{C}^n$ with the standard symplectic form and K is a subgroup of the unitary group $U(n)$. Let $\mathcal{O}$ be a coadjoint orbit of K. We will show that for any $r > 0$ the closed ball

$$\bar{B}(r) = \{\, z \in \mathbf{C}^n : |z|^2 \leq r \,\}$$

intersects $J^{-1}(\mathcal{O})$ in a connected set. Clearly this will prove the theorem.

Note first that the central circle subgroup of $U(n)$,

$$U(1) = \left\{ \begin{pmatrix} e^{i\theta} & & \\ & \ddots & \\ & & e^{i\theta} \end{pmatrix} : \theta \in \mathbf{R} \right\}$$

commutes with K and therefore preserves the level set $J^{-1}(\mathcal{O})$. Consider now the space $N(r)$ obtained from $\bar{B}(r)$ by identifying the points on the boundary that lie in the same $U(1)$-orbit. Let $q : \bar{B}(r) \to N(r)$ denote the quotient map. Since $J^{-1}(\mathcal{O})$ is $U(1)$-invariant and the fibres of q are connected, the set $J^{-1}(\mathcal{O}) \cap \bar{B}(r)$ is connected if and only if its image under q is connected in $N(r)$. We will see shortly that $N(r)$ is K-equivariantly symplectomorphic to $\mathbf{C}P^n(r)$, the complex projective space with the symplectic form equal to the standard one times r. We will also see that under this identification the action of K on $N(r)$ becomes Hamiltonian with the moment map $J_r : N(r) \to \mathfrak{k}^*$ having the property that $J_r^{-1}(\mathcal{O}) = q(J^{-1}(\mathcal{O}) \cap \bar{B}(r))$.

Consider the action of $U(1)$ on $\mathbf{C}^n \times \mathbf{C}$ corresponding to the Hamiltonian $\phi(z, w) = |z|^2 + |w|^2 - r$ for $(z, w) \in \mathbf{C}^n \times \mathbf{C}$. Then

$$\phi^{-1}(0) = \{(z, w) \in \mathbf{C}^n \times \mathbf{C} : |z|^2 + |w|^2 = r\}$$

and $\phi^{-1}(0)/U(1) \simeq \mathbf{C}P^n(r)$. Now, K acts on $\mathbf{C}^n \times \mathbf{C}$ by acting trivally on the second factor. Since the actions of K and $U(1)$ on $\mathbf{C}^n \times \mathbf{C}$ commute, the action of K descends to a Hamiltonian action on the reduced space $\mathbf{C}P^n(r)$. The corresponding moment map J_r is obtained by extending $J : \mathbf{C}^n \to \mathfrak{k}^*$ by zero to a map on $\mathbf{C}^n \times \mathbf{C}$, restricting the extension to the sphere $\phi^{-1}(0)$ and pushing it down to a map on the quotient $\mathbf{C}P^n(r)$.

To get the identification of $N(r)$ with $\mathbf{C}P^n(r)$ we start out by embedding $\bar{B}(r)$ into $\phi^{-1}(0)$ via the map

$$f : z \mapsto \left(z, \sqrt{r - |z|^2}\right).$$

Composing f with the orbit map $\phi^{-1} \to \phi^{-1}/U(1)$ we get a map f' from $\bar{B}(r)$ *onto* $\mathbf{C}P^n(r)$. It is easy to see that f' descends to a homeomorphism f'' from $N(r)$ to $\mathbf{C}P^n(r)$. It is also easy to see that

$$f''(q(J^{-1}(\mathcal{O}) \cap \bar{B}(r))) = f'(q(J^{-1}(\mathcal{O}) \cap \bar{B}(r))) = J_r^{-1}(\mathcal{O}).$$

Obviously, the moment map $J_r : \mathbf{C}P^n(r) \to \mathfrak{k}^*$ is proper. So Remark 2.13 implies that the set $J_r^{-1}(\mathcal{O})$ is connected and we are done. □

5 Reduced Space at Angular Momentum Zero for n Particles in d-space

Let V be the phase space for n particles in d-dimensional Euclidean space;

$$\begin{aligned} V &= T^*\mathbf{R}^d \times T^*\mathbf{R}^d \times \ldots T^*\mathbf{R}^d \quad (n \text{ times}) \\ &= \mathbf{R}^d \times \mathbf{R}^d \times \ldots \mathbf{R}^d \quad (2n \text{ times}). \end{aligned}$$

Take $G = O(d)$ to be the orthogonal group associated to $\mathbf{R}^d$, with $g \in G$ acting on V according to

$$g \cdot (q_1, p^1, q_2, p^2, \ldots, q_n, p^n) = (gq_1, gp^1, gq_2, gp^2, \ldots, gq_n, gp^n).$$

We will use Greek indices, μ, ν, etc. for the particle labels, and Latin indices i, j etc. to index the coordinates on the Euclidean space $\mathbf{R}^d$. So V has coordinates (q^i_μ, p^ν_j), for $\mu, \nu = 1, 2, \ldots, n$ and $i, j = 1, 2, \ldots, d$, which shows that

$$M \cong \mathbf{R}^d \otimes \mathbf{R}^{2n}. \tag{12}$$

Under this isomorphism the G-action becomes $g(x \otimes z) = gx \otimes z$. The symplectic form on V is $\Omega = \Sigma_{i,\mu} dq^i_\mu \wedge dp^\mu_i$. The momentum map for the $O(d)$-action is

$$J(q,p) = \Sigma_\mu q_\mu \wedge p^\mu,$$

where we have used the inner product on $\mathbf{R}^d$ to identify $\Lambda^2\mathbf{R}^d$ with the Lie algebra of $O(d)$ and its dual space. Equation (12) expresses V as the tensor product of the inner product space $\mathbf{R}^d$ with the symplectic vector space $\mathbf{R}^{2n}$. Since $h \in H := Sp(n,\mathbf{R})$ acts by $h(x \otimes z) = x \otimes hz$, it is clear that the actions of $G = O(d)$ and of H commute. The momentum map for the $Sp(n,\mathbf{R})$-action is given by

$$\sigma(q,p) = \begin{pmatrix} q_\mu \cdot q_\nu & q_\mu \cdot p^\nu \\ p^\mu \cdot q_\nu & p^\mu \cdot p^\nu \end{pmatrix}. \tag{13}$$

Here '$\cdot$' denotes the inner product on $\mathbf{R}^d$: $q_\mu \cdot q_\nu = \Sigma_i q^i_\mu q^i_\nu$. Thus $S = \sigma(q,p)$ is a symmetric $2n \times 2n$-matrix which we have written in terms of four $n \times n$-blocks.

In saying that σ is the momentum map we are identifying the space $S^2(\mathbf{R}^{2n})$ of symmetric $2n \times 2n$-matrices on $\mathbf{R}^{2n}$ with the dual of the Lie algebra of $\mathfrak{sp}(n,\mathbf{R})$ since the target of the map σ is $S^2(\mathbf{R}^{2n})$. What is the identification $S^2(\mathbf{R}^{2n}) \cong \mathfrak{sp}(n,\mathbf{R})^*$? The trace pairing (Killing form) $(S_1,S_2) \mapsto \operatorname{trace} S_1S_2$ induces an isomorphism $\mathfrak{sp}(n,\mathbf{R}) \cong \mathfrak{sp}(n,\mathbf{R})^*$. The identification of $S^2(\mathbf{R}^{2n})$ with $\mathfrak{sp}(n,\mathbf{R})$ is described by mapping S to JS where J is the symplectic operator: $J^2 = -1$, $JJ^t = I$, $\Omega(v,w) = \langle v, Jw\rangle$. Composing these identifications yields the desired one: $\mathfrak{sp}(n,\mathbf{R})^* \cong S^2(\mathbf{R}^{2n})$. Under this isomorphism the coadjoint action of $Sp(n,\mathbf{R})$ intertwines with the action $S \mapsto gSg^t$ of $Sp(n,\mathbf{R})$ on $S^2(\mathbf{R}^{2n})$.

The 'first main theorem of invariant theory' (see e.g. [25, Theorem 2.9A]) states that the entries of $S = \sigma(q,p)$ in the formula for σ form a basis for the $O(d)$-invariant polynomials on V. Consequently assumption Q of the previous section holds and so the restriction of $\bar\sigma$ to $J^{-1}(0)/O(d)$ is an isomorphism onto its image. (As in the previous section, $\bar\sigma : M/O(d) \to S^2(\mathbf{R}^2 n)$ is the map induced by σ.) What is its image?

Let $\Sigma \subset S^2(\mathbf{R}^2 n)$ denote the set of *nonnegative* symmetric matrices whose kernel is coisotropic. (This means that the kernel contains its Ω-orthogonal complement.) Let $\Sigma_k \subset \Sigma$ denote the subset of Σ consisting of matrices of rank k, and let

$$\Sigma^k = \cup_{i \leq k} \Sigma_k \tag{14}$$

denote the subset of matrices with rank at most k. As a subset of $\mathfrak{sp}(n,\mathbf{R})$ the set Σ_j is a single coadjoint orbit, and $\Sigma^k = \overline{\Sigma}_k$ is the union of $k+1$ nilpotent orbits, these being the Σ_j, $j \leq k$, with $\Sigma_0 = \{0\}$. These are the strata of Σ^k. We will show that $\sigma(J^{-1}(0)) = \Sigma^k$ where $k = \min(d, n)$. Once this is shown we will have proven:

5.1. THEOREM. *Let V_0 denote the reduced space at angular momentum 0 for the action of $O(d)$ on the phase space V of of n particles in d-space. Then V_0 is isomorphic as a stratified symplectic space to the set Σ^k described in (14), where $k = \min(d, n)$. The isomorphism is the one induced by the $Sp(n, \mathbf{R})$-momentum map, namely the restriction of $\bar{\sigma}$ to $J^{-1}(0)/O(d)$.*

PROOF. We proved in the previous section that σ induces an isomorphism with all the desired properties. It remains to prove that the image of σ restricted to $J^{-1}(0)$ is Σ^k. We have

$$\sigma(q,p) = \begin{pmatrix} q_1 \\ q_2 \\ \vdots \\ q_n \\ p^1 \\ p^2 \\ \vdots \\ p^n \end{pmatrix} \left({q_1}^t \; {q_2}^t \ldots {q_n}^t \; p^{1t} \; p^{2t} \ldots p^{nt} \right).$$

From this expression, it is clear that $\sigma(q,p)$ is nonnegative. Since each of the two factors of $\sigma(q,p)$ is a matrix with rank less than or equal to d, the matrix $\sigma(q,p)$ has rank less than or equal to d.

5.2. REMARK. These are the only constraints on the image of σ. This is the content of the the 'second main theorem of invariant theory' for the orthogonal group [25, Theorem 2.17A]. However, we do not need this result to prove our theorem, as it will in our case follow from equivariance.

Now let us restrict σ to $J^{-1}(0)$. There we have $q_1 \wedge p^1 + q_2 \wedge p^2 + \ldots q_n \wedge p^n = 0$. Let us assume for simplicity that the q^i are linearly independent. Then Cartan's lemma (see e.g. [22, p. 19]) states that we have $p^\mu = \Sigma S^{\mu\nu} q_\nu$ for some symmetric $n \times n$ matrix S. A direct calculation now shows that in this case

$$\sigma(q,p) = \begin{pmatrix} M & MS \\ SM & SMS \end{pmatrix}, \tag{15}$$

where $M_{\mu,\nu} = q_\mu \cdot q_\nu$ is the matrix of inner products. Note that

$$\sigma(q,p) \begin{pmatrix} S \\ -I \end{pmatrix} = 0,$$

from which it follows that the kernel of the map $\sigma(q,p)$ contains the Lagrangian subspace $\{(Sy,-y) : y \in \mathbf{R}^n\}$. But any subspace containing a Lagrangian one is coisotropic, so we have proved our result in this particular case.

In general the q^i are not linearly independent. But a slight variant of the proof of Cartan's lemma shows us that the dimension of the space spanned by $\{q_1, q_2, \ldots, p^1, p^2, \ldots, p^n\}$ is less than or equal to n. It follows from the factorization of σ that the rank of $\sigma(q,p)$ is always less than or equal to n. We have proved the statement regarding the rank of the matrices in $\sigma(J^{-1}(0))$.

A few moments' reflection should convince the reader that each Σ_j is a single orbit of the $Sp(n,\mathbf{R})$-action on $S^2(\mathbf{R}^2 n)$. Hint: Write $\mathbf{R}^{2n} = L_1 \oplus L_2$ where the L_i are Lagrangian subspaces and $A \in \Sigma_j$ annihilates L_2. Note that $Sp(n,\mathbf{R})$ acts transitively on pairs (L_1, L_2) of transverse Lagrangian subspaces, and that, relative to this splitting $g \oplus g^t \in Sp(n,\mathbf{R})$ for any $g \in Gl(L_1)$. (The symplectic form allows us to identify L_2 with the dual of L_1.) Now suppose that we can show that there is *some* matrix $A \in \Sigma_k \cap \sigma(J^{-1}(0))$. Then it follows from the Sp-equivariance of σ and the Sp-invariance of J that $\Sigma_k \subset \sigma(J^{-1}(0))$. It is also clear that the closure of Σ_k is Σ^k. The map σ, being homogeneous and quadratic, is a closed map. It now follows from $\Sigma_k \subset \sigma(J^{-1}(0))$ that $\sigma(J^{-1}(0) = \Sigma^k$ as desired. Thus all we have to do is produce a single matrix A in Σ_k which we can be written in the form $\sigma(q,p)$ for some $(q,p) \in J^{-1}(0)$. Take $(q,p) = (q,0)$. If $d \geq n$, set $q = (e_1, e_2, \ldots e_n)$, the first n elements of an orthonormal basis $\{(e_1, e_2, \ldots, e_d\}$ for $\mathbf{R}^d$. Then (see (15))

$$\sigma(q,p) = \begin{pmatrix} I & 0 \\ 0 & 0 \end{pmatrix},$$

where I is the $n \times n$ identity matrix. This proves the theorem for the case $d \geq n$. In case $d < n$, take $q = (e_1, e_2, \ldots, e_d, 0, \ldots, 0)$. Then $\sigma(q,p)$ again has the above form, except now I is the $d \times d$-identity matrix. □

5.3. REMARK. The dual pair just discussed, $(O(d), Sp(n,\mathbf{R}))$, is the subject of [2]. See also [9] and [10, pp. 501–507].

5.4. REMARK ($O(d)$ versus $SO(d)$). Suppose, in the above discussion, that we replace $O(d)$ by the special orthogonal group $SO(d)$. Then the corresponding reduced space will be a *branched double cover* over the $O(d)$-reduced space. This is because $O(d)/SO(d)$ is the two-element group. Assumption Q fails for the group $SO(d)$. Thus we cannot use dual pairs alone to construct its reduced space. The additional, nonquadratic invariants are the d-ple products $\det[v_1, \ldots, v_d]$, where the v_i are any of the vectors $q_1, \ldots, p^n$. They satisfy the relation $\det[v_1, \ldots, v_d]^2 = \det[v_i \cdot v_j]$. In the special case $d = 2$, the d-ple product is quadratic and we can realize the reduced space via dual pairs. Let us consider the case of our example in Section 1: $d = 2$, $n = 1$. The invariants were written down in Section 1.3 as $(\sigma_1, \sigma_2, \sigma_3, \sigma_4)$. σ_3 is the 2-ple product, i.e., the signed area. The other invariants are $O(2)$-invariants. There is one relation, equation (3). It is quadratic in σ_3, explicitly showing how the $SO(2)$-reduced space is a branched double cover of the $O(2)$-reduced space.

References

[1] R. Abraham and J. Marsden, *Foundations of Mechanics*, second edition, Benjamin/Cummings, Reading, 1978.

[2] J. D. Adams, Discrete Spectrum of the Reductive Dual Pair $(O(p,q), Sp(2m))$, *Invent. Math.* **74** (1983), 449–475.

[3] J. Arms, R. Cushman, M. Gotay, A Universal Reduction Procedure for Hamiltonian Group Actions, preprint 591, University of Utrecht, the Netherlands, 1989.

[4] J. Arms, M. Gotay and G. Jennings, Geometric and Algebraic Reduction for Singular Momentum Maps, *Advances in Mathematics* **79** (1990), 43–103.

[5] R. Cushman and R. Sjamaar, On Singular Reduction of Hamiltonian Spaces, to appear in the proceedings of the Colloque international en l'honneur de Jean-Marie Souriau, 1990.

[6] C. Emmrich, H. Römer, Orbifolds as Configuration Spaces of Systems with Gauge Symmetries, *Commun. Math. Phys.* **129** (1990), 69–94.

[7] M. Goresky and R. MacPherson, Intersection Homology Theory, *Topology* **19** (1980), 135–162.

[8] M. Goresky and R. MacPherson, *Stratified Morse Theory*, Springer-Verlag, New York, 1988.

[9] R. Howe, Remarks on Classical Invariant Theory, *Trans. of the AMS* **313** (1989) no. 2, 539-570.

[10] D. Kazhdan, B. Kostant and S. Sternberg, Hamiltonian Group Actions and Dynamical Systems of Calogero Type, *Comm. Pure Appl. Math.* **31** (1978), 481–507.

[11] F. Kirwan, *Cohomology of Quotients in Symplectic and Algebraic Geometry*, Mathematical Notes **31**, Princeton University Press, Princeton, 1984.

[12] E. Lerman, *Symplectic Fibrations and Weight Multiplicities of Compact Groups*, PhD thesis, MIT, 1989.

[13] J. Marsden and A. Weinstein, Reduction of Symplectic Manifolds with Symmetry, *Rep. Math. Phys.* **5** (1974), 121–130.

[14] J. Mather, *Notes on Topological Stability*, Harvard University, 1970, unpublished.

[15] J. M. Montesinos, *Classical Tesselations and Three-Manifolds*, Springer-Verlag, Berlin, 1987.

[16] M. Otto, A Reduction Scheme for Phase Spaces with Almost Kähler Symmetry. Regularity Results for Momentum Level Sets, *J. Geom. Phys.* **4** (1987), 101–118.

[17] R. S. Palais, On the Existence of Slices for Actions of Non-compact Lie Groups, *Annals of Math.* **73** (1961), 295–323.

[18] I. Satake, On a Generalization of the Notion of Manifold, *Proc. Nat. Acad. Sc.* **42** (1956), 359–363.

[19] G. W. Schwarz, *Generalized Orbit Spaces*, revised version of PhD thesis, MIT, 1972, unpublished.

[20] G. W. Schwarz, Smooth Functions Invariant under the Action of a Compact Lie Group, *Topology* **14** (1975), 63–68.

[21] R. Sjamaar, E. Lerman, Stratified symplectic spaces and reduction, to appear in *Ann. Math.*

[22] S. Sternberg, *Lectures on Differential Geometry*, Chelsea, second edition, 1983.

[23] V. S. Varadarajan, *Harmonic Analysis on Real Reductive Groups*, Lecture Notes in Mathematics **576**, Springer-Verlag, New York, 1977.

[24] C. T. C. Wall, Regular Stratifications, in: *Dynamical Systems, Warwick 1974*, A. Manning, editor, Lecture Notes in Mathematics **468**, Springer-Verlag, New York, 1975.

[25] H. Weyl, *The Classical Groups*, Princeton Univ. Press, second edition, Princeton, 1939.

Remarks on the uniqueness of symplectic blowing up

by Dusa McDuff*

§1 Introduction

In [M2] we began the classification of those symplectic 4-manifolds which satisfy

Hypothesis S: (V, ω) *contains a symplectically embedded 2-sphere with non-negative self-intersection number.*

Our results are as follows: see [M3]. By blowing down "exceptional spheres" (i.e. symplectically embedded 2-spheres with self-intersection -1), every symplectic 4-manifold may be reduced to a minimal manifold (i.e. one which contains no exceptional spheres). Moreoever, if (V, ω) satisfies Hytpothesis S, so does its minimal reduction. (In fact, the class of manifolds satisfying Hypothesis S is closed under blowing up and down.) The minimal manifolds in this class are $\mathbb{C}P^2$ and manifolds which fiber over a Riemann surface M with fiber S^2. By analogy with the complex case, manifolds which satisfy Hypothesis S and have minimal reduction $\mathbb{C}P^2$ or $S^2 \times S^2$ are called *rational*, and those which are S^2-fibrations over a Riemann surface M are said to be *ruled*. In the minimal, rational case on can show the symplectic form is Kähler and is uniquely determined up to diffeomorphism by its cohomology class. In the ruled case, the situation is more complicated, though uniqueness holds when the fiber F is not too large. Unfortunately, as Francois Lalonde has pointed out, Theorem 1.3 in [M2] does not hold as stated. The cohomological restrictions in Theorem 1.3 (i)(b) hold only when the base is a sphere, and the proof of uniqueness for bundles over Riemann surfaces of genus $g > 0$ works only under certain cohomological restrictions. For example, in the case of the trivial bundle, it suffices that the cohomology class a of ω be such that $a([M \times pt]) - [g/2]\, a([F]) > 0$, and in the non-trivial case it is enough to have $a([M_-]) - [(g-1)/2]\, a([F]) > 0$, where M_- is a section with self-intersection number -1. Full details may be found in [M4].

One outstanding question in this classification concerns the uniqueness of blowing up and down. As was pointed out in [M1], blowing up in the

*Partially supported by NSF GRANT DMS 9103033

symplectic category is not a local operation: in order to define a symplectic form $\tilde{\omega}$ on the manifold $\tilde{V}$, obtained by blowing up V at a point, one needs to embed a whole symplectic ball into V. Moreoever, the radius λ of the ball is related to the cohomology class of $\tilde{\omega}$ by the formula: $\tilde{\omega}(S) = \pi\lambda^2$, where S is a copy of $\mathbb{C}P^1$ in the exceptional divisor $\Sigma = \mathbb{C}P^{n-1}$. Conversely, in order to blow down an exceptional divisor Σ in W one cuts W open along Σ and glues in a large ball. We showed in [M2] that the minimal reduction of a symplectic 4-manifold (V, ω) is determined up to isotopy by the homology classes of the blown down spheres. However the extent to which blowing up is unique is not yet clear.

This note begins with a careful discussion (valid in arbitrary dimensions) of the relationship between isotopy classes of symplectic forms on $\tilde{V}$ and the space $\mathrm{Emb}_\omega(B(\lambda), V)$ of symplectic embeddings of the closed 2n-ball of radius λ into V. Specialising to dimension 4, we then give a simpler proof of the main result of [M1] which states that there is a unique way to blow up one point in $\mathbb{C}P^2$. This proof extends to two points. Our result may be stated in several equivalent ways. Note that we will always assume that $\mathbb{R}^4$ and $\mathbb{C}P^2$ have their standard symplectic structures ω_0 and τ_0, where τ_0 is normalised so that $\tau_0(\mathbb{C}P^1) = \pi$.

Theorem 1.1 (i) *The space of symplectic embeddings of two disjoint balls into the open unit 4-ball* Int $B^4(1) \subset \mathbb{R}^4$ *is connected.*
(ii) *The space of symplectic embeddings of one ball into the open polydisc* Int $B^2(r_1) \times B^2(r_2) \subset \mathbb{R}^4$ *is connected, for any* $r_1 \leq r_2$.
(iii) *Let* X_k *denote the manifold obtained by blowing up* k *points in* $\mathbb{C}P^2$. *If* k=1 *or* 2, *any two cohomologous symplectic forms on* X_k, *which are rational, i.e. which satisfy Hypothesis S, are symplectomorphic. In particular, they are Kähler.*

Throughout, we use the language and notation of [M2].

§2 Blowing up and down in the symplectic category

If x is any point in a symplectic manifold (V, ω), there always is an ω-compatible almost complex structure J on V which is integrable near x. (Recall that J is ω-compatible iff $\omega(v, Jv) > 0$ and $\omega(v, w) = \omega(Jv, Jw)$ for all non-zero tangent vectors v and w.) Hence one can blow up V at the point x just as if it were a complex manifold, by replacing x by the copy Σ of the complex projective space $\mathbb{C}P^{n-1}$ formed by all J-complex lines through the origin in the tangent space T_x to V at x. (More precisely, one replaces a neighbourhood of x by a neighbourhood N(Z) of the zero section Z of the canonical line bundle over $\mathbb{C}P^{n-1}$.) Let us call this manifold $\tilde{V}(J, x)$. It is easy to see that if J′ and x′ are another such pair and if γ is a homotopy class of paths in the principal $Sp(2n, \mathbb{R})$-bundle P over V with endpoints lying over x and x′, then there is a diffeomorphism

$$f_\gamma : (\tilde{V}(J,x), \Sigma) \to (\tilde{V}(J',x'), \Sigma')$$

which is unique up to isotopy (Remember that the space of ω-compatible J is contractible.) In particular, if we want to fix an identification of the exceptional divisor Σ with $\mathbb{CP}^{n-1}$ we need to choose an identification of T_x with $\mathbb{C}^n$, i.e. an element p of P which lies over x. We will write $\tilde{V}_p$ for the blow-up of V at x, where Σ is identified with $\mathbb{CP}^{n-1}$ via p, and will write $\tilde{V}$ for any manifold diffeomorphic to some $\tilde{V}_p$.

In order to define a symplectic blow-up one also must put a symplectic form $\tilde{\omega}$ on $\tilde{V}$. It is natural to try to define $\tilde{\omega}$ so that it is standard on the exceptional divisor Σ and equal to $\varphi^*\omega$ outside an ϵ-neighbourhood of Σ, where $\varphi : \tilde{V} \to V$ is the obvious projection taking Σ to x. In fact, we cannot quite do this: as mentioned above, in the symplectic category the operation of blowing up does not just change V at a single point but affects a whole ball around the point.

In order to explain this, we will start with blowing down. Thus we will suppose that Σ is a symplectically embedded copy of $\mathbb{CP}^{n-1}$ in a symplectic 2n-dimensional manifold (W, ω), and that the restriction $\omega|\Sigma$ is $\lambda^2\tau_0$. Further, we will suppose that Σ has the correct normal bundle, i.e. that $c_1(\nu_\Sigma) = -[\tau_0]$. The standard model for Σ is the zero section Z in the canonical line bundle over $\mathbb{CP}^{n-1}$ with appropriate symplectic form ρ_λ. We showed in [M1] that, for each $\epsilon > 0$, there is a symplectomorphism Φ_ϵ between the deleted ϵ-neighbourhood $N_\epsilon(Z) - Z$ of Z and the spherical shell $B(\lambda + \epsilon) - B(\lambda)$ in $\mathbb{R}^{2n}$, where $B(\lambda)$ denotes the closed ball of radius λ and center $\{0\}$ in $\mathbb{R}^{2n}$. Thus, the symplectic neighbourhood theorem implies that:

Lemma 2.1 ([M1]) *For sufficiently small $\epsilon > 0$ there is a neighbourhood* $N_\epsilon(\Sigma)$ *of* Σ *in* W *such that* $N_\epsilon(\Sigma) - \Sigma$ *is symplectomorphic to* $B(\lambda+\epsilon) - B(\lambda) \subset \mathbb{R}^{2n}$.

Definition 2.2: blowing down Suppose that $\Sigma = \mathbb{CP}^{n-1}$ is embedded in (W, ω) with normal bundle as above, and let $\epsilon > 0$ be so small that Lemma 2.1 holds. Then the **blow down** of W along Σ is obtained by cutting out $N_\epsilon(\Sigma)$ from W and sewing $B(\lambda + \epsilon)$ back in. By arguments similar to those in Lemma 2.4 below, one can show that this construction depends only on the choice of symplectomorphism ϕ between (Σ, ω) and $(Z, \lambda^2\tau_0)$. In particular, it is independent of the choice both of ϵ and of the extension of ϕ to $N_\epsilon(\Sigma)$ which is used to define the symplectomorphism between $N_\epsilon(\Sigma) - \Sigma$ and $B(\lambda + \epsilon) - B(\lambda)$. Note that in dimension 4, Σ is just a symplectically embedded 2-sphere with self-intersection number -1: such a sphere is called an **exceptional sphere.**

Definition 2.3: blowing up Conversely, a **blow up** $(\tilde{V}_g, \tilde{\omega}_g)$ of (V, ω) of **weight** λ is obtained from a symplectic embedding g of the ball $B^{2n}(\lambda)$ into V by extending g to a symplectic embedding g_0 of the ball $B^{2n}(\lambda+\epsilon)$, and then replacing the image $g_0(B^{2n}(\lambda+\delta))$ by the standard neighbourhood $N_\delta(Z)$ for some $\delta \in (0, \epsilon)$. Thus the manifold $\tilde{V}_g = \tilde{V}(g_0, \delta)$ is defined to be:

$$\tilde{V}(g_0, \delta) = \big(V - \mathrm{Int}\, g_0(B(\lambda+\delta))\big) \cup N_\delta(Z),$$

where $0 < \delta < \epsilon$ and $N_\delta(Z)$ is the neighbourhood of the zero section Z of γ whose boundary is taken by Φ_ϵ to $S^{2n-1}(\lambda+\delta)$, attached via $g_0 \circ \Phi_\epsilon$. The form ω_g equals ω on V$-$Int $g_0(B(\lambda+\delta))$ and the standard form ρ_λ on $N_\delta(Z)$. It is easy to check that this is independent of the choice of δ. We will denote the exceptional divisor in $\tilde{V}(g_0, \delta)$ by Σ.

Lemma 2.4 (i) *The manifold* $\tilde{V}_g = \tilde{V}(g_0, \delta)$ *is independent of the choice of extension* g_0. *More precisely, if* g_0 *and* g_1 *are two extensions of* g *to* $B(\lambda+\epsilon)$ *there is, for sufficiently small* $\delta > 0$, *a symplectomorphism* $(\tilde{V}(g_0, \delta), \tilde{\omega}_g) \to (\tilde{V}(g_1, \delta), \tilde{\omega}_g)$ *which is unique up to isotopy and which is the identity on* Σ.
(ii) $\tilde{V}_g$ *may be identified with the blow-up* $\tilde{V}_{p(g)}$ *of* V *at* $p(g) = dg(0)$ *by a diffeomorphism which is unique up to isotopy and which is the identity on* Σ.

Proof (i) Without loss of generality, we may suppose that V= $\mathbb{R}^{2n}$ and that g and g_0 are the obvious inclusions. Clearly, it suffices to construct a compactly supported symplectomorphism ϕ of Int $B(\lambda+\epsilon)$ which restricts to g_1 on $B(\lambda+\delta')$ for some $\delta' \in (0, \epsilon)$, and which is unique up to isotopy. To this end, choose a smooth cut off function $\alpha : \mathbb{R} \to [0,1]$ which equals 1 near 0 and 0 near 1, let $\alpha_\delta(x) = \alpha((||x|| - \lambda)/\delta)$, and define $h(t, \delta) : B(\lambda+\epsilon) \to V$ by:

$$h(t, \delta)(x) = x + t\,\alpha_\delta(x)(g_1(x) - x) \quad \text{for} \quad 0 \le t \le 1.$$

Note that $h(t,\delta) - \mathrm{id}$ has support in Int $B(\lambda+\delta)$ for all t, and that $h(1,\delta) = g_1$ near $B(\lambda)$.

Let $\mathcal{C}$ be a convex neighbourhood of ω_0 in the space of all symplectic forms on $B(\lambda+\epsilon)$ which equal ω_0 near $\partial B(\lambda+\epsilon)$, and let $\mathcal{U}$ be a convex neighbourhood of the identity in the group of compactly supported diffeomorphisms of $B(\lambda+\epsilon)$ such that $h^*(\omega_0) \in \mathcal{C}$ for all $h \in \mathcal{U}$. Then, by the convexity of $\mathcal{C}$, the path $t\omega_0 + (1-t)h{*}\omega_0$, $0 \le t \le 1$, lies in $\mathcal{C}$, and Moser's argument implies that there is a retraction R from $\mathcal{U}$ into the group of compactly supported symplectic diffeomorphisms of $B(\lambda+\epsilon)$. Now observe that, because $g_1(x) - x = 0$ on $B(\lambda)$, the maps $h(t,\delta)$, $\delta \to 0$, defined above converge to the identity in the C^1-norm on $B(\lambda+\epsilon)$. Hence there is $\delta_0 > 0$ such that $h(t,\delta) \in \mathcal{U}$ when

$0 \le t \le 1$ and $\delta \le \delta_0$. Put $\phi = R(h(1, \delta_0))$. Because $R(h)-h$ is supported in the closure of the set on which $h^*(\omega_0) \neq \omega_0$, this map ϕ restricts to g_1 on $B(\lambda+\delta')$ for sufficiently small δ'. It remains to check that the isotopy class of ϕ is independent of the choice of α and δ_0. But this follows by the convexity of $\mathcal{U}$.

(ii) Let $Emb(B(\lambda), V)$ denote the space of (not necessarily symplectic) embeddings of $B(\lambda)$ into V, and let E be the total space of the principal $GL(2n, \mathbb{R})$ bundle over V. By using the exponential map associated to some metric on V, one can define an inclusion $\imath: E \to Emb(B(\lambda), V)$ which is homotopy inverse to the map $g \mapsto dg(0)$. Thus $Emb_\omega(B(\lambda), V)$ deformation retracts inside $Emb(B(\lambda), V)$ to $\imath(P)$. The desired conclusion now follows easily. ■

It follows that if we restrict attention to the space $Emb_\omega(B(\lambda), V, p)$ of symplectic embeddings such that $dg(0) = p$, we may identify all the manifolds $\tilde{V}_g$ and call them $\tilde{V}_p$.

Proposition 2.5 *The isotopy class of the form $\tilde{\omega}_g$ on $\tilde{V}_p$ depends only on the isotopy class of g in $Emb_\omega(B(\lambda), V, p)$.*

Proof Let f_t, $0 \le t \le 1$, be a smooth family of symplectic embeddings $B(\lambda) \to V$ such that $df_t(0) = p$ for all t. Extend these to $B(\lambda+\epsilon)$ for some $\epsilon > 0$ and let $\tilde{V}(t) = \tilde{V}(f_t, \epsilon/2)$. By the symplectic isotopy extension theorem, there is an ambient symplectic isotopy F_t of V such that $Ft \circ f_0 = f_t$ on $B(\lambda+\epsilon)$. Clearly, this induces a family of symplectomorphisms $\Psi_t: \tilde{V}(0) \to \tilde{V}(t)$. Further, it is easy to see that one can choose the diffeomorphisms $\phi_t: \tilde{V}_p \to \tilde{V}(t)$ of Lemma 2.4(ii) to be smooth in the sense that the composites $\Phi_t = \phi_t^{-1} \circ \Psi_t \circ \phi_0$ form a smooth family of diffeomorphisms of $(\tilde{V}_p, Z)$. But if $f_0 = g$ and $f_1 = h$, the construction implies that $\tilde{\omega}_g = \Phi_1{}^*(\tilde{\omega}_h)$. Hence $\tilde{\omega}_g$ and $\tilde{\omega}_h$ are isotopic. ■

Note that in the above construction, the path f_t in $Emb_\omega(B(\lambda), V, p)$ gives rise to a family of symplectic forms on $\tilde{V}_p$ whose restrictions to Σ are constant. One could ask whether the converse of Proposition 2.5 holds: that is, if $\tilde{\omega}_g$ and $\tilde{\omega}_h$ are isotopic by an arbitrary isotopy (which need not be constant on Σ), must the embeddings g and h be isotopic? Even if one restricts to dimension 4, it is not clear that this holds in general. However, one can prove the following slightly weaker result. We will say that two embeddings $g, h \in Emb_\omega(B(\lambda), V)$ are *equivalent* (and write $g \approx h$) iff there is a symplectomorphism ϕ of (V, ω) such that $\phi \circ g = h$.

Proposition 2.6 *If* dim V=4, $g \approx h$ *iff the blow-ups* $(\tilde{V}, \tilde{\omega}_g)$ *and* $(\tilde{V}, \tilde{\omega}_h)$ *are symplectomorphic by a symplectomorphism which preserves the homology class* $[\Sigma]$.

Proof Let $\phi : (\tilde{V}, \tilde{\omega}_g) \to (\tilde{V}, \tilde{\omega}_h)$ be a symplectomorphism which preserves the homology class $[\Sigma]$. Then, as in [M2] (3.3), because for generic J there is a unique J-holomorphic representative of the class $[\Sigma]$, there is a $\tilde{\omega}_h$-symplectic isotopy of $\tilde{V}$ which takes $\phi(\Sigma)$ to Σ. Hence we may suppose that $\phi(\Sigma) = \Sigma$, and it easily follows that $g \approx h$. The converse is obvious. ∎

Proposition 2.7 *If* $V = \mathrm{Int}\, B(1)$, $\mathbb{C}P^2$ *or* $S^2 \times S^2$, *the blow-ups* $(\tilde{V}, \tilde{\omega}_g)$ *and* $(\tilde{V}, \tilde{\omega}_h)$ *are symplectomorphic by a map which induces the identity on homology iff the embeddings* g *and* h *are isotopic.*

Proof By Proposition 2.6, it suffices to show that for each of the three listed manifolds the group of compactly supported symplectomorphisms which are the identity on homology is connected. Gromov showed in [G1] that this group is contractible when $V = \mathrm{Int}\, B(1)$, and that it deformation retracts onto $SU(3)$ when $V = \mathbb{C}P^2$. When $V = S^2 \times S^2$, one must distinguish two cases depending on the cohomology class of ω. Let σ_1 (resp σ_2) denote the pull-back to $S^2 \times S^2$ of the standard area form on S^2 with total area π via projection onto the first (resp. second) factor. If $\omega = \lambda(\sigma_1 \oplus \sigma_2)$, Gromov showed that the group of symplectomorphisms under consideration deformation retracts onto $SO(3) \times SO(3)$: see [G1] 0.3.C. (Note that in this case the whole group of symplectomorphisms is disconnected, since it is possible to interchange the two factors.) If $\omega = \lambda\sigma_1 \oplus \mu\sigma_2$ where $\lambda \neq \mu$, this is no longer true. However, his proof shows that the group is connected, which is all we need here. ∎

Appropriate versions of the above propositions apply to embeddings of a finite union of disjoint balls. In particular, there is a unique way of blowing up $\mathbb{C}P^2$ at k points with weights $\lambda_1, \dots, \lambda_k$ iff the space $\mathrm{Emb}_\omega(\coprod B(\lambda_i), \mathbb{C}P^2)$ is path-connected. (Here $\coprod B(\lambda_i)$ denotes the disjoint union of the balls $B(\lambda_1), \dots, B(\lambda_k)$.) We now show that $\mathrm{Emb}_\omega(\coprod B(\lambda_i), \mathbb{C}P^2)$ is path-connected iff $\mathrm{Emb}_\omega(\coprod B(\lambda_i), \mathrm{Int}\, B(1))$ is also, thereby reproving Proposition 1.4 of [M1]. Let ı be the standard symplectic embedding of Int B(1) onto $\mathbb{C}P^2 - \mathbb{C}P^1$ given by

(2.7.1) $$\iota(z, w) = [\sqrt{1 - |z|^2 - |w|^2} : z : w].$$

Note that $\iota^*\tau_0 = \omega_0$ because ι factors though the obvious projection $\pi : S^5 \to \mathbb{C}P^2$, and $\pi^*\tau_0 = \omega_0|S^5$.) Further, define $\kappa : \mathrm{Int}\, (B^2(r_1) \times B^2(r_2)) \to S^2 \times S^2$ to be a product of embeddings $B^2(r_i) \to S^2-$pt., where we take the symplectic form on $S^2 \times S^2$ to be a product of embeddings $B^2(r_i) \to S^2-$pt., where we take the symplectic form on $S^2 \times S^2$ to be ${r_1}^2\sigma_1 \oplus {r_2}^2\sigma_2$ so that κ may be chosen to be symplectic.

Proposition 2.8 (i) ι *induces an isomorphism between* $\pi_0\big(\mathrm{Emb}_\omega(\coprod B(\lambda_i), \mathrm{Int}\, B(1))\big)$ *and* $\pi_0((\mathrm{Emb}_\omega(\coprod B(\lambda_i), \mathbb{C}P^2)))$.

(ii) *κ induces an isomorphism between* $\pi_0\big(\mathrm{Emb}_\omega(\coprod B(\lambda_i), \mathrm{Int}\,(B^2(r_1)\times B^2(r_2))\big)$ *and* $\pi_0(\mathrm{Emb}_\omega(\coprod B(\lambda_i), S^2\times S^2))$.

Proof (i) Let $g_t, 0 \le t \le 1$, be a path in $\mathrm{Emb}_\omega(\coprod B(\lambda_i), \mathbb{C}P^2)$ with g_0 and g_1 in Im $\imath$, and let S denote the 2-sphere in $\mathbb{C}P^2$ whose complement is Im ι. Suppose that we can find a smooth path S_t of symplectically embedded 2-spheres with $S_0 = S_1 = S$ such that S_t is disjoint from Im g_t for each t. The isotopy extension theorem then implies that there is a path ϕ_t of symplectic diffeomorphisms of $\mathbb{C}P^2$ such that $\phi_0 = $ id. and $\phi_t(S) = S_t$. Further, because the group of symplectomorphisms ϕ of $\mathbb{C}P^2$ such that $\phi(S) = S$ is connected, we may also suppose that $\phi_1 = $ id. Thus ${\phi_t}^{-1} \circ g_t$ is a path in Im $\imath$ joining g_0 to g_1.

To find the S_t, let X_k denote $\mathbb{C}P^2$ blown up at k points, and let $\tilde{S}$ be the lift of S. (This is disjoint from the exceptional spheres.) X_k has a path σ_t of symplectic forms given by the embeddings g_t. Note that both σ_0 and σ_1 are non-degenerate on $\tilde{S}$. Now choose a path J_t of almost complex structures on X_k so that J_t is σ_t-compatible for each t, the exceptional spheres are J_t-holomorphic for each t and $\tilde{S}$ is J_0-holomorphic. Then, as in [M1], there is a unique J_t-holomorphic sphere through every pair of points outside the exceptional spheres, and so one can pick out a smooth family of embedded spheres $\tilde{S}_t$ starting at $\tilde{S}$ such that $\tilde{S}_t$ is J_t-holomorphic for each t. These spheres cannot meet the exceptional spheres in X_k since the relevant homology classes have zero intersection. Further, the space of σ_1-symplectically embedded 2-spheres in class $[\tilde{S}]$ is connected, because each such sphere is J-holomorphic for some σ_1-tame J. Thus, we may suppose that $\tilde{S}_1 = \tilde{S}$, and take the S_t to be the images of the $\tilde{S}_t$ in the blow down $\mathbb{C}P^2$ of X_k.

(ii) The proof of (ii) is similar and will be left to the reader. Here, because the complement of Im κ is $S^2 \times pt \cup pt \times S^2$, one had to find two families $\tilde{S}_t$ and $\tilde{S}_t'$ of curves, one in class $A = [S^2\times pt]$ and the other in class $B = [pt \times S^2]$. ■

§3 Uniqueness of blow ups of $\mathbb{C}P^2$

Our argument is based on the following easy lemma.

Lemma 3.1 *Suppose that* U *is a star-shaped open subset of* $\mathbb{R}^4$ *and let* σ *be any symplectic form on* $\mathbb{R}^4$ *which equals the standard form outside some compact subset of* U. *Then there is a diffeomorphism* ϕ *of* $\mathbb{R}^4$ *with support in* U *such that* $\phi{*}\sigma = \omega_0$.

Proof We will suppose that U is star-shaped about the origin. Define $\sigma_t = t^2.{\mu_t}^*(\sigma)$ for $0 < t \le 1$, where $\mu_t : \mathbb{R}^4 \to \mathbb{R}^4$ is the expansion $x \mapsto x/t$. Then

$\sigma_1 = \sigma$ and, for all t, $\sigma_t = \omega_0$ outside some compact subset of t·U. We may choose $t > 0$ and $\lambda > 0$ so that $t \cdot U \subset \mathrm{Int}\, B^2(\lambda) \times B^2(\lambda) \subset U$. Then σ_t is standard near the boundary of the polydisc $B^2(\lambda) \times B^2(\lambda)$, and so, by [G1] 2.4.$A_1$, there is a diffeomorphism ϕ with support in $B^2(\lambda) \times B^2(\lambda)$ such that $\phi * \sigma = \omega_0$. ■

Next, consider $\mathbb{C}P^2$ with its standard Kähler metric μ_0, normalised as before so that the area of a complex line is π. Thus each complex projective line is isometric to the sphere $S^2(1/2)$ of radius 1/2 in $\mathbb{R}^3$. Fix a complex projective line S, identify $\mathbb{C}P^2 - S$ with $\mathrm{Int}\, B(1) \subseteq \mathbb{C}^2$ via the map ι of (2.7.1) and let $B = B_g(x, r)$ be a metric ball in $\mathbb{C}P^2$ with center on S.

Lemma 3.2 *The set* $U \subset \mathrm{Int}\, B(1)$ *which is mapped by* ι *to* $\mathbb{C}P^2 - (S \cup B)$ *is star-shaped.*

Proof This holds because the rays through $\{0\}$ in B(1) are mapped by $\imath$ to geodesics from $x_0 = \imath(0)$ to S, and metric balls are geodesically convex. ■

Lemma 3.3 *The set* $B(\lambda_1) \amalg B(\lambda_2)$ *embeds symplectically onto a pair of metric balls in* $\mathbb{C}P^2$ *iff* ${\lambda_1}^2 + {\lambda_2}^2 < 1$.

Proof Using the fact that $\iota\, (B(\lambda))$ is a metric ball, one easily checks that $B(\lambda)$ is symplectomorphic to $B_g(x,r)$ iff $\lambda = \sin(r)$. Therefore the condition ${\lambda_1}^2 + {\lambda_2}^2 < 1$ is equivalent to $r_1 + r_2 < \pi/2$. But because the diameter of the projective lines in $(\mathbb{C}P^2, g)$ is $\pi/2$, this is exactly the condition which allows one to place the metric balls so that they have disjoint intersections with the projective line S through their centers. But then they are disjoint by Lemma 3.2. ■

We will call a symplectic embedding of $B(\lambda_1) \amalg B(\lambda_2)$ onto a pair of metric balls in $(\mathbb{C}P^2, \mu_0)$ a "standard embedding". Note that these are all isotopic, and that they give rise to Kähler forms on X_2. Further, by [G1] 0.2.B (see also [M1]), $B(\lambda_1) \amalg B(\lambda_2)$ embeds symplectically into $\mathbb{C}P^2$ iff ${\lambda_1}^2 + {\lambda_2}^2 < 1$. Hence there is a standard embedding whenever there is a symplectic embedding. We now show:

Theorem 3.4 *Any symplectic embedding of* $B(\lambda_1) \amalg B(\lambda_2)$ *into* $\mathbb{C}P^2$ *is isotopic to a standard embedding. Thus,* $\mathrm{Emb}_\omega(B(\lambda_1) \amalg B(\lambda_2), \mathbb{C}P^2)$ *is path-connected (provided it is non-empty).*

Proof Let σ be the symplectic form on X_2 constructed from the symplectic embedding $g: B(\lambda_1) \amalg B(\lambda_2) \to \mathbb{C}P^2$. By Proposition 2.7 it suffices to show that it is symplectomorphic to a standard form τ, i.e. one constructed from a standard embedding. Let A denote the homology class of the projective lines in $\mathbb{C}P^2$ and also of their lifts to X_2, and let E_1, E_2 be the homology classes of the exceptional curves Σ_1, Σ_2. Choose a σ-tame almost complex structure J_1 on X_2 such that the two exceptional curves in X_2 are J_1-holomorphic. Using the fact that for each τ_0-tame J on $\mathbb{C}P^2$ there is a unique J-holomorphic A-curve through each pair of points, and arguing as in [M1] Lemma 3.3, one can easily show that there is a unique J_1-holomorphic curve C in X_2 in the class $A - E_1 - E_2$. (In fact, this curve is a component of the unique J_1-holomorphic A-cusp-curve which meets Σ_1 and Σ_2.) It follows from positivity of intersections (see [M2](2.5)) that C is embedded and meets each exceptional curve once transversally. Therefore, by the symplectic neighbourhood theorem, there is a symplectomorphism from a neighbourhood N of $Y = C \cup \Sigma_1 \cup \Sigma_2$ to a neighbourhood of the corresponding curves in the standard model (X_2, τ). It follows that the deleted neighbourhood $N - Y$ is symplectomorphic to a neighbourhood of the boundary of the star-shaped set $U \subset \text{Int } B(1)$, where $\iota(U) = \mathbb{C}P^2 - (S \cup B_1 \cup B_2)$. The result now follows from Lemma 3.1. ■

Theorem 1.1 follows by combining Theorem 3.4 with Propositions 2.6 and 2.7.

Note 3.5 The proof of Theorem 3.4 clearly applies to one ball, and so gives a new proof of the main result in [M1].

(3.6) Embeddings of more than two balls.

Note that, given any two symplectic embeddings $g_1, g_2 : \coprod B(\lambda_i) \to V$, one can always decrease the λ_i to a set of radii λ_i' such that the restrictions of g_1 and g_2 to $\coprod B(\lambda_i')$ are isotopic. (This holds because the group of symplectomorphisms of (V, ω) acts k-transitively on V, and so is true for any symplectic manifold V.) Of course, the size of the λ_i' depends on g in general: in fact, there are no cases besides those mentioned above when the λ_i' may be chosen independently of g_1 and g_2.

However, there is a little more to be said about > 2 balls in $\mathbb{C}P^2$. First, observe that the proof of Theorem 3.4 applies to any symplectic embedding $g : \coprod B(\lambda_i) \to \mathbb{C}P^2$ such that there is a J-holomorphic $(A - \sum_i E_i)$-curve in $(X_k, \tilde{\omega}_g)$ for some $\tilde{\omega}_g$-tame J. Hence these embeddings are isotopic to standard embeddings, i.e. embeddings whose image is a union of metric balls. This condition can be satisfied only if $\tilde{\omega}_g(A - \sum_i E_i) > 0$, i.e. $\sum_i \lambda_i^2 < 1$, and so is more restrictive than the symplectic packing inequalities given in [G1] 0.2.B. (For example, if $k \leq 5$ there is a symplectic embedding of $\coprod B(\lambda_i)$ into $\mathbb{C}P^2$ iff $\lambda_1^2 + \ldots + \lambda_k^2 < 2$ and $\lambda_i^2 + \lambda_j^2 < 1$ for all pairs $i \neq j$.) However, it is not clear that the numerical condition $\sum_i \lambda_i^2 < 1$ is sufficient to guarantee that there is a J-holomorphic $(A - \sum_i E_i)$-curve since, when $k > 2$, these curves have negative index and so are never generic. Equivalently, one needs to find a symplectically embedded 2-sphere C in $\mathbb{C}P^2$ which meets each ball $g(B(\lambda_i))$

in the image under g of a flat disc through the center of $B(\lambda_i)$. By elementary arguments (see [G2] 3.4.2 B′) one can show that the space of symplectic embeddings of a 2-disc into any symplectic 4-manifold is connected. Hence one can arrange that the intersection of C with Im g contains these flat discs, but it is not clear that one can isotop away all the other parts of $C \cap \mathrm{Im}\, g$.

We end by proving a result stated in [M1]. We pointed out there that all cohomolgous Kähler structures on X_k are symplectomorphic. We will call the corresponding component of $\mathrm{Emb}_\omega(\coprod B(\lambda_i), \mathbb{C}P^2)$ the "standard component".

Proposition 3.7 *If g is an embedding of $\coprod B(\lambda_i)$ into* Int B(1) *which is isometric with respect to the usual flat metric on $\mathbb{C}^n$, then $\iota \circ g$ belongs to the standard component.*

Proof Let J_0 be the standard complex structure on $\mathbb{C}^2$. We first claim that it suffices to show that there is an integrable τ_0-tame almost complex structure J on $\mathbb{C}P^2$ which extends the complex structure $(\iota \circ g)_* J_0$ on $\mathrm{Im}\, \iota \circ g$. To see this, observe that such a J induces a complex structure $\tilde{J}$ on the blow-up X_k which is tamed by the form $\tilde{\omega}_g$ The set $\mathcal{S}$ of forms cohomologous to $\tilde{\omega}_g$ and tamed by $\tilde{J}$ contains forms which are compatible with $\tilde{J}$ and hence $\tilde{J}$-Kähler. (Recall that ω tames J iff $\omega(v,Jv) > 0$ for $v \neq 0$, and is compatible with J if, in addition, $\omega(v,w) = \omega(Jv,Jw)$.) Since $\mathcal{S}$ is convex, $\tilde{\omega}_g$ is symplectomorphic to a Kähler form, as required.

By hypothesis, the mapping given by g of each ball $B(\lambda_i)$ may be described as a translation followed by a unitary rotation. In particular $g_*(J_0) = J_0$. Therefore, it suffices to show that, if $0 < \epsilon_2 < \epsilon_1 < 1$, there is a ω_0-tame complex structure J on Int B(1) which equals the usual complex structure J_0 in $B(1-\epsilon_1)$ and equals J_1 outside $B(1-\epsilon_2)$ where J_1 denotes the pull-back by $\imath$ of the usual complex structure $\mathbb{C}P^2$. But, it is easy to see that $J_1 = \Psi^* J_0$ where $\Psi : \mathrm{Int}\, B(1) \to \mathbb{C}^2$ is a map of the form $x \mapsto r(||x||)x$. (In fact, Ψ is essentially stereographic projection.) Thus, for all $\lambda < 1$, J_0 and J_1 agree on the complex part of the tangent bundle of the sphere $\partial B(\lambda)$, and differ by a scalar factor (depending only on λ) on the vector field ξ which generates the Hopf flow $(z, w) \mapsto (e^{i\theta}z, e^{i\theta}w)$. Since both are tamed by ω_0, it is not hard to find an integrable ω_0-tame J on the annulus $B(1-\epsilon_2) - B(1-\epsilon_1)$ which has the form $\phi^* J_0$ for suitable ϕ and which interpolates between them. ■

We claimed in Note 2.5 of [M1] that the form $\tilde{\omega}_g$ on X_k is Kähler iff the complex structure $g_*(J_0)$ on Im g extends to an ω_0-tame complex structure J on B(1) which equals J_0 near $\partial B(1)$. The above argument shows that one can ask equivalently that the extension J equal J_1 near $\partial B(1)$, and it also proves the "if" part of the claim. To prove the "only if" part one can argue as follows. By construction, the form $\tilde{\omega}_g$ is non-degenerate on the exceptional spheres $\Sigma_1, \ldots, \Sigma_k$ in X_k, and it is Kähler iff there is a compatible complex structure

$\tilde{J}$ on X_k. Now, the exceptional spheres Σ_i need not be $\tilde{J}$-holomorphic, but, by the argument in Proposition 2.6 above, there is always a $\tilde{\omega}_g$-symplectic isotopy Ψ_t of X_k such that the Σ_i are $(\Psi_1)_*\tilde{J}$-holomorphic. Therefore, by replacing $\tilde{J}$ by $(\Psi_1)_*\tilde{J}$, we may suppose that the Σ_i are $\tilde{J}$-holomorphic. Further, since the complex structure in the neighbourhood of a rational holomorphic curve in a complex surface is uniquely determined by the topological type of the normal bundle, $\tilde{J}$ is standard near the Σ_i and near any complex projective line and so descends to give the required complex structure on Int B(1).

References

G1. Gromov, M.: Pseudo-holomorphic curves in almost complex manifolds, Invent. Math 82 (1985), 307–347.

G2. Gromov, M.: *Partial Differential Relations*, Spring-Verlag 1986.

M1. McDuff, D.: Blow ups and symplectic embeddings in dimension 4, Topology, 1991.

M2. McDuff, D.: The Structure of Rational and Ruled Symplectic 4-manifolds, Journ. Amer. Math. Soc. 3 (1990), 679–712.

M3. McDuff, D.: Symplectic 4-manifolds, to appear in Proceedings of International Congress of Mathematics, Kyoto, 1990.

M4. McDuff, D.: Notes on Rules symplectic manifolds, preprint, January 1992.

Stony Brook, NY 11794
January 1992

The 4-Dimensional Symplectic Camel and Related Results

DUSA MCDUFF AND LISA TRAYNOR

January 1991, revised January 1992

§1 INTRODUCTION

Consider $\mathbf{R}^4$ with the standard symplectic structure ω_0. The open set $E_\lambda \subset (\mathbf{R}^4, \omega_0)$ is defined to be the union of the two half spaces $\{y_1 < 0\}$, $\{y_1 > 0\}$, and the open 3-ball $U(\lambda) := \{(x_1, 0, x_2, y_2)\colon x_1^2 + x_2^2 + y_2^2 < \lambda^2\} \subset \{y_1 = 0\}$. The 4-dimensional symplectic camel problem, first posed by Gromov, asks the following question [A]: Using a 1-parameter family of symplectic diffeomorphisms, can a closed ball $B(R)$ of radius $R \geq \lambda$ be moved from one half space of E_λ to the other? It is not hard to see that it is possible to move such a ball from one half space to the other via a 1-parameter family of volume-preserving diffeomorphisms. However, using tools developed in Gromov's 1985 paper [G], it has been proven that it is impossible to carry out this construction using a 1-parameter family of symplectic diffeomorphisms. Eliashberg and Gromov sketch a proof in [EG, 3.4], which is valid in all dimensions and uses Eliashberg's technique of filling by holomorphic discs.

Rephrased, this result says the space $\mathcal{E}(R, \lambda)$ of symplectic embeddings of $B(R)$ into E_λ has at least two components when $R \geq \lambda$. Viterbo then conjectured that E_λ and E_μ are not symplectomorphic when $\lambda \neq \mu$. This conjecture could immediately be proven if one also knew that $\mathcal{E}(R, \lambda)$ has exactly one component when $R < \lambda$. However there might be an embedding whose image is so tangled up in the "hole" $U(\lambda)$ of E_λ that it is impossible to isotop it to one side or the other. Thus it is not known whether an arbitrary embedding is symplectically isotopic to one of the standard inclusions. As we will see below, an embedding which has an extension to the ball of radius $R + \lambda$ is symplectically isotopic to an inclusion. Using this, we will show:

MAIN THEOREM. *E_λ is symplectomorphic to E_μ if and only if $\lambda = \mu$.*

Our proof uses techniques specific to dimension 4, but we conjecture the result is true in higher dimensions as well.

In order to explain the proof, it is convenient to introduce the following terminology. Let g_R^+ and g_R^- be elements of $\mathcal{E}(R, \lambda)$ which embed $B(R)$ linearly into the spaces $\{y_1 < 0\}$ and $\{y_1 > 0\}$, respectively. Further, let $\mathcal{E}(R, R_1, \lambda)$ denote the subset of $\mathcal{E}(R, \lambda)$ consisting of those embeddings which extend to $B(R_1)$. Then we have:

The first author is partially supported by NSF grant no. DMS-9103033. The second author is supported by an Alfred P. Sloan Doctoral Dissertation Fellowship.

EXTENDABLE EMBEDDINGS LEMMA. *Every element in $\mathcal{E}(R, R+\lambda, \lambda)$ is symplectically isotopic either to g_R^+ or to g_R^-.*

PROOF OF MAIN THEOREM: Suppose $\psi: E_\lambda \to E_\mu$ is a symplectomorphism with $\lambda < \mu$, and R is chosen so that $\lambda < R < \mu$. Then ψ_* induces an isomorphism

$$\pi_0(\mathcal{E}(R,\lambda)) \longrightarrow \pi_0(\mathcal{E}(R,\mu)).$$

By the symplectic camel theorem, the embeddings $g_R^\pm$ represent distinct elements in $\pi_0(\mathcal{E}(R,\lambda))$. But, because they extend over $B(R+\mu)$, the above Lemma implies that their composites with ψ are isotopic to standard linear embeddings into $\mathcal{E}(R,\mu)$, and hence, since $R < \mu$, represent the same element of $\pi_0(\mathcal{E}(R,\mu))$. This contradiction proves the theorem. ∎

The Extendable Embeddings Lemma is proved by using the special properties of fillings in dimension 4. Recall that a filling of a 2-sphere S is a 1-parameter family of disjoint, embedded J-holomorphic discs whose boundaries foliate S and whose union $F(J)$ is homeomorphic to a 3-ball. Eliashberg has proven, although not published all details of his proof, that an embedded copy S of S^2 in $\mathbf{R}^4$ can be filled as long as it sits inside a J-convex hypersurface $\partial\Omega$, for some integrable, ω_0-tame J (see [E], [K]). In §2-§4, we write out the details of his proof in the special case when S, J and $\partial\Omega$ are all "standard" near S. In this case, because there are no hyperbolic points, we do not need to assume that J is everywhere integrable. For completeness, the proof of the Camel Theorem is sketched in §5. In fact, in order to prove the Camel Theorem, one does not need the discs in the filling to be disjoint and embedded. All that matters is that the filling, $F(J)$, of the boundary sphere of the "hole" $U(\lambda)$ in E_λ separates E_λ. Thus it is easy to extend the proof to higher dimensions. However, to prove the Extendable Embeddings Lemma we need these special properties, since the main step is to join a suitable filling to the "wall" $\mathbf{R}^4 - E_\lambda$ to create a hypersurface Q in $\mathbf{R}^4$ such that the pair $(\mathbf{R}^4, Q)$ is symplectomorphic to $(\mathbf{R}^4, \mathbf{R}^3)$. The details are given in §6.

We would like to thank Ya. Eliashberg and W. Klingenberg for explaining the technique of filling by J-holomorphic discs to us.

§2 BASIC DEFINITIONS

Given a Riemann surface (Σ, i) and an almost complex manifold (V, J), a *J-holomorphic map of Σ into V* is a (C^1) map $f: \Sigma \to V$ with J-linear derivative, i.e.

$$df \circ i = J \circ df.$$

Recall, also, that an almost complex structure J is said to be *ω-tame* if $\omega(v, Jv) > 0$ for all non-zero tangent vectors v, and *ω-compatible* (or *ω-calibrated*) if $\omega(v,w) = \omega(Jv, Jw)$. We will use the coordinates $z_1 = x_1 + iy_1$ and $z_2 = x_2 + iy_2$ on $\mathbf{C}^2 = \mathbf{R}^4$ and will write J_0 for the standard ω_0-compatible almost complex structure on $\mathbf{R}^4$.

Let $\partial\Omega$ be an oriented hypersurface in an almost complex 4-manifold (V, J), and, for each $x \in \partial\Omega$, let $\xi_{x,J}$ be the maximal J-invariant subspace of the tangent space $T_x\partial\Omega$. $\partial\Omega$ is said to be *J-convex* if for one (and hence any) defining 1-form α of ξ we have $d\alpha(v, Jv) > 0$ for all non-zero $v \in \xi_{x,J}$. Since $\xi = ker\ \alpha$, it follows that ξ is a contact structure on $\partial\Omega$.

Throughout this paper, $\Omega \subset \mathbf{R}^4$ will denote a domain bounded by a smooth hypersurface $\partial\Omega$. For our applications, we will be able to choose Ω so that $\partial\Omega$ is transverse to the radial vector field

$$\partial = x_1\frac{\partial}{\partial x_1} + y_1\frac{\partial}{\partial y_1} + x_2\frac{\partial}{\partial x_2} + y_2\frac{\partial}{\partial y_2}.$$

Consider $\alpha = i(\partial)\omega_0$. Then since $d\alpha = 2\omega_0$ and ∂ is transverse, it follows that α is a contact form on $\partial\Omega$. Moreover, $\partial\Omega$ is J-convex for any ω_0-tame J such that $ker\ \alpha|_{\partial\Omega}$ is J-invariant. It will be convenient to assume further that $\Omega \cap \{|y_1| < \beta\} = B^4 \cap \{|y_1| < \beta\}$, for some $\beta > 0$, where B^4 is the unit ball centered at the origin. In particular, notice that the standard unit 2-sphere $S \subset \{y_1 = 0\}$ sits inside $\partial\Omega$.

With Ω as described above, it is possible to construct a smooth function φ on $\mathbf{R}^4$ such that $\Omega = \{\varphi \leq 1\}$ and $\partial\Omega = \{\varphi = 1\}$. Let $\mathcal{J}$ be the set of smooth, ω_0-compatible almost complex structures on $\mathbf{R}^4$ and let $\mathcal{C}_\Omega$ be the open subset of $\mathcal{J}$ consisting of those J which make $\partial\Omega$ J-convex. Fix $1 > \varepsilon > 0$ and let

$$\mathcal{J}_\Omega = \{J \in \mathcal{C}_\Omega \colon J = J_0 \text{ on } \{\varphi > 1 - \varepsilon\} \cap \{|y_1| < \beta\}\}.$$

It follows from the above remarks that $\mathcal{J}_\Omega$ is nonempty. Next, given δ such that $1 - \varepsilon < \delta < 1$, put

$$S_\delta = \{(x_1, 0, x_2, y_2) \in S \colon |x_1| < \delta\}.$$

Then S_δ is diffeomorphic to an open annulus and, since the elements of $\mathcal{J}_\Omega$ are standard near S, for each $J \in \mathcal{J}_\Omega$, S_δ is a totally real surface sitting inside a J-convex $\partial\Omega$. (Recall that a submanifold $M \in (V, J)$ is said to be *totally real* if $TM \cap JTM = \{0\}$.) Let D be the unit disc in $\mathbf{C}$ with almost complex structure i and let A be the generator of $\pi_2(\mathbf{R}^4, S_\delta)$ which is represented by

$$(D, \partial) \to (\mathbf{R}^4, S_\delta)$$
$$(x, y) \mapsto (0, 0, x, y).$$

Notice that by our construction of $\mathcal{J}_\Omega$, for each c such that $1 - \varepsilon < |c| < 1$,

$$f_c(x, y) = (c, 0, \sqrt{1 - c^2}x, \sqrt{1 - c^2}y)$$

is a J-holomorphic map representing A for all $J \in \mathcal{J}_\Omega$. We denote the images of these flat maps at height c by D_c. Finally, for $J \in \mathcal{J}_\Omega$, we define a ***J-holomorphic A-disc*** to be the image of a J-holomorphic map

$$f: (D, \partial) \to (\mathbf{R}^4, S_\delta),$$

which represents $A \in \pi_2(\mathbf{R}^4, S_\delta)$ and is in the connected component of the flat discs. Thus each J-holomorphic A-disc, *Im* f, may be joined to a flat disc, D_c, by a path *Im* f_t of J_t-holomorphic A-discs, where $J_t \in \mathcal{J}_\Omega$.

§3 Properties of J-Holomorphic A-Discs

Technically, we only require a J-holomorphic map of the disc, f, to have J-linear derivative on the closed disc. However, since the boundary is sent to S_δ, we can say more because of the following.

LEMMA 3.1. *For some neighborhood U of S_δ in $(\mathbf{R}^4, J_0)$ there is an anti-holomorphic reflection $\sigma: U \to U$ in S_δ.*

PROOF: Consider the cylinder $C = \{(x_1, 0, x_2, y_2): |x_1| < 1 - \delta, x_2^2 + y_2^2 = 1\}$. There exists a neighborhood U_1 of C in $\mathbf{C}^2$ and $\sigma_1: U_1 \to U_1$ an antiholomorphic reflection in C given by $\sigma(z_1, z_2) = (\overline{z_1}, \frac{1}{\overline{z_2}})$. Choosing U_1 sufficiently small, let U be the image of U_1 under the biholomorphism $\varphi(z_1, z_2) = (z_1, \sqrt{1 - z_1^2}z_2)$ and define σ on U by $\sigma = \varphi \circ \sigma_1 \circ \varphi^{-1}$. ∎
Thus, by the Schwarz reflection principle, f has a unique J-holomorphic extension to a neighborhood of the closed disc.

PROPOSITION 3.2. *All interior points of a J-holomorphic A-disc are contained in Int Ω.*

PROOF: Suppose there exists an f such that *Int* f is not contained in *Int* Ω. Since f, by definition, is in the connected component of the flat discs, there exists a family f_t of J_t-holomorphic discs such that, for some t_0, we have *Im* $f_t \subset \Omega$, for $t \leq t_0$, and *Im* f_{t_0} is tangent to $\partial\Omega$ from the inside at $p = f_{t_0}(z_0)$. We can break this into two cases.
(i) $z_0 \in$ *Int* D. This is impossible by an application of the maximum principle. See [M5, Lemma 2.4].
(ii) $z_0 \in \partial D$. By construction $J = J_0$ near $p \in S_\delta$, and $\partial\Omega$ coincides with S^3 near S_δ. Let $T_p^{\mathbf{C}} S^3$ be the complex part of the tangent space to $\partial\Omega$ at p and $(T_p^{\mathbf{C}} S^3)^\perp$ its ω_0 perpendicular. Then, if π_p is the orthogonal projection of $\mathbf{R}^4$ onto the complex line $(T_p^{\mathbf{C}} S^3)^\perp$, the composite $\pi_p \circ f_{t_0}$ is holomorphic in a neighbourhood of ∂D in $\mathbf{C}$. If it had a zero of infinite order at z_0, it would be constant near ∂D, which would contradict the fact that f_{t_0} represents A. Thus it must have a zero of finite order at z_0 and, by hypothesis, must have a zero of order at least two. Hence it maps a neighborhood of $z_0 \in \partial D$ onto a neighborhood of $0 \in T_p^{\mathbf{C}} S^3 \equiv \mathbf{C}$. But this is impossible, since *Im* $f_{t_0} \subset \Omega$ and π_p takes a neighborhood of p in Ω into a half space in $\mathbf{C}$. ∎

Because S_δ is totally real, $TS_\delta \cap T^{\mathbf{C}}S^3$ is a real, orientable line field on S_δ. It is not hard to check that this line field is vertical, i.e. that its integral curves are the restrictions to S_δ of the great arcs on S through the poles $\{(\pm 1, 0, 0, 0)\}$. The above proof tells us the following.

COROLLARY 3.3. *$f|_{\partial D}$ is never vertical, i.e.*

$$T_p(f|_{\partial D}) \neq T_p S_\delta \cap T_p^{\mathbf{C}} S^3.$$

We will derive further important properties of our discs by a "doubling argument." Let $N = U \cup Int\ \Omega$ where U is as in Lemma 3.1. Let N_1, N_2 be two copies of N and let U_i be the copy of U in N_i. Form the double

$$V = N_1 \cup_\sigma N_2 = \frac{N_1 \amalg N_2}{p \sim \sigma(p)} \quad \text{for } p \in U_1.$$

One can check V is a manifold and further that

$$T_pV = \begin{cases} T_pN_i, & p \notin U_i \\ \dfrac{T_pN_1 \amalg T_{\sigma(p)}N_2}{(p,v) \sim (\sigma(p), \sigma_* v)}, & p \in U_1. \end{cases}$$

Consider

$$J^V(p) = \begin{cases} J(p), & p \in N_1 \\ -J(p), & p \in N_2. \end{cases}$$

Since σ is anti-holomorphic and J is smooth, it follows that J^V is a well-defined, smooth almost complex structure on V.

Given a J-holomorphic map $f\colon (D, \partial) \to (\Omega, S_\delta)$, a J^V-holomorphic map $f^V\colon (S^2, i) = (\mathbf{C} \cup \{\infty\}, i) \to (V, J^V)$ can be constructed as follows. Let f_1, f_2 be copies of f into N_1, N_2 respectively and let $r\colon \mathbf{C}^2 \cup \{\infty\} \to \mathbf{C}^2 \cup \{\infty\}$ be the antiholomorphic reflection through S^1 given by $r(z) = 1/\bar{z}$. Define $f^V\colon \mathbf{C} \cup \{\infty\} \to V$ by

$$f^V(z) = \begin{cases} f_1(z), & |z| \leq 1 \\ f_2 \circ r(z), & |z| > 1. \end{cases}$$

It is not hard to verify that f^V is a smooth, J^V-holomorphic map of the sphere.

PROPOSITION 3.4. *The J-holomorphic A-discs are embedded and disjoint.*

PROOF: Let f be a J-holomorphic map representing A, and Z the image of its double. We will show that Z is embedded and is disjoint from the image Z' of the double of any other J-holomorphic A-disc. Let D_c be a flat J-holomorphic A-disc and S_c its double. Then Z and S_c are homologous J^V-holomorphic 2-spheres and S_c is embedded. But it is shown in [M2, theorem 1.3] that there is a homological criterion for a J-holomorphic 2-sphere C to be embedded: namely, C is embedded iff its virtual genus $g(C)$ is 0, where $g(C) = 1 + \frac{1}{2}(C \cdot C - c(C))$. (Here $c \in H^2(V;\mathbf{Z})$ is the first Chern class of the complex vector bundle (TV, J).) Since $g(Z) = g(S_c) = 0$, it follows that Z is embedded. Further

$$Z \cdot Z' = S_c \cdot S_c = 0.$$

But, by [**M2**, theorem 1.1], every intersection point of Z with Z' contributes positively to the algebraic intersection number $Z \cdot Z'$. Hence Z and Z' are disjoint. ∎

COROLLARY 3.5. *For all J-holomorphic A maps f,*

$$|\omega_0\text{-area of } f| = |\int_{f(D)} \omega_0| \leq \pi.$$

PROOF: By an easy calculation, integrating ω_0 over a portion of the top hemisphere of the 2-sphere S $(x_1 > 0)$ gives a positive value while integrating over a portion of the bottom hemisphere produces a negative value. The result now follows from Stokes' theorem and the fact that $Im\ f$ is an embedded disc with boundary on $S_\delta \subset S$. ∎

§4 FILLING THE SPHERE

In this section we show that the unit 2-sphere S in the hyperplane $\{y_1 = 0\}$ has a J-filling $F(J)$ for all $J \in \mathcal{J}_\Omega$. More precisely:

(4.1) SPHERE FILLING THEOREM. *For all $J \in \mathcal{J}_\Omega$, there is a 1-parameter family of disjoint, J-holomorphic discs whose boundaries foliate S and whose union $F(J)$ is diffeomorphic to B^3.*

Note that $F(J)$ contains two degenerate discs at the poles $(\pm 1, 0, 0, 0)$ of S.

We will just give a sketch of the proof since it follows by standard Fredholm theory from the results of §3. The first step is to define suitable Banach manifolds of almost complex structures and maps.

The results used in Proposition 3.4 require that J be smooth. Therefore, we will use the following procedure of Floer [**F**] to construct a Banach manifold $N(J) \subset \mathcal{J}_\Omega$ containing a given J. The tangent space $T_J\mathcal{J}_\Omega$ to $\mathcal{J}_\Omega$ at J consists of the space $C^\infty(S_J)$ of smooth sections j of the bundle $End\ (T_x\mathbf{R}^4)$ such that $jJ + Jj = 0$, $\langle j\alpha, \beta\rangle + \langle \alpha, j\beta\rangle = 0$ (where $\langle\ ,\ \rangle$ is the standard inner product), and $j(x) = 0$ on $\{\varphi > 1 - \varepsilon\} \cap \{|y_1| < \beta\}$. Let $\bar{\varepsilon} = (\varepsilon_i)_{i\in\mathbf{N}}$ be any sequence of positive real numbers. Then

$$\|j\|_{\bar{\varepsilon}} = \sum_{k\in\mathbf{N}} \varepsilon_k \max_{x\in\mathbf{R}^4} |D^k j(x)|$$

is a norm on the linear space

$$C^{\bar{\varepsilon}}(S_J) = \{j \in C^\infty(S_J) : \|j\|_{\bar{\varepsilon}} < \infty\}.$$

Further, one can check $C^{\bar{\varepsilon}}(S_J)$ is a Banach space. As Floer has shown [**F**, Lemma 5.1], it is possible to choose $\bar{\varepsilon}$ so that $C^{\bar{\varepsilon}}(S_J)$ is dense in $T_J\mathcal{J}_\Omega$ with respect to the L^2-norm. Choose $r > 0$ small enough so that, for $\|j\|_{\bar{\varepsilon}} < r$, the exponential map is injective. Let $N'(J) = \{j \in C^{\bar{\varepsilon}}(S_J) : \|j\|_{\bar{\varepsilon}} < r\}$. Then $N'(J)$ is an open set of a Banach space and thus a Banach manifold. Define $N(J)$ to be the image of $N'(J)$ under the exponential map diffeomorphism.

Now, fix $s > 1$ and let $\mathcal{F} = \mathcal{F}^\delta_{A,s}$ be the Sobolev space $H^{s+1}(D, \partial; \mathbf{R}^4, S_\delta)$ of all maps $f : (D, \partial) \to (\mathbf{R}^4, S_\delta)$ whose $(s+1)^{th}$ derivative is L^2 and which represent $A \in \pi_2(\mathbf{R}^4, S_\delta)$.

For $J_1 \in \mathcal{J}_\Omega$, define

$$\mathcal{M} = \mathcal{M}(N(J_1)) = \{(f, J) \in \mathcal{F} \times N(J_1) : \bar{\partial}_J f = df + J \circ df \circ i = 0\}.$$

PROPOSITION 4.2. *$\mathcal{M}$ is a Banach submanifold of $\mathcal{F} \times N(J_1)$.*

PROOF: This can be proved locally using the procedure of McDuff [**M1**, §4]. There are two points to note. First observe that, because S_δ is totally real, the boundary conditions imposed on our maps are elliptic (see [**BB**]). Thus, in the notation of [**M1**, §4], L_J is still Fredholm and $Im\ d\Phi$ is closed and of finite dimension. The proof that 0 is a regular value of $d\Phi_{(f,J)}$ goes through as before, provided that $Im\ f$ intersects the part of $\mathbf{R}^4$ where the elements of $\mathcal{J}_\Omega$ are allowed to vary. However, one must also consider the possibility that $Im\ f$ is contained in the region $\{\varphi \geq 1 - \varepsilon\} \cap \{|y_1| \leq \beta\}$ where J is constrained to equal J_0. But then the integrability tensor of J vanishes on $Im\ f$. Thus L_J is just the usual $\bar{\partial}$-operator and so is surjective, as required. ∎

It follows as in McDuff [**M1**, Proposition 4.2] that the projection map

$$P_A = P_{A,J_1} : \mathcal{M}(N(J_1)) \to N(J_1)$$

is Fredholm. By the Sard-Smale theorem, there is a subset of second category $N(J_1)_{reg} \subset N(J_1)$ which consists of regular elements. For these J, the inverse image

$$P_A^{-1}(J) =: \mathcal{M}_p(J, A, \delta)$$

is a manifold. Let $(\mathcal{J}_\Omega)_{reg}$ be the union of all the $N(J)_{reg}$, $J \in \mathcal{J}_\Omega$. Since this is an uncountable union, the set $(\mathcal{J}_\Omega)_{reg}$ need not be of second category. However, it is dense.

Since the index of $P_{A,J}$ is determined by the symbol of the elliptic operator L_J, one easily sees that it is independent of $J \in \mathcal{J}_\Omega$, and so may be calculated by considering a specially nice J. Thus, suppose that $J = J_0$

near the unit ball $B^3 \subset \{y_1 = 0\}$. Then the flat maps f_c, for $|c| < 1 - \varepsilon$ are J-holomorphic. Since the boundaries of these flat discs completely fill S_δ and, by Proposition 3.4, we know that distinct discs must be disjoint, it follows that these flat discs are the only J-holomorphic A-discs. Thus, there is a 4-parameter family of flat J-holomorphic A-discs. (Note that three of these dimensions correspond to the reparametrization group $PSL(2, \mathbf{R})$.) Further, since J is integrable near the image of each flat disc, L_J is the usual $\bar{\partial}$ operator and it is easy to see that (f_c, J) is a regular point for all c. It follows that J is a regular value for $P_{A,J}$. Hence the index of $P_{A,J}$ is 4 and we have shown:

PROPOSITION 4.3. *For all J in a dense set $(\mathcal{J}_\Omega)_{reg}$ of $\mathcal{J}_\Omega$, $\mathcal{M}_p(J, A, \delta)$ is a smooth 4-manifold.*

The next step is to compactify the manifolds $\mathcal{M}_p(J_i, A, \delta)$. It is not enough simply to quotient out by the noncompact reparametrization group $G = PSL(2, \mathbf{R})$ of all biholomorphisms of the disc, because S_δ is itself not compact. To fix this, we will consider the closure $\overline{S}_\delta$ of S_δ. Let $\overline{\mathcal{M}}_p(J, A, \delta)$ consist of all J-holomorphic discs with image in $(\Omega, \overline{S}_\delta)$.

LEMMA 4.4. *$\overline{\mathcal{M}}_p(J, A, \delta) = \mathcal{M}_p(J, A, \delta) \cup (f_\delta)_G \cup (f_{-\delta})_G$, where $(f_\delta)_G$ is the G-orbit of the flat disc at height δ and similarly for $(f_{-\delta})_G$.*

PROOF: The inclusion $\supset$ is clear. Conversely, suppose there exists an f such that $Im\ f$ is a non-flat A-disc which intersects one of the circles of $\overline{S}_\delta \setminus S_\delta$. By construction of $\mathcal{J}_\Omega$ and definition of δ, there are flat J-holomorphic discs whose boundaries lie on these circle. Thus, $Im\ f$ must intersect one of these flat discs, which contradicts Proposition 3.4. ∎

PROPOSITION 4.5. *For all $J \in \mathcal{J}_\Omega$, the space of unparametrized J-holomorphic A-discs, $\overline{\mathcal{M}}_p(J, A, \delta)/G$ is diffeomorphic to a compact interval.*

PROOF: First suppose $J \in (\mathcal{J}_\Omega)_{reg}$. It is easy to see that in this case $\overline{\mathcal{M}}_p(J, A, \delta)/G$ is a 1-dimensional manifold with two boundary points given by $[f_{\pm\delta}]$. Therefore, it suffices to prove that it is compact and connected.

To see that this space is compact, double all the discs so that they become J^V-holomorphic A^V-spheres. Construct a metric g_J^V on V by piecing together $g_1(v, w) = \omega_0(v, J^V w)$ on N_1 and $g_2(v, w) = -\omega_0(v, J^V w)$ on N_2, and observe that, by Corollary 3.5, all the J^V-holomorphic A^V-spheres are uniformly bounded in the associated H^1 Sobolev norm. Thus one can apply the standard compactness theorem for closed J-holomorphic spheres. (See, for example, [M4, (2.4)].) Clearly, any limiting cusp-curve is the double of a J-holomorphic "cusp-disc" in Ω. The most general form for such a cusp-disc is a connected union of J-holomorphic components, some of which are discs and some of which are "bubbles" (i.e. J-holomorphic spheres). In our case, Corollary 3.3 implies that the boundary of the cusp-disc is transverse to the vertical, and so only one of its components can be a disc. Further, there can be no bubbles since $\pi_2(\Omega) = 0$. Hence, the limit must consist of a single disc. This proves compactness. (An alternative proof, which does not use doubling, may be constructed by the methods of [**O**].)

It follows that one connected component of $\overline{\mathcal{M}}_p(J, A, \delta)$ is diffeomorphic to a compact interval. Since this component is non-empty and contains the flat discs $f_{\pm\delta}$, the boundaries of the discs in this component must fill out the whole of S_δ. But distinct discs are disjoint. Therefore there can be no other discs and $\overline{\mathcal{M}}_p(J, A, \delta)$ is connected, as claimed.

Now consider an arbitrary J. We first claim that there is a J-holomorphic disc through each point of S_δ. This follows from the compactness theorem, because it is true for regular J and because regular elements are dense in $\mathcal{J}_\Omega$. Next observe that, if α is a vertical arc in S going from one pole to the other, Corollary 3.3 implies that each disc intersects α exactly once. The result now follows easily. ∎

(4.6) PROOF OF THEOREM 4.1.

Let

$$\mathcal{M}_p(J, A) = \overline{\mathcal{M}}_p(J, A, \delta) \cup \{f_c \circ \gamma \colon \delta < |c| \leq 1, \gamma \in G\},$$

and let $F(J)$ be the image of the evaluation map

$$e_D(J) \colon \mathcal{M}_p(J, A) \times_G (D, \partial) \to (\mathbf{R}^4, S).$$

We showed that $\partial F(J) = S$ in the proof of Proposition 4.5. Since G acts freely away from the degenerate discs at the poles, $\overline{\mathcal{M}}_p(J, A, \delta) \times_G (D, \partial)$ is a fiber bundle over the interval $\overline{\mathcal{M}}_p(J, A, \delta)/G$ with fiber (D, ∂). Thus $\mathcal{M}_p(J, A) \times_G (D, \partial)$, with its obvious smooth structure, is diffeomorphic to a 3-ball. By Proposition 3.4, $e_D(J)$ is injective and restricts to an embedding on each disc. The fact that it is a diffeomorphism onto its image $F(J)$ may be proved by the argument in [**M3**, Lemma 3.5]. ∎

§5 THE CAMEL THEOREM

We start by constructing particular regions Ω to which we can apply our Sphere Filling result. The regions will be constructed by centering large 4-balls of radius κ at $(0, \pm\kappa, 0, 0)$ and smoothing the union via a "solid cylinder" which agrees with $B^4(\lambda)$ near $\{y_1 = 0\}$. More precisely, choose $\kappa > \lambda > 0$ and consider $\Omega(\kappa, \lambda) = \{\varphi_{\kappa,\lambda} \leq 1\}$ for $\varphi_{\kappa,\lambda} = \frac{1}{\kappa^2}(x_1^2 + \rho(y_1) + x_2^2 + y_2^2)$ where $\rho = \rho_{\kappa,\lambda}$ is a smooth function such that

$$\rho(y_1) = \begin{cases} (y_1 + \kappa)^2, & y_1 < -\alpha_{\kappa,\lambda} \\ (y_1 - \kappa)^2, & y_1 > \alpha_{\kappa,\lambda} \\ y_1^2 + \kappa^2 - \lambda^2, & |y_1| < \beta_{\kappa,\lambda} \end{cases} \tag{1}$$

and

$$\rho(y_1) \leq min((y_1+\kappa)^2, (y_1-\kappa)^2). \tag{2}$$

Here we assume that $0 < \beta_{\kappa,\lambda} < \alpha_{\kappa,\lambda}$, and that, for each λ, $\alpha_{\kappa,\lambda} \to 0$ as $\kappa \to \infty$. Then, $E_\lambda = \cup_\kappa Int(\Omega(\kappa,\lambda))$. Also notice these conditions guarantee that $\Omega(\kappa,\lambda)|_{|y_1|<\beta_{\kappa,\lambda}} = B^4(\lambda)|_{|y_1|<\beta_{\kappa,\lambda}}$. An easy calculation shows:

LEMMA 5.1. *It is possible to choose ρ so that it satisfies (1) and (2) and so that $\partial\Omega(\kappa,\lambda) = \{\varphi_{\kappa,\lambda} = 1\}$ is transverse to the radial vector field ∂.*

We can therefore apply our filling result to $S^2(\lambda) \subset \partial\Omega(\kappa,\lambda)$.

(5.2) THE CAMEL THEOREM. *If $R \geq \lambda$, a standard linear embedding g_R^- of $B(R)$ into $E_\lambda \cap \{y_1 < 0\}$ is not symplectically isotopic to the corresponding embedding g_R^+ into $E_\lambda \cap \{y_1 > 0\}$.*

PROOF: Any isotopy $g_t, 1 \leq t \leq 2$ from g_R^- to g_R^+ is contained in a compact subset of E_λ and hence in $Int\ \Omega = Int\ \Omega(\kappa,\lambda)$ for some κ. Let $J_t \in \mathcal{J}_\Omega$ be a smooth family of almost complex structures chosen so that $J_t = (g_t)_* J_0$ on $Im\ g_t$. Because each filling $F(J_t)$ disconnects Ω, the set

$$X = \{(t,x) : x \in F(J_t), 1 \leq t \leq 2\}$$

disconnects $[1,2] \times \Omega$. Further, since $(g_t)_* J_0 = J_0$ when $t = 1,2$, we may suppose that J_1 and J_2 equal J_0 near the 3-ball $B^3(\lambda) \subset \{y_1 = 0\}$. Then $F(J_t) = B^3(\lambda)$ when $t = 1,2$ and so the points $(1, g_1(0))$ and $(2, g_2(0))$ lie in distinct components of $([1,2] \times \Omega) \setminus X$. Therefore, the path $(t, g_t(0))$, $1 \leq t \leq 2$, must intersect X. In other words, for some t, there is a J_t-holomorphic A-disc C through $g_t(0)$. The argument is now completed in the usual way: see [G] or [M4, (2.5.2)]. Let C' be the pull-back of the disc C by g_t. Then C' is a J_0-holomorphic curve through the center of the ball $B(R)$ with boundary on $\partial B(R)$. Thus C' is a minimal surface and, by the monotonicity theorem, must have area $\geq \pi R^2$ with respect to the standard metric. On the other hand, this area can be calculated by integrating ω_0 over C' and hence is strictly bounded above by the integral of ω_0 over C. Yet we know, by corollary 3.5, that $\omega_0(C) \leq \pi\lambda^2$. ∎

§6 EMBEDDINGS OF BALLS

In this section we prove the Extendable Embeddings Lemma which states that any embedding g of $B(R)$ into E_λ which extends to $B(R+\lambda)$ is isotopic to one of the standard linear embeddings $g_R^\pm$. The following Lemma shows why we require that g extend to $B(R+\lambda)$. We will suppose that Ω is one of the regions $\Omega_{\kappa,\lambda} \subset E_\lambda$ defined in §5.

LEMMA 6.1. *If $g(B(R+\lambda)) \subset Int\ \Omega$, then for every $J \in \mathcal{J}_\Omega$ which equals $g_*(J_0)$ on $g(B(R+\lambda))$, the filling $F(J)$ does not intersect $g(B(R))$.*

PROOF: If $g(B(R))$ did intersect $F(J)$, there would be a J_0-holomorphic curve C' of area $< \pi\lambda^2$ through a point $z \in B(R) \subset B(R+\lambda)$. Since C' goes through the center of the ball $B(z,\lambda) \subset B(R+\lambda)$, this leads to a contradiction as in the proof of the Camel Theorem (5.2). ∎

PROPOSITION 6.2. *Let $W = \mathbf{R}^3 - B^3(\lambda) \subset \{y_1 = 0\}$. For all $J \in \mathcal{J}_\Omega$, the filling $F(J)$ may be ε-perturbed so that it fits together with W to form a 3-manifold Q such that $(\mathbf{R}^4, Q, \omega_0)$ is symplectomorphic to $(\mathbf{R}^4, \mathbf{R}^3, \omega_0)$.*

Before proving Proposition 6.2, we show how Lemma 6.1 and Proposition 6.2 together prove our Extendable Embeddings Lemma. If $g \in \mathcal{E}(R, R+\lambda, \lambda)$, Lemma 6.1 tells us that $g(B(R))$ must lie on one of the sides of $W \cup F(J)$. Since we may choose ε sufficiently small so that $Q \cap g(B(R)) = \emptyset$, this means $g(B(R))$ lies in a subset of E_λ which is symplectomorphic to a half-space and hence to the whole of $\mathbf{R}^4$. However, the space of symplectic embeddings of a ball into $\mathbf{R}^4$ is connected. In fact, it has the same homotopy type as the symplectic linear group $Sp(4,\mathbf{R})$. Hence g is isotopic to $g_R^\pm$ as claimed.

In the course of proving Proposition 6.2, we will use a "parameterized family of fillings." Consider the following 2-spheres and 4-balls of radius γ:

$$S(\gamma, s) = \{x_1^2 + x_2^2 + y_2^2 = \gamma^2, y_1 = s\},$$
$$B(\gamma, s) = \{x_1^2 + (y_1 - s)^2 + x_2^2 + y_2^2 \leq \gamma^2\}.$$

If we choose an almost complex structure J such that $J = J_0$ outside the 4-ball $B(\mu) = B(\mu, 0)$ and choose $\gamma > 2\mu$ so that $B(\mu) \subset B(\gamma, s)$ for $-\mu \leq s \leq \mu$ then, for each s, we may let $B(\gamma, s)$ play the role of Ω, and so may fill $S(\gamma, s)$ with respect to J. Denote the filling by $F^s(J)$.

LEMMA 6.3. *$F^r(J) \cap F^s(J) = \emptyset$ when $r \neq s$.*

PROOF: Notice that $F^r(J) \subset B(\gamma, r)$ for all r by Proposition 3.2, and that $B(\gamma, r) \cap S(\gamma, s) = \emptyset$ when $r \neq s$. Thus if the discs $D^r \in F^r(J)$ and $D^s \in F^s(J)$ intersect, they must do so at a point which is in the interior of each disc. But this is impossible by positivity of intersections: see Proposition 3.4. ∎

PROOF OF PROPOSITION 6.2: We first construct the hypersurface Q. It will be defined as the union of a 1-parameter family of symplectic 2-manifolds, each of which consists of a disc in the filling $F(J)$ extended so as to join together with a flat 2-plane $\{x_1 = \text{constant},\ y_1 = 0\}$ outside some large ball.

By Corollary 3.3, the discs in $F(J)$ meet any vertical arc α on $S = S^2(\lambda)$ once transversally. Let $D_\nu = Im\ f_\nu$ be the disc in $F(J)$ which intersects this arc α at $x_1 = \nu$. Then

$$\begin{aligned} f_\nu^*\omega_0 &= f_\nu^*(dx_1 \wedge dy_1 + dx_2 \wedge dy_2) \\ &= (\pi_1 \circ f_\nu)^*(dx_1 \wedge dy_1) + (\pi_2 \circ f_\nu)^*(dx_2 \wedge dy_2), \end{aligned}$$

where π_i denotes the projection onto the (x_i, y_i) plane. Both terms here are ≥ 0 near ∂D since $\pi_i \circ f_\nu$ is J_0-holomorphic near ∂D. Moreover, the second term is strictly positive, because the boundary of D_ν is transverse to the vertical arcs. Thus, by flattening the y_1 coordinate of f_ν near ∂D, we can perturb the disc D_ν to a disc $\tilde{D}_\nu = Im\ \tilde{f}_\nu$ which is infinitely tangent to the hyperplane $\{y_1 = 0\}$ along its boundary and which is still symplectically embedded. Clearly, we may suppose that this perturbation is smooth with respect to ν, and that all the discs $\tilde{D}_\nu, -\lambda < \nu < \lambda$, are disjoint. Note that this perturbation does not alter the flat discs D_ν for ν near $\pm\lambda$.

Now, let P_ν be the portion of the flat plane $\{(x_1, y_1) = (\nu, 0)\}$ which lies outside $B^3(\lambda)$, i.e. $P_\nu = \{(\nu, 0, x_2, y_2) \notin B^3(\lambda)\}$. Perturb each P_ν inside the hyperplane $\{y_1 = 0\}$ to a 2-dimensional space $\tilde{P}_\nu$ which is still a graph over the (x_2, y_2) plane and which joins smoothly with $\partial\tilde{D}_\nu$. Then $L_\nu = \tilde{D}_\nu \cup \tilde{P}_\nu$ is a smoothly embedded 2-manifold which equals P_ν outside a compact set. It is symplectic since both $\omega_0|_{\tilde{D}_\nu}$ and $\omega_0|_{\tilde{P}_\nu}$ are positive. Again, we may suppose that the L_ν vary smoothly with respect to ν, and that they coincide with the flat planes $\{(x_1, y_1) = (\nu, 0)\}$ outside some compact subset X of the strip $|x_1| < \lambda \in \mathbf{R}^4$.

Let $Q = \cup_\nu L_\nu$. By construction, Q is foliated by the symplectic 2-manifolds L_ν.

We now construct a symplectomorphism from $(\mathbf{R}^4, Q)$ to $(\mathbf{R}^4, \mathbf{R}^3)$. Choose a J' which equals J_0 outside some large 4-ball $B(\mu)$ containing the compact subset X defined above and such that the leaves of the foliation of Q are J'-holomorphic. With γ as in Lemma 6.3, let $S(\gamma, s)$ be the 2-sphere in $\{y_1 = s\}$ with filling $F^s(J')$. Note that the discs in $F^0(J')$ are just the intersections of the leaves L_ν with the ball $B(\gamma, 0)$. By Lemma 6.3, we know that the fillings $F^s(J')$ are mutually disjoint. Thus, by repeating the above argument, we can perturb the $F^s(J')$ so they fit together with the hyperplanes $\{y_1 = s\}$ to form a foliation of $\mathbf{R}^4$ with leaves Q^s, each of which is foliated by symplectic 2-planes. Further, outside of a compact subset of $\mathbf{R}^4$, Q^s coincides with $\{y_1 = s\}$ and is foliated by the planes $\{x_1 = \text{constant}\}$.

Now consider the characteristic foliation $\Lambda^s = ker\ \omega_0|_{Q^s}$ of Q^s. This is transverse to the symplectic leaves and points in the direction $-\frac{\partial}{\partial x_1}$ outside a large ball. Therefore, we may parametrize the characteristic flow ϕ_t^s on Q^s smoothly in s so that it preserves the foliation of each Q^s and has tangent $-\frac{\partial}{\partial x_1}$ outside a compact set. Choose K large enough so that the plane $\{x_1 = K, y_1 = s\}$ is a symplectic leaf in Q^s for all s. Then define $\Phi\colon \mathbf{R}^4 \to \mathbf{R}^4$ by

$$\Phi(p) = (K - t, s, x_2, y_2)$$

where $p \in Q^s$ is the image of the point (K, s, x_2, y_2) under the map ϕ_t^s. This takes Q^s to the hyperplane $\Pi^s = \{y_1 = s\}$, for all s, but is not quite symplectic. Thus to complete the proof of Proposition 6.2, it remains to show

LEMMA 6.4. *There exists a diffeomorphism h of $(\mathbf{R}^4, Q)$ such that $h^*\Phi^*\omega_0 = \omega_0$.*

PROOF: Let ξ^s be the vector field on Q^s whose flow is ϕ_t^s. By construction, $\Phi_*(\xi^s) = -\frac{\partial}{\partial x_1}$. It follows that if $q = (x_1, s, x_2, y_2)$ where $x_1 > K$, then for all t, $\Phi \circ \phi_t^s(q) = \tau_t \circ \Phi(q)$ where $\tau_t(x_1, y_1, x_2, y_2) = (x_1 - t, y_1, x_2, y_2)$. Since the restriction of ω_0 to Q^s is invariant under ϕ_t^s, we can then conclude $\Phi^*\omega_0|_{Q^s} = \omega_0|_{Q^s}$ for all s. It follows that the forms $\omega_r = r\Phi^*\omega_0 + (1-r)\omega_0$ are non-degenerate on $\mathbf{R}^4$ when $0 \le r \le 1$. Suppose we can find a 1-form β which vanishes at all points of Q and satisfies $d\beta = \Phi^*\omega_0 - \omega_0$. Then if the vector field v_r defined by

$$i(v_r)\omega_r + \beta = 0$$

is integrable, the standard Moser method argument shows that the integral h_r of v_r satisfies $h_r^*\omega_r = \omega_0$. Since $h_r = id$ on Q the desired diffeomorphism is then given by h_1.

Consider $\beta = \Phi^*(x_1 dy_1 + x_2 dy_2) - (x_1 dy_1 + x_2 dy_2)$. Outside a compact set K_1,

$$\Phi(x_1, y_1, x_2, y_2) = (x_1, y_1, \psi_{y_1}(x_2, y_2))$$

where $\psi_{y_1} : \mathbf{R}^2 \to \mathbf{R}^2$ is independent of x_1, and equals the identity outside a compact set K_2 in the (y_1, x_2, y_2)-space. It follows that outside K_1,

$$\omega_r = dx_1 \wedge dy_1 + dx_2 \wedge dy_2 + r(h_1\ dy_1 \wedge dx_2 + h_2\ dy_1 \wedge dy_2)$$

where $|h_1|$ and $|h_2|$ are bounded functions of y_1, x_2, y_2. It is then easy to check that given any bounded 1-form β, the associated vector field v_r defined by the equation $(v_r)\omega_r + \beta = 0$ has bounded growth and is thus integrable. The β which we constructed above is in fact bounded. However since it does not vanish on Q we make the following modifications.

$\Phi^*\omega_0|_Q = \omega_0|_Q$ implies that $\beta|_Q$ is closed. From the description of Φ outside the compact set K_1, it is easy to check $\beta|_Q$ is bounded. Thus $\beta|_Q = df$ where f is a function on Q and we can extend f to a function $\tilde{f}$ on $\mathbf{R}^4$ so that $d\tilde{f}$ is bounded. Then $\beta' = \beta - d\tilde{f}$ is cohomologous to β, is still bounded, and vanishes for all vectors in TQ. Lastly we modify β' to a form β'' which equals zero at all points of Q. To do this, first choose coordinates x_1', y_1', x_2', y_2' on $\mathbf{R}^4$ so that $Q = \{y_1' = 0\}$ and so that these are the standard coordinates on $\mathbf{R}^4$ outside a compact set. Suppose

$$\beta' = \sum f_i dx_i' + g_i dy_i'.$$

Then f_1, f_2 and g_2 vanish on Q by construction. Further, outside a compact set, $g_1 = \psi^1 \frac{\partial \psi^2}{\partial y_1}$ where $\psi_{y_1}(x_2, y_2) = (\psi^1, \psi^2)$. Hence g_1 is independent of x_1 and has compact support with respect to the other variables. Therefore $d(g_1 y_1')$ is bounded and we may take $\beta'' = \beta' - d(g_1 y_1')$. ∎

REFERENCES

[A] Arnol'd, V.I., *First steps in symplectic topology*, Russian Math. Surveys **41** 6 (1986), 1-21.

[BB] Booss, B. and Bleecker, D. D., "Topology and Analysis: The Atiyah-Singer index formula and gauge theoretic physics," Springer-Verlag, New York, 1985.

[E] Eliashberg, Ya., *Filling by holomorphic discs and its applications*, "Geometry of low dimensional manifolds," London mathematical society lecture note series no. 152, Cambridge Univ. Press, 1990.

[EG] Eliashberg, Y. and Gromov, M., *Convex symplectic manifolds*, preprint (1990).

[F] Floer, A., *The unregularized gradient flow of the symplectic action*, Comm. Pure Appl. Math. **XLI** (1988), 775-813.

[G] Gromov, M., *Pseudo-holomorphic curves on almost complex manifolds*, Invent. Math. **82** (1985), 307-347.

[K] Klingenburg, W., *Filling a 2-sphere in* $\mathbf{C}^2$ *by holomorphic discs*, in preparation.

[M1] McDuff, D., *Examples of symplectic structures*, Invent. Math. **89** (1987), 13-36.

[M2] ———, *The local behaviour of holomorphic curves in almost complex 4-manifolds*, Jour. Diff Geom. **34** (1991), 143-164.

[M3] ———, *Blow ups and symplectic embeddings in dimension 4*, Topology **30** (1991), 409-421.

[M4] ———, *Elliptic methods in symplectic geometry*, Bull. Amer. Math. Soc. **23** (1990), 311-358.

[M5] ———, *Symplectic manifolds with contact type boundaries*, Invent. Math. **23** (1990), 311-358.

[O] Oh, Y.-G., *Removal of boundary singularities of pseudo-holomorphic curves with lagrangian boundary conditions*, preprint (1990).

Department of Mathematics, SUNY at Stony Brook, Stony Brook, NY 11794-3651, USA

Differential forms and connections adapted to a contact structure, after M. Rumin[1]

Pierre Pansu

U.R.A. 169 du C.N.R.S.
Centre de Mathématiques
Ecole Polytechnique
F-91128 Palaiseau

U.R.A. 1169 du C.N.R.S.
Mathématiques
Université Paris-Sud
F-91405 Orsay

pansu@cmep.polytechnique.fr

Michel Rumin is a student of Mikhael Gromov, who asked him the following question : Let M be a manifold with contact structure ξ, E a vector bundle over M. A *partial connection* on E is a covariant derivative $\nabla_v e$ defined for smooth sections e of E but only for vectors v in ξ. In particular, parallel translation is defined only along Legendrian curves, that is curves which are tangent to ξ. Can one define the curvature of such a connection ?

Gromov provided the following hint : For an ordinary connection A, curvature arises in the asymptotics of holonomy around short loops. A loop encompasses a certain "span" (a 2-vector, see below), quadratic in length, and holonomy deviates from the identity by an amount proportional to curvature times span, that is, quadratic in length. In case M has dimension 3 and carries a contact structure, then every Legendrian loop has essentially zero area. Gromov conjectured that, in this case, curvature should arise as the cubic term in the asymptotic expansion of holonomy.

Michel Rumin has found a notion of curvature for partially defined connections along the above lines. The point is to understand the exterior differential for a partially defined 1-form. In fact, M. Rumin constructs a substitute for the de Rham complex : a locally exact complex of hypoelliptic operators naturally attached to a contact manifold (M, ξ) of dimension $2m+1$. The operator which sends m-forms to m+1-forms is new. It is of second order.

---

[1]Research partially supported by the EEC under contract SC1–0105–C, GADGET

In this lecture, after some comments on the asymptotics of holonomy, I explain M. Rumin's construction. Then I describe an application of M. Rumin's ideas to analysis on complex hyperbolic space : boundary values of L^2 harmonic forms. I learned most of the material presented here thanks to P.Y. Gaillard, V. Goldshtein, J. Heber, P. Julg and M. Rumin.

1 Asymptotic expansion of holonomy

In this section, I explain how the asymptotic expansion of the holonomy of a smooth connection specializes to Legendrian curves in a contact manifold.

1.1 The classical formula.

1.1 The classical formula.– I define the "span", i.e., the algebraic area spanned by a loop c in a vector space $V = \mathbf{R}^n$. Let $\alpha \in \Lambda^2 V^*$ be a 2-form, viewed as a translation invariant differential form on V. Then $d\alpha = 0$, and there exists a 1-form β such that $d\beta = \alpha$. The linear functional

$$\alpha \mapsto \int_c \beta$$

on $\Lambda^2 V^*$ corresponds to a unique 2-vector span(c).

Now let M be a differentiable manifold, $x \in M$, $c \subset M$ a small loop through x. Using coordinate charts ϕ we define various covectors

$$\text{span}_\phi(c) = \text{span}(\phi(c)) \in \Lambda^2 T_x M$$

which coincide up to an error of size $o(\text{area}(c))$, where area(c), the "geometric area", is the least area of a surface spanned by c. Thus it makes sense to state

1.2 Fact .–*Let D be a connection on some vector bundle E over a manifold M. For $x \in M$, c a short loop through x, let $Hol(D, c)$ denote the holonomy of the connection D along c (an endomorphism of the fibre E_x). Then, as the length of c tends to 0,*

$$Hol(D, c) = 1 + \langle F^D, \text{span}(c)\rangle + o(\text{area}(c)),$$

where F^D is the curvature of D at x, an $End(E_x)$-valued 2-form on $T_x M$.

1.3 The case of contact manifolds.– Geometric areas will be measured relative to Carnot-Caratheodory metrics. Given a norm on the plane field ξ, a Carnot-Caratheodory metric is defined by minimizing the length of Legendrian curves between two points.

In dimension $2m+1 \geq 5$, every Legendrian curve bounds a Legendrian surface, which has Hausdorff dimension 2. Thus we define the geometric area area(c) to be the infimum of the 2-dimensional Hausdorff measures of surfaces whose boundary is c. However, in dimension 3, all smooth surfaces have Hausdorff dimension 3, and we take 3-dimensional Hausdorff measure in the definition of geometric area. Note that contact transformations are Lipschitz with respect to Carnot-Caratheodory metrics, so they preserve the rough size of geometric areas.

Recall that the Heisenberg group N of dimension $2m+1$ is the simply connected group attached to the following Lie algebra $\mathcal{N}$:

$$\mathcal{N} = \mathbf{R}^{2m} \oplus \mathbf{R}$$

with center $\mathbf{R}$ and the Lie bracket is given by the symplectic form $\mathbf{R}^{2m} \times \mathbf{R}^{2m} \rightarrow \mathbf{R}$. The left-invariant plane field generated by the factor $\mathbf{R}^{2m}$ is a contact structure ξ_0. Every contact manifold is locally isomorphic to the Heisenberg group (N, ξ_0).

We define the "algebraic area" spanned by a Legendrian loop c in the Heisenberg group as before, but replacing the symplectic vector space, which is the local model for manifolds, by the Heisenberg group, which is the local model for contact manifolds.

In dimensions $2m+1 \geq 5$, the closed left invariant 2-forms on N all are pulled-back from $N/[N,N] = \mathbf{R}^{2m} = \xi_0$. For such an $\alpha = d\beta$, the formula

$$\langle \mathrm{span}\,(c), \alpha \rangle = \int_c \beta$$

defines a 2-covector span$\,(c) \in \Lambda^2 \xi_0$, equivariantly with respect to Heisenberg automorphisms, and also, up to an error controlled by geometric area, with respect to contact diffeomorphisms.

Note that, if τ denotes some left-invariant 1-form whose kernel is the canonical contact structure ξ_0, then

$$\langle \mathrm{span}\,(c), d\tau \rangle = 0.$$

In dimension 3, all left-invariant 2-forms are closed, and one gets a 2-covector span$\,(c) \in \Lambda^2 \mathcal{N}$.

In both cases, one has

1.4 FACT .– *Let D be a connection on some vector bundle E over a contact manifold (M,ξ). Let $x \in M$, and c be a short Legendrian loop through x. Then, as the length of c tends to 0,*

$$Hol(D,c) = 1 + \langle F^D, \mathrm{span}\,(c)\rangle + o(\mathrm{area}(c)).$$

In dimensions $2m+1 \geq 5$, geometric area is quadratic in length (Y. Eliashberg, unpublished) and we see that the projection of F^D in

$$\Lambda^2\xi^*/\mathbf{R}d\tau$$

only depends on the restriction of D to the ξ directions, i.e., it is an invariant attached to the partial connection $\nabla = D_{|\xi}$. We call it the *curvature* of ∇. It has $(m(m-1)/2\ -1)(\mathrm{rk}\,E)^2$ independant components.

In dimension 3, the geometric area is cubic in length, and we cannot ignore the vertical components of span (c). We define the *curvature* of $\nabla = D_{|\xi}$ to be the projection of F^D mod multiples of the symplectic structure $d\tau$. It has $2(\mathrm{rk}\,E)^2$ independant components.

In the next section, these definitions will be shown to fit into the formalism of Rumin differential forms.

2 RUMIN'S COMPLEX

It is a substitute for the de Rham complex, where 1-forms are replaced by *partial* 1-forms, i.e., sections of the dual of a contact structure.

Let (M,ξ) be a $2m+1$-dimensional contact manifold. Let Ω^* denote the graded algebra of smooth differential forms. Let $\mathcal{I}^*$ be the graded differential ideal generated by contact forms (i.e., 1-forms τ whose kernel is the contact hyperplane ξ) and $\mathcal{J}^*$ the annihilator of $\mathcal{I}^*$ (i.e., forms α such that $\alpha \wedge \beta = 0$ for all $\beta \in \mathcal{I}^*$). It is again a graded differential ideal.

Once a contact form τ is chosen, the elements of $(\Omega^*/\mathcal{I}^*)\oplus\mathcal{J}^*$ identify with sections of subbundles or quotients of $\Lambda^*\xi^*$. Indeed, let

$$L\ : \Lambda^k\xi^* \to \Lambda^{k+2}\xi^*$$

denote exterior multiplication by $d\tau_{|\xi}$,

$$E^k = \Lambda^k\xi^*/\mathrm{im}\,L,$$

for $k \leq m$, and

$$E^k = \ker L \subset \Lambda^k\xi^*,$$

for $k \geq m$. Then $(\Omega^k/\mathcal{I}^k)\oplus\mathcal{J}^k$ coincides with smooth sections of E^k if $k \leq m$, and with smooth sections of E^{k-1} multiplied by τ if $k \geq m+1$. In particular, the elements in $\Omega^1/\mathcal{I}^1$ are sections of E^1, i.e., partial 1-forms.

There is an induced complex

$$d_\xi \ : (\Omega^*/\mathcal{I}^*) \oplus \mathcal{J}^* \to (\Omega^*/\mathcal{I}^*) \oplus \mathcal{J}^*.$$

One easily checks that d_ξ is locally exact at $\Omega^k/\mathcal{I}^k$ (resp. at $\mathcal{J}^k$) if $k < m$ (resp. if $k > m+1$). This has been observed independandly by V. Ginzburg [5], and generalized by Zhong Ge, [15].

2.1 THEOREM (M. Rumin, [14]).– *There exists a linear second order operator*

$$d_R \ : \Omega^m/\mathcal{I}^m \to \mathcal{J}^{m+1}$$

such that the sequence

$$0 \to \mathbf{R} \to \Omega^0 \xrightarrow{d_\xi} \Omega^1/\mathcal{I}^1 \xrightarrow{d_\xi} \cdots \xrightarrow{d_\xi} \Omega^m/\mathcal{I}^m \xrightarrow{d_R} \mathcal{J}^{m+1} \xrightarrow{d_\xi} \cdots \xrightarrow{d_\xi} \mathcal{J}^{2m+1} \to 0$$

is a locally exact complex, i.e., a resolution of the constant sheaf $\mathbf{R}$.

We explain this in 3 dimensions. If $\alpha \in \Omega^1$ is a 1-form, there is a unique choice of a function f so that $d(\alpha + f\tau)$ vanishes on ξ. Take

$$d_R(\alpha \ mod \ \mathcal{I}^1) = d(\alpha + f\tau) \in \mathcal{J}^2.$$

The function f is determined by the equation

$$d\alpha_{|\xi} + f \, d\tau_{|\xi} = 0$$

and depends on first derivatives of α (in the ξ directions only), thus d_R involves two derivatives in the ξ directions.

Given a metric on ξ, there is a normalization of the contact form τ so that $|d\tau_{|\xi}| = 1$. One gets a pointwise inner product on $\Omega^*/\mathcal{I}^* \oplus \mathcal{J}^*$, and a Hodge operator

$$* \ : \Omega^k/\mathcal{I}^k \to \mathcal{J}^{2m+1-k}$$

such that $*1 = \tau \wedge (d\tau)^m$ and

$$\langle \alpha, \beta \rangle * 1 = \alpha \wedge *\beta.$$

2.2 THEOREM (M. Rumin, [14]).– *Given a metric on* ξ, *put* $d_\xi^* = - * d_\xi *$ *(resp.* $d_R^* = - * d_R *$*). This is a formal adjoint to* d_ξ *(resp.* d_R*).*

The laplacians

$$(n-k)d_\xi d_\xi^* + (n-k-1)d_\xi^* d_\xi \quad \text{on} \quad (\Omega^k/\mathcal{I}^k) \oplus \mathcal{J}^k \quad \text{for} \quad k \neq m, m+1,$$

$$(d_\xi d_\xi^*)^2 + d_R^* d_R \quad \text{on} \quad \Omega^m/\mathcal{I}^m,$$

$$(d_\xi^* d_\xi)^2 + d_R d_R^* \quad \text{on} \quad \mathcal{J}^{m+1},$$

are maximally hypoelliptic.

Maximal hypoellipticity means that, given vector fields X_i tangent to ξ, one locally has estimates of the form

$$\| X_1 X_2 \alpha \|_2 \leq C(\| \Delta\alpha \|_2 + \| \alpha \|_2)$$

for the second order Laplacian, and similarly for the fourth order Laplacian.

There is an analogue in contact geometry of the principal symbol, which gives a criterion for hypoellipticity, see [11]. However this criterion becomes effective only when combined with clever Bochner type formulas, see [14]. Simultaneously, various vanishing theorems are obtained. They include the following important feature : on a CR manifold whose Webster torsion vanishes (these are integrability conditions on the metric, analogous to the Kähler condition for Hermitian metrics), the above Laplacians preserve the bidegree (the unusual choice of coefficients is essential).

Back to connections : A partial connection is a Lie algebra valued partial 1-form A and one can make sense of the curvature $dA + A \wedge A$ as was done for dA. In dimensions $2m+1 \geq 5$, it is a Lie algebra valued 2-form on ξ mod $d\tau$, in dimension 3, it is a Lie algebra valued 2-form vanishing on ξ, as announced in the previous section.

3 L^2-HARMONIC FORMS ON COMPLEX HYPERBOLIC SPACE

Complex hyperbolic $m+1$-space is a complete symmetric Kähler manifold, isometric to the unit ball of $\mathbf{C}^{m+1}$ equipped with its Bergman metric. It is the symmetric space of the simple Lie group $SU(m+1,1)$. It is a generalization of the unit disk in $\mathbf{C}$, equipped with its Poincaré metric, which is the symmetric space of $SU(1,1) = PSL(2,\mathbf{R})$.

The new feature when $m \geq 1$ is that the boundary S^{2m+1} inherits a canonical contact structure. At a point $x \in S^{2m+1}$, the contact plane ξ_x is the maximal complex vector subspace in $T_x S^{2m+1} \subset T_x \mathbf{C}^{m+1}$.

Complex hyperbolic $m+1$-space has L^2-cohomology in the middle dimension $m+1$ and in each type (p,q), $p+q = m+1$. Following recent work by Pierre Julg and Michel Rumin ([9]), we explain that L^2-harmonic forms have boundary values, which realize an isomorphism with an explicit space of Rumin differential forms on the boundary, the sphere S^{2m+1} equipped with its canonical contact structure.

3.1 The case of real hyperbolic space.– Let us first explain the corresponding theory for real hyperbolic $2m+2$-space, i.e., the space form of constant sectional curvature -1. L^2-harmonic forms in the middle dimension are conformally invariant, so we can replace hyperbolic space minus one point with $S^{2m+1} \times \mathbf{R}_+$ in a product metric, a situation which has been studied by Atiyah-Patodi-Singer, [1].

Since the Hodge $*$ commutes with the Laplacian and $*^2 = (-1)^{m+1}$, harmonic forms split into self-dual and anti-self-dual forms ($*\alpha = \pm i_m \alpha$ where $i_m = 1$ if $m+1$ is even, $i_m = i$ if $m+1$ is odd). The equations for a closed self dual (resp. anti-self-dual) form α on $S^{2m+1} \times \mathbf{R}_+$ can be viewed as an ODE in the $y \in \mathbf{R}_+$ variable. Splitting $\alpha = a \pm i_m(*a) \wedge dy$, it reads

$$da = 0 \quad \text{and} \quad \frac{\partial}{\partial y} a = - \pm i_m d * a. \tag{1}$$

This equation has constant operator coefficients and explicit solutions in terms of data at $y = 0$, i.e., along the boundary, are easily found. This leads to

3.2 Proposition .– *A closed, self-dual (resp. anti-self-dual) $m+1$-form α on real hyperbolic $2m+2$-space has a boundary value $BV(\alpha)$, which is a closed $m+1$-form on S^{2m+1}, $\pm$-invariant under the operator*

$$F = \operatorname{sign} A, \quad A = i_m d *_{|\ker d} \,.$$

The L^2 norm translates into a Sobolev norm on the boundary :

$$\| \alpha \|_2 = \| \, |A|^{-1/2} \alpha|_\partial \|_2 \,.$$

The L^2 norm of harmonic forms is recovered as follows (P. Julg) : Forms α smooth up to the boundary are dense in L^2 solutions of (1). Choose a smooth form β such that $d\beta = \alpha$ on $S^{2m+1} \times \mathbf{R}_+$ and $d * \beta = 0$ on S^{2m+1}, then

$$\begin{aligned} \| \alpha \|_2^2 &= \int_{S^{2m+1} \times \mathbf{R}_+} \pm i_m \alpha \wedge \alpha \\ &= \int_{S^{2m+1}} \beta \wedge \pm i_m \alpha \\ &= \int_{S^{2m+1}} \langle \alpha, \pm i_m (-1)^{m+1} * \beta \rangle \\ &= \int_{S^{2m+1}} \langle \alpha, \pm A^{-1} \alpha \rangle \\ &= \| \, |A|^{-1/2} (\alpha_{|S^{2m+1}}) \|_2^2 \,. \end{aligned}$$

3.3 Problem.– Since conformal mappings of S^{2m+1} extend as isometries of real hyperbolic space, we observe that both the operator F and the norm

$$\| \, |A|^{-1/2}\alpha \, \|_2$$

on closed $m+1$-forms are Möbius invariants. More generally, since every quasiconformal mapping of S^{2m+1} extends to a quasiisometry of real hyperbolic space, the norm on closed $m+1$-forms is quasiinvariant under quasiconformal mappings. To what extent is the operator F invariant under quasiconformal mappings ?

3.4 The case of complex hyperbolic space.– A similar computation can be done in the complex case. Harmonic $m+1$-forms split into types and primitive components. Only primitive forms can be in L^2. A conformal change leads to a metric on $S^{2m+1} \times \mathbf{R}_+$ of the form

$$g_{|\xi} + y^{-2} g_{|\xi^\perp} + dy^2$$

where ξ, the complex tangent to S^{2m+1} in the embedding of S^{2m+1} in $\mathbf{C}^{m+1}$, is the canonical contact structure. The ODE for ∂ and $\overline{\partial}$-closed primitive forms does not have constant, nor even commuting coefficients. Splitting forms on S^{2m+1} according to $\xi \oplus \xi^\perp$ looks hopeless since the splitting is not invariant under the exterior differential. Nevertheless, Rumin's complex precisely extracts the part of d that preserves the splitting.

It turns out that the ODE, when rephrased in terms of Rumin's d_R and $*$ operators, can be reduced to scalar equations. These equations are singular at $y = 0$. Still, their L^2 solutions are determined by their values at $y = 0$. One concludes

3.5 Theorem (P. Julg,[9]).– *There exists a boundary value operator BV on L^2 harmonic $m+1$-forms on complex hyperbolic $m+1$-space, with values in (non smooth) closed partial $m+1$-forms on S^{2m+1} (i.e., elements of $\mathcal{J}m+1 \cap \ker d_\xi$). It is an isometry for the norm*

$$\| \, |A|^{-1/2}\alpha \, \|_2$$

where $A = i_m d_R *_{|\ker d_\xi}$.

The boundary value operator BV sends the Hodge i_m* to the operator $F = \operatorname{sign} A$.

The finer splitting of L^2 harmonic forms into complex types $\mathcal{H}^{p,q}$ seems to translate as follows. Since the contact hyperplane ξ carries a complex structure, Rumin forms of degree $k \geq m+1$ split into types,

$$\mathcal{J}^k = \bigoplus_{p+q=k-1} \mathcal{J}^{p,q}.$$

Then

- for $p,\ q \geq 2$, $BV(\mathcal{H}^{p,q})$ consists of closed forms in $\mathcal{J}^{p-1,q} \oplus \mathcal{J}^{p,q-1}$;
- $BV(\mathcal{H}^{m+1,0})$ (resp. $BV(\mathcal{H}^{0,m+1})$) consists of closed forms in $\mathcal{J}^{m,0}$ (resp. in $\mathcal{J}^{0,m}$) ;
- $BV(\mathcal{H}^{m,1})$ (resp. $BV(\mathcal{H}^{1,m})$) is the L^2-orthogonal complement of $BV(\mathcal{H}^{m+1,0})$ (resp. of $BV(\mathcal{H}^{0,m+1})$) inside closed forms in $\mathcal{J}^{m,0} \oplus \mathcal{J}^{m-1,1}$ (resp. $\mathcal{J}^{0,m} + \mathcal{J}^{1,m-1}$), unless $m = 1$;
- when $m = 1$, $BV(\mathcal{H}^{1,1})$ is the orthogonal complement of $BV(\mathcal{H}^{2,0}) \oplus BV(\mathcal{H}^{0,2})$ in $\mathcal{J}^2 = \mathcal{J}^{1,0} \oplus \mathcal{J}^{0,1}$.

3.6 The ring of representations of $SU(m+1,1)$.– The ring $R(G)$ consists of equivalence classes of formal differences of G-modules with a finite difference of dimension, i.e., of Fredholm G-modules.

P. Julg and G. Kasparov ([10]) prove that $R(SU(m+1,1)) = R(U(m+1))$.

Theorem 3.5 is a crucial tool in the proof. Indeed, it allows them to construct a representative of an important element γ of $R(SU(m+1,1))$ as a representation of $SU(m+1,1)$ on a module over the algebra of continuous functions on the compactification $\overline{X}$ of complex hyperbolic space – forms on X plus Rumin forms on ∂X – which implies that $\gamma = 1$.

3.7 Poisson transform.– The results 3.2 and 3.5 provide us with a Poisson transform for closed middle degree forms, whose inverse is given by taking boundary values, in an L^2 setting.

More generally, one may naively wonder wether there is a Poisson transform for differential forms on symmetric spaces G/K with the following properties :

- it is G-equivariant,
- it commutes with the exterior differential,
- it coincides with the ordinary Poisson transform for functions,
- its inverse amounts to take some kind of boundary value,
- its image consists of all harmonic forms on G/K.

P.Y. Gaillard has studied in [4] the case of real hyperbolic space, i.e., $G = SO(n,1)$ (see also [8]). The Poisson transform takes forms on the boundary isomorphicly onto coclosed harmonic forms, and commutes with exterior derivative. (There is however an exception : in dimension $2m+1$, the Poisson transform kills coclosed m-forms on the boundary, and thus reaches only closed and coclosed m-forms on hyperbolic space). In general, Poisson transforms have boundary values only in degrees strictly less than half the dimension.

It is likely that there is an analogous Poisson transform for differential forms on complex hyperbolic space. Obviously, Rumin differential forms and modified exterior differential should be used. Also, Poisson transforms are probably automaticly primitive, ∂ and $\overline{\partial}$-coclosed.

4 L^p-COHOMOLOGY

The proof of P. Julg's theorem involves several magic identities satisfied by special functions. We present now a direct argument that shows that L^2-harmonic forms on complex hyperbolic space are representable by partial boundary values. It turns out that the method applies to L^p-cohomology as well.

Recall that the L^p-cohomology $H^*_p(X)$ of a Riemannian manifold X is the cohomology of the complex $(\Omega^*_{(p)}(X), d)$ where $\Omega^*_{(p)}(X)$ is the space of differential forms α with $|\alpha| \in L^p$ and $|d\alpha| \in L^p$. In general, the image

$$d(\Omega^{k-1}_{(p)}(X)) \subset \Omega^k_{(p)}(X)$$

is not closed, and one defines *reduced* L^p-cohomology as the quotient

$$\overline{H}^k_p(X) = \Omega^k_{(p)}(X)/\overline{d\Omega^{k-1}_{(p)}(X)}$$

by the closure of the image of d.

If $H^*_p(X) = \overline{H}^*_p(X)$, i.e., if the image $d(\Omega^{k-1}_{(p)}(X))$ is closed in $\Omega^k_{(p)}(X)$, we say that X *has only reduced L^p-cohomology in degree k*. This property is invariant under coarse quasiisometries, like those arising from isomorphisms of cocompact isometry groups.

For $p = 2$, there is exactly one L^2-harmonic form in each reduced L^2-cohomology class, i.e., the space of L^2-harmonic forms is isomorphic to $\overline{H}^*_2(X)$.

We explain next that on a negatively curved manifold, a closed form in L^p often admits a boundary value. We start again with the easier case of real hyperbolic space, which has been computed independently by V. Goldshtein, V. Kuz'minov and I. Shvedov, [6].

4.1 L^p-cohomology of real hyperbolic space.–

One uses the decomposition of real hyperbolic n-space X as a warped product

$$X = \mathbf{R}_+ \times_{\sinh r} S^{n-1}.$$

Split a k-form α as

$$\alpha = a + b \wedge dr$$

where a and b are viewed as functions on $\mathbf{R}_+$ with values in L^p-differential forms on the sphere S^{n-1}. The L^p norm of α is roughly the norm of a in $L^p(e^{(n-1-kp)r}\,dr)$ plus the norm of b in $L^p(e^{(n-1-(k-1)p)r}\,dr)$. The form α is closed iff a is closed and $\frac{\partial}{\partial r}a = \pm db$, which can be written

$$\frac{\partial}{\partial r} d^{-1}a = \pm d^{-1}db$$

where d^{-1} takes exact k-forms to coexact $k-1$-forms. Thus d^{-1} denotes the pseudo-inverse of d.

If $p < n-1/k-1$, $L^p(e^{(n-1-(k-1)p)r}\,dr) \subset L^1(dr)$ so $d^{-1}\alpha$ converges in L^p as $r \to +\infty$, and a converges to a distribution $a(\infty) = BV(\alpha)$.

If $\alpha \in d(\Omega^{k-1}_{(p)})$, or if $p \leq n-1/k$, then $BV(\alpha) = 0$.

Conversely, if $BV(\alpha) = 0$, then α admits a primitive β in L^p. Indeed, the Poincaré homotopy formula

$$\beta(r) = -\int_0^{+\infty} b(r+s)\,ds$$

(no dr component) solves $d\beta = \alpha$ and is in L^p (Hardy inequality) for $p < n-1/k-1$.

In conclusion, for real hyperbolic n-space, L^p-cohomology in degree k vanishes for $p \leq n-1/k$, and, for $n-1/k < p < n-1/k-1$, it is isomorphic to a certain function space of closed k-forms on S^{n-1}. In particular, it is a Hausdorff Banach space, thus, for such values of p and k, real hyperbolic space has only reduced cohomology. The L^p norm can be recovered in terms of boundary values - up to a constant, see [12] for the case when $k=1$.

For $p = n-1/k-1$, reduced cohomology vanishes but L^p-cohomology is huge.

The same argument applies to manifolds with variable curvature. Indeed, what matters is the Lie derivative of the metric on forms under the radial vector field $\frac{\partial}{\partial r}$. This is controlled by sectional curvature. This leads to the following comparison result (Jens Heber's help was instrumental in obtaining the sharp curvature assumption).

4.2 THEOREM [13].– *Let X be a complete simply connected Riemannian manifold of dimension n with negatively δ-pinched sectional curvature, i.e., $-1 \leq K \leq \delta < 0$. For all*

$$p < 1 + \frac{n-k}{k-1}\sqrt{-\delta},$$

an L^p closed k-form admits a boundary value, which determines its cohomology class. In particular, X has only reduced L^p-cohomology in degree k.

4.3 L^2-cohomology of complex hyperbolic plane.– We now check that the L^2-cohomology of complex hyperbolic plane in degree 2 is a limiting case of the above comparison theorem. Indeed, the theorem applies to L^p closed 2-forms on complex hyperbolic plane, for all $p < 2$: there exists a boundary value, which determines the L^p-cohomology class.

For $p \geq 2$, the boundary value does not exist any more, but a partial boundary operator will replace it, at least when $p < 4$

The complex hyperbolic plane in polar coordinates is not a warped product : the metric on spheres increases at different speeds along the factors of the splitting

$$TS^3 = \xi \oplus \xi^{\perp}.$$

Accordingly, a 2-form has to be split into four components

$$\alpha = a + a' d\tau + b \wedge dr + b' \tau \wedge dr.$$

The L^p norm of α in roughly the sum of the norms of $a \in L^p(e^{(4-3p)r}\, dr)$, $a' + b' \in L^p(e^{(4-2p)r}\, dr)$, $b \in L^p(e^{(4-p)r}\, dr)$. For $p = 2$, the limiting exponent 0 for b' prevents one from having an ordinary boundary value as in the preceding paragraph.

If we view the forms $a + da' \wedge \tau \in \mathcal{J}^2$ and $b \in \Omega^1/\mathcal{I}^1$ as elements of Rumin's complex, the equation $d\alpha = 0$ implies

$$d_{\xi}(a + da' \wedge \tau) = 0$$

and

$$\frac{\partial}{\partial r}(a + da' \wedge \tau) = d_R b$$

which implies that $a + da' \wedge \tau$ converges (when $p < 4$), this is our partial boundary value $BV(\alpha)$. It factors through reduced L^p-cohomology , and is injective on the reduced cohomology.

It turns out that the complex hyperbolic plane has only reduced cohomology in degree 2. This is a special case of a theorem of A. Borel, [2]. It also follows from estimates on the spectrum of the Laplacian. Indeed, for L^2-functions and 1-forms, the spectrum of the Laplacian is bounded below, [3]. This implies an estimate of the form

$$\| \beta \|_2^2 \leq C(\| d\beta \|_2^2 + \| \delta\beta \|_2^2)$$

for compactly supported 1-forms β, which therefore implies that the image $d(\Omega^1_{(2)}(\mathbf{C}H^2))$ is closed in $\Omega^2_{(2)}(\mathbf{C}H^2)$. We conclude that our partial boundary value BV is injective on L^2-cohomology .

This elementary approach cannot give the finer information on complex types contained in theorem 3.5.

References

[1] M. ATIYAH, V. PATODI, I. SINGER, *Spectral assymetry and Riemannian geometry* , Math. Proc. Cambridge Philos. Soc. **78**, 405-432 (1975).

[2] A. BOREL, *The L^2-cohomology of negatively curved Riemannian symmetric spaces,* Ann. Acad. Sci. Fennicae **10**, 95-105 (1985).

[3] H. DONNELLY, CH. FEFFERMAN, *L^2-cohomology and index theorem for the Bergman metric,* Annals of Math. **118**, (1983),593-618.

[4] P.Y. GAILLARD, *Transformation de Poisson de formes différentielles,* Comment. Math. Helvetici **61**, 581- (1986).

[5] V. GINZBURG, *On closed characteristics of 2-forms,* PhD Thesis, Berkeley (1990).

[6] V. GOL'DSHTEIN, V. KUZ'MINOV, I. SHVEDOV, *L^p-cohomology of noncompact riemannian manifolds,* to appear in Sib. Mat. Zh.

[7] M. GROMOV, *Asymptotic properties of discrete groups,* Preprint I.H.E.S. (1992).

[8] A. JUHL, *On the Poisson transformation for differential forms on hyperbolic spaces,* Seminar Analysis der K. Weierstrass Institut 1987/88, p. 224-236, Teubner, Leipzig (1988).

[9] P. JULG, *K-théorie des C^*-algèbres associées à certains groupes hyperboliques,* Thèse d'Etat, Université de Strasbourg (1991).

[10] P. JULG, G. KASPAROV, *L'anneau $KK_G(\mathbf{C},\mathbf{C})$ pour $G = SU(n,1)$,* to appear in C. R. Acad. Sci. Paris.

[11] B. HELFFER, J. NOURRIGAT, *Hypoellipticité maximale pour des opérateurs polynomes de champs de vecteurs,* Progress in Math. **58**, Bikhäuser (1985).

[12] P. PANSU, *Cohomologie L^p des variétés à courbure négative, cas du degré un,* in "P.D.E. and Geometry 1988", Rend. Sem. Mat. Torino, Fasc. spez., 95-120 (1989).

[13] P. PANSU, *Cohomologie L^p et pincement,* in preparation.

[14] M. RUMIN, *Un complexe de formes différentielles sur les variétés de contact,* C. R. Acad. Sci. Paris **310**, 401-404 (1990).

[15] ZHONG GE, *Generalized characteristics for Carnot-Caratheodory metrics,* Preprint Univ. Arizona, Tuczon (1991).

The Maslov class rigidity and non-existence of Lagrangian embeddings

Leonid Polterovich
Institut des Hautes Etudes Scientifiques
35, route de Chartres
91440-BURES sur YVETTE (France)*

October 1990

1 Introduction and main results

1. An important question of symplectic topology is the following: given a manifold, does it admit a Lagrangian embedding into $\mathbf{C}^n$? A series of obstructions to existence of such embeddings arises due to pure topological reasons (see M. Audin's paper [1] for a detailed discussion). However in [3] M. Gromov discovered an obstruction of another nature. Using infinite-dimensional analysis he showed that on every embedded Lagrangian submanifold of $\mathbf{C}^n$ there exists a cycle with positive symplectic area. Thus the first Betti number of a manifold admitting a Lagrangian embedding into $\mathbf{C}^n$ does not vanish.

Besides the symplectic area there is another remarkable first cohomology class on Lagrangian submanifold of $\mathbf{C}^n$ – the Maslov class. Recently different restrictions on the Maslov class of Lagrangian embeddings were discovered (see [7], [5], [6]). It is natural to suppose that they also lead to an obstruction to existence of Lagrangian embeddings. In the present paper we construct such obstruction (see theorem 1 below). We use it in order to show that certain flat manifolds do not admit Lagrangian embeddings into $\mathbf{C}^n$ (see theorems 2, 3 below). Our approach is based on the Maslov class rigidity phenomenon for manifolds of non-positive curvature which was discovered by C. Viterbo (see [7] and section 2.2 below).

I am deeply grateful to I.H.E.S. for hospitality and to B. Bowditch, M. Gromov, M. Kapovich, J.-C. Sikorav and C. Vitero for numerous useful consultations and discussions.

* *Permanent address:* School of Mathemtical Sciences, Tel-Aviv University, Ramat-Aviv, Tel-Aviv 69978, Israel.

2. We describe our obstruction in the following:

Theorem 1. *Let X and Y be closed manifolds of dimension n admitting Riemannian metrics of non-positive sectional curvature. Suppose that there exists a Lagrangian embedding $f : Y \to T^*X$ such that the following conditions hold:*

(c1) *the Maslov class of f vanishes;*

(c2) *the composition of f with the natural projection $T^*X \to X$ induce monomorphism $A : H^1(X;\mathbf{Z}) \to H^1(Y;\mathbf{Z})$ which "expands" in the following sense:*

$$A(H^1(X;\mathbf{Z})) \subset mH^1(Y;\mathbf{Z})$$

where m is a positive integer such that $m > (n+1)/2$ if X is orientable and $m > n+1$ if X is non-orientable.

Then X does not admit a Lagrangian embedding into $\mathbf{C}^n$.

Here the Maslov class $\mu_f \in H^1(Y,\mathbf{Z})$ of a Lagrangian embedding $f : Y \to T^*X$ is defined as follows. If $\gamma : \mathbf{R}/\mathbf{Z} \to V$ is a loop then the vector bundle $V = (f \circ \gamma)^* T(T^*X)$ over $\mathbf{R}/\mathbf{Z}$ with fibers $V(t) = T_{f(\gamma(t))}T^*X$ has two Lagrangian subbundles $\Lambda_0(t) = T^*_{f(\gamma(t))}X$ and $\Lambda_1(t) = \text{range df}(\gamma(t))$. The integer $\langle \mu_f, \gamma \rangle$ is the relative Maslov index of Λ_0 and Λ_1.

Theorem 1 is proved in § 2.

3. We will now give an application of Theorem 1. Let $\mathbf{T}^n$ be a torus with coordinates $x_1, \ldots, x_n \pmod 1$.

Consider a map $\alpha : \mathbf{T}^n \to \mathbf{T}^n$, given by

$$(x_1, x_2, \ldots, x_{n-1}, x_n) \mapsto (x_1 + 1/(2n-2), x_3, \ldots, x_n, -x_2).$$

Note that α generates a group, say G, of transformations of the torus which is isomorphic to $\mathbf{Z}_{2n-2}$ and acts freely. Denote by K^n the quotient $\mathbf{T}^n/G$.

Theorem 2. *If $n \geq 3$ then K^n does not admit a Lagrangian embedding into $\mathbf{C}^n$.*

Remark 1. The manifold K^2 is the Klein bottle. Unfortunately our obstruction does not work in this case.

Remark 2. The manifold K^3 is described in the book [8], p. 117 (there it is called $\mathcal{G}_4$). One can check that K^3 is orientable. Thus there is no "soft" obstructions to Lagrangian embeddings of K^3 into $\mathbf{C}^3$ (see [1], 4.3.2.). This means that K^3 admits a Lagrangian immersion into $\mathbf{C}^3$ as well as a totally real embedding. Moreover $H^1(K^3, \mathbf{Z}) = \mathbf{Z}$. Thus the symplectic area control does not give an obstruction in this case.

The proof of Theorem 2 and its generalisation for certain flat manifolds one can find in §3.

2 Proof of theorem 1

1. For an element $\mu \in H^1(W, \mathbf{Z})$ denote by $\| \mu \|$ the non-negative integer such that

$$\mu = \| \mu \| \cdot (\text{primitive element}).$$

2. The Maslov class rigidity for manifolds of non-positive curvature is established in the following:

Proposition 1. (C. Viterbo, [7]). *Let W be an embedded closed Lagrangian submanifold of $\mathbf{C}^n$ with the Maslov class μ. Suppose that W admits a Riemannian metric of non-positive sectional curvature. Then the following estimate holds:*

$$1 \leq \| \mu \| \leq n+1$$

Remark 3. Note that the Maslov class of each cycle is even if W is orientable. Thus in the orientable case one can improve the previous estimate: $\| \mu \| \geq 2$.

3. In this section we prove Theorem 1. We use essentially a construction invented by F. Lalonde and J.-C. Sikorav in [4].

Suppose that the manifold X admits a Lagrangian embedding, say g into $\mathbf{C}^n$. Due to Weinstein's theorem one can extend g to a symplectic embedding

$$\varphi : \mathcal{U} \to \mathbf{C}^n$$

where $\mathcal{U}$ is a tubular neighbourhood of the zero section in T^*X. Let f' be a composition of f with a suitable homothety along the fibres of T^*X such that $f'(Y) \subset \mathcal{U}$. Denote by $h : Y \to \mathbf{C}^n$ the composition $\varphi \circ f'$. One can easily check that the following relation holds:

$$\mu_h = \mu_f + A\mu_g$$

Thus $\mu_h = A\mu_g$ due to condition c1. Recall that due to Proposition 1 and Remark 3 the following inequalities hold: $\| \mu_g \| \geq 1$ if X is non-orientable and $\| \mu_g \| \geq 2$ if X is orientable. Thus $\| \mu_h \| \geq m \| \mu_g \| > n+1$ due to condition c2. We get a contradiction with Proposition 1. □

3 An obstruction to Lagrangian embeddings of certain flat manifolds

1. Let W be a closed connected flat manifold of dimension n. Consider the torus $\mathbf{T}^n = \mathbf{R}^n/\mathbf{Z}^n$. It is well known (see [8], chapter 3) that W can be represented as a quotient $\mathbf{T}^n/G$ where G is a finite group with the following properties:

(p1) G acts freely on $\mathbf{T}^n$;

(p2) Each element $G \in G$ can be written as a transformation $g(x) = Qx + p$, where $x \in \mathbf{R}^n, Q \in \mathrm{SL}(n; \mathbf{Z}), p \in \mathbf{R}^n$.

Denote by $\tau : G \to \mathrm{SL}(n, \mathbf{Z})$ the projection $g \mapsto Q$.

(p3) τ^{-1} (id)=id.

Denote by θ the natural covering $\mathbf{T}^n \to W$. Using another language one can say that θ is the covering corresponding to the maximal normal free abelian subgroup of the fundamental group of W.

It is easy to check that the induced map $\theta^* : H^1(W; \mathbf{Z}) \to H^1(\mathbf{T}^n; \mathbf{Z})$ is a monomorphism (use transfer homomorphism, see chapter 3 written by E. Floyd in [2]). Denote by $m(W)$ the maximal positive integer such that

$$\theta^* : H^1(W, \mathbf{Z}) \subset m(W) \cdot H^1(\mathbf{T}^n, \mathbf{Z}).$$

We assume be definition that $m(W) = 0$ if $H^1(w, \mathbf{Z}) = 0$.

2. Theorem 3. *Let W be a closed connected flat manifold of dimension n. suppose that $m(W) > n+1$ if W is non-orientable and $m(W) > (n+1)/2$ if W is orientable. Then W does not admit a Lagrangian embedding into $\mathbf{C}^n$.*

Theorem 3 follows from Theorem 1 (see §1) and the following:

Proposition 2. *Let W be a closed connected flat manifold. Let $\theta : \mathbf{T}^n \to W$ be the covering constructed in section 1. Then there exists a Lagrangian embedding $f : \mathbf{T}^n \to T^*W$ such that the following conditions hold:*

(a) *the Maslov class of f vanishes;*

(b) *$\pi \circ f = \theta$, where π is the natural projection $T^*W \to W$.*

Proposition 2 is proved in the next section.

3. Denote by $\bar{\theta}$ the induced sympletic covering $T^*\mathbf{T}^n \to T^*W$. For a translation invariant 1-form λ on $\mathbf{T}^n$ denote by L_λ its graph. By definition L_λ lies in $T^*\mathbf{T}^n$. Moreover L_λ is Lagrangian since λ is closed.

Proposition 3. *There exists a translation invariant 1-form λ on $\mathbf{T}^n$ such that the map $\bar{\theta} : L_\lambda \to T^*W$ is an embedding.*

Proof of Proposition 2. Set $f = \bar{\theta} \circ \kappa^{-1}$ where $\kappa : L^\lambda \to \mathbf{T}^n$ is the natural projection and λ is chosen according to Proposition 3. Obviously $\pi \circ f = \theta$. Moreover the Maslov class of f vanishes since f is transversal to fibers of the bundle $T^*W \to W$. □

4. Proof of Proposition 3

Identify each tangent space to the torus with $\mathbf{R}^n$ and the set of translation invariant 1-forms on $\mathbf{T}^n$ with $(\mathbf{R}^n)^*$. In fact we have to prove that there exists covector $\xi \in (\mathbf{R}^n)^*$ such that $g^*\xi \neq \xi$ for every $g \in G\backslash\{\mathrm{id}\}$. This fact follows immediately from finiteness of G and properties p2, p3 of the group G (see Section 1). □

5. Proof of Theorem 2

(1) We claim that $m(K^n) = 2n - 2$. Indeed consider the universal covering $\mathbf{R}^n \to K^n$. The group of Deck transformation is generated by $\alpha, t_2, \ldots, t_n)$ where

$$t_j(x_1, \ldots, x_j, \ldots, x_n) = (x_1, \ldots, x_{j-1}, x_j + 1, x_j + 1, x_{j+1}, \ldots, x_n).$$

One can easily check that the following relations hold:

$$\begin{aligned} \alpha^{2n-2} &= t_1; \\ t_j\alpha &= \alpha t_{j+1} \quad \text{for } j = 2, \ldots, n-1; \\ t_n\alpha &= \alpha t_2^{-1}. \end{aligned}$$

Now one can show that $H_1(W, \mathbf{Z}) = \mathbf{Z} \oplus \mathbf{Z}_2$ and find the homomorphism $H_1(\mathbf{T}^n, \mathbf{Z}) \to H_1(W, \mathbf{Z})$. A direct computation gives us that $m(K^n) = 2n - 2$. Our claim is proved.

(2) Note that $2n - 2 > n + 1$ if $n > 3$. Thus K^n does not admit a Lagrangian embedding into $\mathbf{C}^n$ due to Theorem 3.

(3) The last observation is that K^3 is orientable and $m(K^3) = 4 > (3 + 1)/2$. Thus K^3 does not admit a Lagrangian embedding into $\mathbf{C}^n$ due to Theorem 3. □

References

[1] Audin, M., Fibres normaux d'immersions en dimension double, points doubles d'immersions Lagrangiennes et plongements totalement réels, Comment, Math. Helv. 63 (1988), 583–623.

[2] Borel, A., Seminar on transformation groups, Annals of Math. Studies, 46, Princeton University Press, 1960.

[3] Gromov, M., Pseudo-holomorphic curves in sympletci manifolds, Invent. Math., 82 (1985), 307–347.

[4] Lalonde, F. and Sikorav, J.-C., Sous-variétés Lagrangiennes et Lagrangiennes exactes des fibres contangentes, Comment. Math. Helv. (to appear).

[5] Polterovich, L., The Maslov class of Lagrange surfaces and Gromov's pseudoholomorphic curves, Trans. A.M.S. (to appear).

[6] Polterovich, L., Monotone Lagrange submanifolds of linear spaces and the Maslov class in cotangent bundles, Math. Zeitschrift (to appear).

[7] Viterbo, C., A new obstruction to embedding Lagrangian tori, Invent. Math., 100 (1990), 301–320.

[8] Wolf, J., Spaces of constant curvature, McGraw-Hill, Inc. 1967.

Phase Functions and Path Integrals

Joel Robbin† and Dietmar Salamon*
Mathematics Department†
University of Wisconsin
Madison, WI 53706 USA
and
Mathematics Institute*
University of Warwick
Coventry CV4 7AL Great Britain

March 12, 1993

This note is an introduction to our forthcoming paper [17]. There we show how to construct the metaplectic representation using Feynman path integrals. We were led to this by our attempts to understand Atiyah's explanation of topological quantum field theory in [2].

Like Feynman's original approach in [9] (see also [10]) an action integral plays the role of a phase function. Unlike Feynman, we use paths in phase space rather than configuration space and use the symplectic action integral rather than the (classical) Lagrangian integral. We eventually restrict to (inhomogeneous) quadratic Hamiltonians so that the finite dimensional approximation to the path integral is a Gaussian integral. In evaluating this Gaussian integral the signature of a quadratic form appears. This quadratic form is a discrete approximation to the second variation of the action integral.

For Lagrangians of the form kinetic energy minus potential energy, evaluated on curves in configuaration space, the index of the second variation is well-defined and, via the Morse Index Theorem,[1] related to the Maslov Index of the corresponding linear Hamiltonian system. The second variation of the symplectic action has both infinite index and infinite coindex. However, this second variation does have a well-defined signature via the aforementioned discrete approximation. This signature can be expressed in terms of the Maslov index of the corresponding linear Hamiltonian system. This is a symplectic analog of the Morse Index Theorem.

*This research has been partially supported by the SERC.
[1]See [8] for example.

Our treatment is motivated by the formal similarity between Feynman path integrals and the Fourier integral operators of Hörmander [13]. A key point of Hörmander's theory is that the phase function which appears in the expression for a Fourier integral operator can be replaced by another phase function which defines the same symplectic relation. This is how Feynman path integrals can be evaluated: one replaces the symplectic action by the generating function of the corresponding symplectic relation.

In sections 1 and 2 we review how to use phase functions to construct Lagrangian manifolds and symplectic relations. These generalities are motivated by the examples in section 3 where the phase function is the action integral.

Our topic has a vast literature. Our formula for the metaplectic representation appears in [16] where it is obtained by other arguments. Souriau [26] found an explicit solution for the quantum harmonic oscillator involving the Maslov index (thus correcting Feynman's original formula which is valid only for short times). Keller [14] first noticed the phase shift due to the Maslov index in Theorem 5.2 below and for this reason the Maslov index is sometimes called the *Keller-Maslov index*. Duistermaat's article [8] explains how to interpret the Morse index in terms of the Maslov index but in the situation studied here the Morse index is undefined. The article [1] explains how Feynman and Dirac [7] were motivated by using the method of stationary phase to obtain classical mechanics as the limit (as $\hbar \to 0$) of quantum mechanics. Daubechies and Klauder [5] (see also [6]) have formulated a theory of path integrals on phase space where the Hamiltonian function can be any polynomial. They remark that the 'time slicing' construction used by Feynman does not generalize. However, our Hamiltonians are at worst quadratic and Feynman's original method is adequate.

1 Lagrangian manifolds

A **variational family** is a pair

$$\pi : P \to X, \qquad \phi : P \to \mathbf{R}$$

consisting of a surjective submersion π between manifolds P and X, and a smooth function ϕ on P. Each choice of $x \in X$ determines a constrained variational problem

$$\textit{extremize } \phi(c) \quad \textit{subject to} \quad \pi(c) = x.$$

We call a critical point of $\phi|\pi^{-1}(x)$ a **fiber critical point** of ϕ. Denote by $C(\pi, \phi) \subset P$ the set of all fiber critical points $c \in P$ of ϕ. At a fiber critical point c the differential $d\phi(c)$ vanishes on the vertical tangent space

$\ker d\pi(c) = T_c\pi^{-1}(x)$. This means that there exists a **Lagrange multiplier** $y \in T_x^*X$ such that

$$d\phi(c)\gamma = \langle y, d\pi(c)\gamma\rangle \qquad (1)$$

for every $\gamma \in T_cP$. The Lagrange multiplier y is uniquely determined since $d\pi(c)$ is surjective. Consider the map

$$\lambda_{\pi,\phi} : C(\pi, \phi) \to T^*X$$

defined by $\lambda_{\pi,\phi}(c) = (x, y)$ where $x = \pi(c)$ and y is given by (1). Denote its image by

$$\Lambda(\pi, \phi) = \{(x, y) \in T^*X : \exists c \in \pi^{-1}(x) \text{ such that } (1)\}.$$

If $\Lambda = \Lambda(\pi, \phi)$ is a set of this form then we say that (π, ϕ) *defines* Λ and call ϕ a **phase function** for Λ with respect to π. An extreme case is where $P = X$ and $\pi : P \to X$ is the identity so that ϕ is a function on X and $\Lambda = \mathrm{Gr}(d\phi)$. In this case ϕ is called a **generating function** for Λ.

Let $N_\pi \subset T^*P$ denote the **fiber normal bundle:**

$$N_\pi = \{(c, b) \in T^*P : b \in \ker(d\pi(c))^\perp\}.$$

This is a co-isotropic submanifold of T^*P and its symplectic quotient is T^*X. In the lingo of [28] $\Lambda(\pi, \phi)$ is the **symplectic reduction** of the Lagrangian manifold $\mathrm{Gr}(d\phi)$. Recall that two submanifolds G and N of a manifold W are said to intersect

- **transversally** in W iff $T_zW = T_zG + T_zN$, and

- **cleanly** in W iff $T_z(G \cap N) = T_zG \cap T_zN$

for $z \in G \cap N$. (For clean interesection impose the condition that the intersection $G \cap N$ be a submanifold; for transverse intersections this follows. A transversal intersection is automatically clean.) We call the variational family (π, ϕ) **tranversal** (resp. **clean**) iff $\mathrm{Gr}(d\phi)$ intersects N_π transversally (resp. cleanly) in P.

Proposition 1.1 *If (π, ϕ) is a clean variational family, then $\Lambda(\pi, \phi)$ is an immersed Lagrangian manifold. If (π, ϕ) is a transversal variational family, then $\lambda_{\pi,\phi}$ is a Lagrangian immersion.*[2]

Proof: Localize and choose co-ordinates so that $P = X \times U$ where $X \subset \mathbf{R}^n$ and $U \subset \mathbf{R}^N$ and that $\pi : X \times U \to X$ is the projection

$$\pi(c) = x, \qquad c = (x, u).$$

[2] In the transversal case this is due to Hörmander [13]. The clean case is folklore.

Then $C(\pi,\phi)$ is defined by the equation $\partial_u\phi = 0$ and $\Lambda(\pi,\phi)$ is defined by eliminating u from the equations

$$\partial_u\phi = 0, \qquad y = \partial_x\phi.$$

The family is transversal iff 0 is a regular value of $\partial_u\phi$ and clean iff $C(\pi,\phi)$ is a manifold and the tangent space at a point $c = (x,u) \in C(\pi,\phi)$ is given by

$$T_cC(\pi,\phi) = \{(\xi,v) \in \mathbf{R}^n \times \mathbf{R}^N : \partial_u\partial_x\phi(x,u)\xi + \partial_u\partial_u\phi(x,u)v = 0\}.$$

To prove Proposition 1.1 fix $c = (x,u) \in C(\pi,\phi)$ and apply the next lemma with $A = \partial_x\partial_x\phi(x,u)$, $B = \partial_x\partial_u\phi(x,u)$, $B^T = \partial_u\partial_x\phi(x,u)$, $C = \partial_u\partial_u\phi(x,u)$, $d\partial_u\phi(x,u) = (B^T, C)$, $T = T_cC(\pi,\phi)$, $\ell = d\lambda_{\pi,\phi}(c)$.

Lemma 1.2 *Suppose that $A \in \mathbf{R}^{n\times n}$ and $C \in \mathbf{R}^{N\times N}$ are symmetric and that $B \in \mathbf{R}^{n\times N}$. Let*

$$T = \{(\xi,v) : B^T\xi + Cv = 0\} \subset \mathbf{R}^n \times \mathbf{R}^N$$

and $\ell : T \to \mathbf{R}^n \times \mathbf{R}^n$ by

$$\ell(\xi,v) = (\xi, A\xi + Bv).$$

Then $\ell(T) \subset \mathbf{R}^n \times \mathbf{R}^n$ is a Lagrangian subspace.

Proof: Note that $(\xi,\eta) \in \ell(T)$ iff the inhomogeneous system

$$\begin{aligned} Bv &= -A\xi + \eta \\ Cv &= -B^T\xi \end{aligned}$$

has a solution v. Hence $(\xi,\eta) \in \ell(T)$ iff

$$B^T\xi' + Cv' = 0 \Longrightarrow \langle \xi', -A\xi + \eta\rangle + \langle v', -B^T\xi\rangle = 0.$$

On the other hand $(\xi,\eta) \in \ell(T)^\omega$ iff

$$B^T\xi' + Cv' = 0 \Longrightarrow \langle \xi, A\xi' + Bv'\rangle - \langle \eta, \xi'\rangle = 0.$$

Hence $\ell(T) = \ell(T)^\omega$. □

At a critical point of a function, the Hessian is a well-defined quadratic form on the tangent space; at a fiber critical point c the **vertical Hessian** is defined on the vertical tangent space. By Taylor's theorem the vertical Hessian Φ is characterized by the equation

$$\phi(c+\gamma) = \phi(c) + \tfrac{1}{2}\Phi(\gamma) + O(\|\gamma\|^3)$$

for $d\pi(c)\gamma = 0$. Here $c + \gamma = \exp_c(\gamma) \in P$ where exp is an exponential map which carries vertical tangent vectors to the fiber; Φ is independent of the choice.

Proposition 1.3 *Assume that (π, ϕ) is a transversal variational family, and $c \in C(\pi, \phi)$. Then Φ is non-degenerate iff $d\pi(c) : T_cC(\pi, \phi) \to T_xX$ is an isomorphism.*

Proof: In local coordinates $c = (x, u)$ the tangent space $T_cC(\pi, \phi)$ is defined by the equation

$$\partial_u\partial_x\phi(c)\hat{x} + \partial_u\partial_u\phi(c)\hat{u} = 0.$$

Hence the projection $(\hat{x}, \hat{u}) \mapsto \hat{x}$ is an isomorphism on $T_cC(\pi, \phi)$ if and only if the Hessian matrix $\Phi = \partial_u\partial_u\phi(c)$ is invertible. □

Definition 1.4 A fiber critical point $c \in C(\pi, \phi)$ is called **nondegenerate** if the fiber Hessian Φ is nondegenerate. This implies that

(1) $\mathrm{Gr}(d\phi)$ and N_π intersect transversally at c,

(2) $d\pi(c) : T_cC(\pi, \phi) \to T_xX$ is invertible, and

(3) $T_x^*X \cap T_{(x,y)}\Lambda(\pi, \phi) = 0$.

In (3) $T_x^*X \subset T_{(x,y)}T^*X$ is the vertical tangent space of the cotangent bundle. The inverse

$$G = d\pi(c)^{-1} : T_xX \to T_cC(\pi, \phi)$$

of the projection in (2) is called the **Green's function** of ϕ at c. By the implicit function theorem $\pi|C(\pi, \phi)$ is a diffeomorphism in a neighborhood of $x = \pi(c)$: we denote the local inverse by g and call it the **nonlinear Green's function**. Clearly

$$G = dg(x).$$

Remark 1.5 If the projection $\pi : C(\pi, \phi) \to X$ is a diffeomorphism, there is a global nonlinear Green's function $g : X \to P$. Its image is the set $g(X) = C(\pi, \phi)$ of fiber critical points. In this case $\Lambda(\pi, \phi)$ admits a generating function $\phi \circ g : X \to \mathbf{R}$.

2 Symplectic relations

A **symplectic relation** from a symplectic manifold M_0 to a symplectic manifold M_1 is a Lagrangian submanifold of $\bar{M}_0 \times M_1$. The bar indicates that the sign of the symplectic form in the first factor has been reversed. We do not assume that M_0 and M_1 have the same dimension. A clean variational family (π_{01}, ϕ_{01}) with $\pi_{01} : P_{01} \to X_0 \times X_1$ determines a symplectic relation from T^*X_0 to T^*X_1. The construction is as in the previous section but with a sign change. As before at a critical point c there is a Lagrange multiplier $(-y_0, y_1) \in T_{x_0}^*X_0 \times T_{x_1}^*X_1$ such that

$$d\phi_{01}(c)\gamma = \langle y_1, \xi_1\rangle - \langle y_0, \xi_0\rangle \tag{2}$$

for every $\gamma \in T_cP$ where $d\pi(c)\gamma = (\xi_0, \xi_1)$. Denote by

$$R_{01} = \{(x_0, y_0, x_1, y_1) \in T^*X_0 \times T^*X_1 : \exists c \in \pi_{01}^{-1}(x_0, x_1) \text{ such that (2)}\}$$

the relation induced by ϕ_{01}. We call ϕ_{01} a **phase function** for R_{01} and call R_{01} the relation defined by (π_{01}, ϕ_{01}). If (π_{01}, ϕ_{01}) is a clean variational family then R_{01} is a symplectic relation. In the extreme case where $P_{01} = X_0 \times X_1$ and π_{01} is the identity we call ϕ_{01} a **generating function** for R_{01}. If $\psi_{10} : T^*X_0 \to T^*X_1$ is a symplectomorphism then its graph $R_{01} = \mathrm{Gr}(\psi_{10})$ is a symplectic relation. In this case we call ϕ_{01} a phase function (or generating function) for ψ_{10}.

The **composition** of two relations $R_{01} \subset \bar{M}_0 \times M_1$ and $R_{12} \subset \bar{M}_1 \times M_2$ is the relation

$$R_{01}\#R_{12} = \{(z_0, z_2) : (z_0, z_1) \in R_{01}, (z_1, z_2) \in R_{12}\}.$$

Note that by our conventions the graph operation is a contravariant functor:

$$\mathrm{Gr}(\psi_{21} \circ \psi_{10}) = \mathrm{Gr}(\psi_{10})\#\mathrm{Gr}(\psi_{21}).$$

A Lagrangian manifold is a special case of a symplectic relation (take M_0 to be a point) and we have the formula

$$\psi_{21}(\Lambda_1) = \Lambda_1\#\mathrm{Gr}(\psi_{21})$$

if $\Lambda_1 \subset M_1$ is Lagrangian.

Let (π_{01}, ϕ_{01}) and (π_{12}, ϕ_{12}) be variational families with

$$\pi_{01} : P_{01} \to X_0 \times X_1, \qquad \pi_{12} : P_{12} \to X_1 \times X_2.$$

Define another variational family (π_{02}, ϕ_{02}) by

$$P_{02} = \{(c_{01}, c_{12}) \in P_{01} \times P_{12} : \pi_{01}(c_{01}) = (x_0, x_1), \pi_{12}(c_{12}) = (x_1, x_2)\}$$

with $\pi_{02} : P_{02} \to X_0 \times X_2$ given by

$$\pi_{02}(c_{01}, c_{12}) = (x_0, x_2)$$

and $\phi_{02} : P_{02} \to \mathbf{R}$ by

$$\phi_{02}(c_{01}, c_{12}) = \phi_{01}(c_{01}) + \phi_{12}(c_{12}).$$

Let $R_{jk} \subset T^*X_j \times T^*X_k$ be the relation defined by (π_{jk}, ϕ_{jk}).

Proposition 2.1 $R_{02} = R_{01}\#R_{12}$.

Proof: Fix $c_{02} = (c_{01}, c_{12}) \in P_{02}$ and

$$\gamma_{02} = (\gamma_{01}, \gamma_{12}) \in T_{c_{01}}P_{01} \times T_{c_{12}}P_{12}$$

and introduce the notations $(x_j, x_k) = \pi_{jk}(c_{jk})$ and

$$d\pi_{01}(c_{01})\gamma_{01} = (\xi_0, \xi_1), \quad d\pi_{12}(c_{12})\gamma_{12} = (\xi_1', \xi_2).$$

The tangent space to P_{02} at c_{02} is the set of all pairs γ_{02} such that $\xi_1 = \xi_1'$. The tangent space to the fiber of P_{02} is defined by the three constraints $\xi_0 = 0$, $\xi_1 = \xi_1'$, and $\xi_2 = 0$. Then c_{02} is a fiber critical point iff there are Lagrange multipliers $y_0 \in T^*_{x_0}X_0$, $y_1 \in T^*_{x_1}X_1$, and $y_2 \in T^*_{x_2}X_2$ such that

$$d\phi_{01}(c_{01})\gamma_{01} + d\phi_{12}(c_{12})\gamma_{12} = \langle y_2, \xi_2\rangle + \langle y_1, \xi_1 - \xi_1'\rangle - \langle y_0, \xi_0\rangle$$

for *all* $(\gamma_{01}, \gamma_{12}) \in T_{c_{01}}P_{01} \times T_{c_{12}}P_{12}$. In particular

$$d\phi_{01}(c_{01})\gamma_{01} = \langle y_1, \xi_1\rangle - \langle y_0, \xi_0\rangle$$

(take $\gamma_{12} = 0$) and

$$d\phi_{12}(c_{12})\gamma_{12} = \langle y_2, \xi_2\rangle - \langle y_1, \xi_1'\rangle$$

(take $\gamma_{01} = 0$). This shows that $R_{02} \subset R_{01}\#R_{12}$. For the reverse inclusion argue backwards. □

The composition operation has the following interpretation. For $x \in X$ we identify T^*_xX with the vertical tangent space

$$V = T^*_xX \subset T_{(x,y)}T^*X.$$

It is a Lagrangian submanifold of T^*X. Now fix $x_0 \in X_0$ and $x_2 \in X_2$. The goal is to find all pairs $y_0 \in T^*_{x_0}X_0$ and $y_2 \in T^*_{x_2}X_2$ such that

$$(x_0, y_0, x_2, y_2) \in R_{02}.$$

These points correspond to Lagrangian intersections of the image of $T^*_{x_0}X_0$ under R_{01} with the preimage of $T^*_{x_2}X_2$ under R_{12}. For every point

$$(x_1, y_1) \in \left(T^*_{x_0}X_0\#R_{01}\right) \cap \left(R_{12}\#T^*_{x_2}X_2\right) \tag{3}$$

in this intersection there exist points $y_0 \in T^*_{x_0}X_0$ and $y_2 \in T^*_{x_2}X_2$ such that $(x_0, y_0, x_1, y_1) \in R_{01}$ and $(x_1, y_1, x_2, y_2) \in R_{12}$ and hence $(x_0, y_0, x_2, y_2) \in R_{02}$ as required. In the special case where R_{jk} is the graph of a symplectomorphism each intersection point (x_1, y_1) determines y_0 and y_2 uniquely. Thus given x_0 and x_2 there is a one-to-one correspondence of Lagrangian intersection points in T^*X_1 with points in $R_{02} \cap \left(T^*_{x_0}X_0 \times T^*_{x_2}X_2\right)$.

Now assume that R_{jk} is a manifold and fix

$$(x_j, y_j, x_k, y_k) \in R_{jk}.$$

Let $c_{02} = (c_{01}, c_{12}) \in P_{02}$ be the corresponding fiber critical point so that

$$\pi_{jk}(c_{jk}) = (x_j, x_k)$$

With this notation we have (3). In the tangent space $T_{(x_1,y_1)}T^*X_1$ there are three interesting Lagrangian subspaces:

$$L_0 = T_{(x_1,y_1)}\left(T^*_{x_0}X_0 \# R_{01}\right), \qquad L_2 = T_{(x_1,y_1)}\left(R_{12} \# T^*_{x_2}X_2\right),$$

and the vertical space

$$V = T^*_{x_1}X_1.$$

We assume that L_0 and L_2 are transverse to V. Then the pair (L_0, L_2) determines a quadratic form on the quotient space $T_{x_1}X_1 = T_{(x_1,y_1)}T^*X_1/T^*_{x_1}X_1$. To define it choose a Lagrangian complement H of V:

$$T_{(x_1,y_1)}T^*X_1 = H \oplus V.$$

Identify H with the dual space V^* using the symplectic form on $T_{(x_1,y_1)}T^*X_1$. Since L_0 and L_2 are transverse to V there exist quadratic forms $Q_j : H \to V = H^*$ such that

$$L_0 = \mathrm{Gr}(Q_0), \qquad L_2 = \mathrm{Gr}(Q_2).$$

There is a natural projection (isomorphism) $H \to T_{x_1}X_1$ and the difference

$$Q = Q_0 - Q_2 \tag{4}$$

descends to a quadratic form $T_{x_1}X_1 \to T^*_{x_1}X_1$. The result is independent of the choice of H. We abuse language and identify Q with a form on $T_{x_1}X_1$. The form Q is called the **composition form** of (L_0, L_2). Denote by Φ_{jk} the fiber Hessian at c_{jk}. Assume that the fiber critical points $c_{jk} \in C_{\phi_{jk}}$ are nondegenerate and denote the Green's function by $G_{jk} : T_{x_j}X_j \times T_{x_k}X_k \to T_{c_{jk}}C_{\phi_{jk}}$. Define

$$G_0\xi_1 = G_{01}(0, \xi_1), \qquad G_2\xi_1 = G_{12}(\xi_1, 0).$$

Proposition 2.2 *The linear map*

$$T_{c_{02}}\pi_{02}^{-1}(x_0, x_2) \to T_{c_{01}}\pi_{01}^{-1}(x_0, x_1) \times T_{x_1}X_1 \times T_{c_{12}}\pi_{12}^{-1}(x_1, x_2)$$

given by

$$\gamma_{02} = (\gamma_{01}, \gamma_{12}) \mapsto (\gamma_{01} - G_0\xi_1, \xi_1, \gamma_{12} - G_2\xi_1)$$

where $\xi_1 = d\pi_1(c_{01})\gamma_{01}$ *is an isomorphism. Moreover,*

$$\Phi_{02}(\gamma_{02}) = \Phi_{01}(\gamma_{01} - G_0\xi_1) + \Phi_{12}(\gamma_{12} - G_2\xi_1) + \langle Q\xi_1, \xi_1 \rangle. \tag{5}$$

Proof: In local coordinates we have

$$\phi_{02}(x_0, u_{01}, x_1, u_{12}, x_2) = \phi_{01}(x_0, u_{01}, x_1) + \phi_{12}(x_1, u_{12}, x_2).$$

The relation R_{01} is defined by eliminating u_{01} from the nonlinear system

$$\begin{aligned} -y_0 &= \partial_{x_0}\phi_{01}(x_0, u_{01}, x_1) \\ y_1 &= \partial_{x_1}\phi_{01}(x_0, u_{01}, x_1) \\ 0 &= \partial_{u_{01}}\phi_{01}(x_0, u_{01}, x_1). \end{aligned}$$

The last equation defines the set $C_{\phi_{01}}$. The Lagrangian manifold $T^*_{x_0}X_0\#R_{01}$ is defined by fixing x_0 and eliminating u_{01} from the last two. The tangent space $T_{c_{01}}R_{01}$ is defined by eliminating v_{01} from

$$\begin{aligned} -\eta_0 &= (\partial_{x_0}\partial_{x_0}\phi_{01})\xi_0 + (\partial_{x_0}\partial_{x_1}\phi_{01})\xi_1 + (\partial_{x_0}\partial_{u_{01}}\phi_{01})v_{01} \\ \eta_1 &= (\partial_{x_1}\partial_{x_0}\phi_{01})\xi_0 + (\partial_{x_1}\partial_{x_1}\phi_{01})\xi_1 + (\partial_{x_1}\partial_{u_{01}}\phi_{01})v_{01} \\ 0 &= (\partial_{u_{01}}\partial_{x_0}\phi_{01})\xi_0 + (\partial_{u_{01}}\partial_{x_1}\phi_{01})\xi_1 + (\partial^2_{u_{01}}\phi_{01})v_{01}. \end{aligned}$$

The last equation defines the tangent space to $C_{\phi_{01}}$ and the Green's function G_{01} is given by solving for v_{01}. Thus

$$G_0\xi_1 = (0, \Gamma_0\xi_1, \xi_1), \qquad \Gamma_0 = -\left(\partial^2_{u_{01}}\phi_{01}\right)^{-1}\partial_{u_{01}}\partial_{x_1}\phi_{01}.$$

To define the Lagrangian subspace $L_0 = T_{(x_1,y_1)}\left(T^*_{x_0}X_0\#R_{01}\right)$, set $\xi_0 = 0$ and eliminate v_{01} in the last two equations. Hence L_0 is the graph of the symmetric matrix

$$Q_0 = (\partial^2_{x_1}\phi_{01}) + (\partial_{x_1}\partial_{u_{01}}\phi_{01})\Gamma_0.$$

Similarly, L_2 is the graph of Q_2 where

$$-Q_2 = (\partial^2_{x_1}\phi_{12}) + (\partial_{x_1}\partial_{u_{12}}\phi_{12})\Gamma_2, \qquad \Gamma_2 = -\left(\partial^2_{u_{12}}\phi_{12}\right)^{-1}\partial_{u_{12}}\partial_{x_1}\phi_{12}.$$

Now the tangent vector γ_{02} is in local co-ordinates given by

$$\gamma_{02} = (0, v_{01}, \xi_1, v_{12}, 0)$$

with $\gamma_{01} = (0, v_{01}, \xi_1)$ and $\gamma_{12} = (\xi_1, v_{12}, 0)$. Hence

$$\gamma_{01} - G_0\xi_1 = (0, v_{01} - \Gamma_0\xi_1, 0)$$

and

$$\begin{aligned} \Phi_{01}(\gamma_{01} - G_0\xi_1) &= \langle(\partial^2_{u_{01}}\phi_{01})v_{01}, v_{01}\rangle + 2\langle(\partial_{u_{01}}\partial_{x_1}\phi_{01})\xi_1, v_{01}\rangle \\ &\quad -\langle\xi_1, (\partial_{x_1}\partial_{u_{01}}\phi_{01})\Gamma_0\xi_1\rangle. \end{aligned}$$

Similarly,

$$\begin{aligned} \Phi_{12}(\gamma_{12} - G_2\xi_1) &= \langle(\partial^2_{u_{12}}\phi_{12})v_{12}, v_{12}\rangle + 2\langle(\partial_{u_{12}}\partial_{x_1}\phi_{12})\xi_1, v_{12}\rangle \\ &\quad -\langle\xi_1, (\partial_{x_1}\partial_{u_{12}}\phi_{12})\Gamma_2\xi_1\rangle. \end{aligned}$$

Subtract these two identities from the Hessian

$$\begin{aligned}\Phi_{02}(\gamma_{02}) &= \langle(\partial^2_{u_{01}}\phi_{01})v_{01}, v_{01}\rangle + \langle(\partial^2_{u_{12}}\phi_{12})v_{12}, v_{12}\rangle \\ &\quad +2\langle(\partial_{u_{01}\partial_{x_1}}\phi_{01})\xi_1, v_{01}\rangle + 2\langle(\partial_{u_{12}\partial_{x_1}}\phi_{12})\xi_1, v_{12}\rangle \\ &\quad +\langle(\partial^2_{x_1}\phi_{01})\xi_1, \xi_1\rangle + \langle(\partial^2_{x_1}\phi_{12})\xi_1, \xi_1\rangle.\end{aligned}$$

and use the above formulae for Q_0 and Q_2 to prove the proposition. □

3 Examples

In our examples, except for the last one, the space P is a space of paths and the work of the previous section can be interpreted formally. Alternatively one can introduce Hilbert manifold structures and generalize the previous work to the infinite dimensional case.

Example 3.1 Let X be a manifold and $L : \mathbf{R} \times TX \to \mathbf{R}$ be a function. Fix $t_0, t_1 \in \mathbf{R}$ and take $P_{01} = P(t_0, t_1)$ to be the space of all paths $c : [t_0, t_1] \to X$. Take $X_0 = X_1 = X$ and define the projection $\pi = \pi_{01}$ by evaluation at the endpoints:

$$\pi(c) = (c(t_0), c(t_1)).$$

The phase function $\phi = \phi_{01}$ is the Lagrangian action integral

$$\phi(c) = \int_{t_0}^{t_1} L(t, c(t), \dot{c}(t))\, dt$$

from the calculus of variations. A tangent vector $\gamma \in T_cP_{01}$ is a vectorfield along c and it is vertical iff it vanishes at the endpoints. A curve c is a fiber critical point iff it satisfies the Euler-Lagrange equations

$$\dot{y} = \partial_x L, \qquad y = \partial_{\dot{x}} L$$

where $L = L(t, x, \dot{x})$. The right side of Equation (2) consists of the boundary terms which result from the integration by parts in the derivation of the Euler-Lagrange equations. If t_1 is replaced by t and allowed to vary between t_0 and t_1, the restriction $c|[t_0, t]$ is a fiber critical point of the new problem on $P(t_0, t)$ and the Lagrange mutiplier $y(t)$ is the $y(t)$ which appears in the Euler-Lagrange equations. When the Legendre transformation

$$TX \to T^*X : (x, \dot{x}) \mapsto (x, y), \qquad y = \partial_{\dot{x}} L$$

is a diffeomorphism, the Euler-Lagrange equations take the form of Hamilton's equations

$$\dot{x} = \partial_y H, \qquad \dot{y} = -\partial_x H$$

where $H : T^*X \to \mathbf{R}$ is defined by eliminating $\dot{x}$ (via the Legendre transformation) from

$$H(t, x, y) = \langle y, \dot{x}\rangle - L(t, x, \dot{x}).$$

Example 3.2 Specialize the previous example by taking X a Riemannian manifold with energy function $L(x,\dot{x}) = \frac{1}{2}|\dot{x}|_x^2$. Then the fiber critical points are the *geodesics*. The Hessian $\Phi = \Phi_{01}$ is defined by

$$\Phi(\gamma) = \int_{t_0}^{t_1} \langle (W\gamma)(t), \gamma(t) \rangle_{c(t)} \, dt$$

where the operator W is given by

$$W\gamma = \frac{D^2\gamma}{dt^2} + R(\dot{c},\gamma)\dot{c}.$$

Here D/dt denotes the covariant derivative along c and $\gamma(t) \in T_{c(t)}X$. The linear second order differential equation $W\gamma = 0$ is the *Jacobi equation.* The geodesic is non-degenerate when its end points are not *conjugate* and the Green's function G has its usual interpretation of assigning to the boundary conditions $\xi_0 \in T_{x(t_0)}X$ and $\xi_1 \in T_{c(t_1)}X$ the unique solution $\gamma = G(\xi_0, \xi_1)$ of the boundary value problem

$$W\gamma = 0, \qquad \gamma(t_0) = \xi_0, \qquad \gamma(t_1) = \xi_1.$$

Example 3.3 Again take $X_0 = X_1 = X$ but now take $\mathcal{P}(t_0, t_1)$ the space of all curves

$$c = (x,y) : [t_0, t_1] \to T^*X$$

with projection $\pi = \pi_{t_0 t_1} : \mathcal{P}(t_0, t_1) \to X \times X$ given by

$$\pi(c) = (x(t_0), x(t_1)).$$

A time-dependent Hamiltonian $H : \mathbf{R} \times T^*X \to \mathbf{R}$ determines a one-form σ_H on $\mathbf{R} \times T^*M$ via

$$\sigma_H = \langle y, dx \rangle - H\,dt$$

called the **action form** of H. Define the phase function $\phi = \phi_{t_0 t_1}$ to be the integral

$$\phi(c) = \int_c \sigma_H$$

of the action form along c is called the **action integral**. A more explicit formula is

$$\phi(c) = \int_{t_0}^{t_1} \Big(\langle y, \dot{x} \rangle - H(t,x,y) \Big) \, dt.$$

where $c(t) = (x(t), y(t))$. As before the vertical critical points of (π, ϕ) are the solutions of the Euler-Lagrange equations of this functional. They are Hamilton's equations

$$\dot{x} = \partial_y H, \qquad \dot{y} = -\partial_x H.$$

The Lagrange multipliers in Equation (2) are given by $y_0 = y(t_0)$ and $y_1 = y(t_1)$. Assume for simplicity that these differential equations are complete. Then the symplectic relation determined by $(\pi_{t_0t_1}, \phi_{t_0t_1})$ is $\mathrm{Gr}(\psi_{t_0}^{t_1})$ where $t \mapsto \psi_{t_0}^{t}(x,y)$ is the solution of Hamilton's equations satisfying the initial condition $\psi_{t_0}^{t_0}(x,y) = (x,y)$. These symplectomorphisms define an **evolution system** meaning that

$$\psi_{t_1}^{t_2} \circ \psi_{t_0}^{t_1} = \psi_{t_0}^{t_2}, \qquad \psi_{t_0}^{t_0} = 1\!\!1.$$

Remark 3.4 Assume that c_0 is a nondegenerate fiber critical point in the previous example. Then there is a local nonlinear Green's function which assigns to every point (x_0, x_1) near $\pi(c_0)$ the unique solution $c(t) = (x(t), y(t))$ near c_0 of Hamilton's equation which satisfies $x(t_0) = x_0$ and $x(t_1) = x_1$. Let

$$S_{t_0t_1}(x_0, x_1) = \int_{t_0}^{t_1} \Big(\langle y, \dot{x} \rangle - H(t, x, y) \Big)\, dt$$

denote the action integral of this solution. This is a local generating function of the symplectomorphism $\psi_{t_0}^{t_1}$.

Remark 3.5 Now fix t_0 and x_0. Then the generating function $S(t,x) = S_{t_0t}(x_0, x)$ satisfies the **Hamilton-Jacobi equation**

$$\partial_t S + H(t, x, \partial_x S) = 0.$$

To prove this differentiate the identity

$$S(t, x(t)) = \int_{t_0}^{t} \Big(\langle y(s), \dot{x}(s) \rangle - H(s, x(s), y(s)) \Big)\, ds$$

with respect to t and use $y = \partial S / \partial x$.

Example 3.6 By a **partition** of **R** we mean an infinite discrete subset $\mathcal{T} \subset \mathbf{R}$ extending to infinity in both directions. Every $t \in \mathcal{T}$ has a unique **successor** $t^+ \in \mathcal{T}$ and **predecessor** $t^- \in \mathcal{T}$ defined by

$$t^- = \sup \mathcal{T} \cap (-\infty, t), \qquad t^+ = \inf \mathcal{T} \cap (t, \infty).$$

Denote the **mesh** of $\mathcal{T}$ by

$$|\mathcal{T}| = \sup_{t \in \mathcal{T}} |t^+ - t|.$$

Now take $X = \mathbf{R}^n$, $T^*X = \mathbf{R}^n \times \mathbf{R}^n$, and fix a time dependent Hamiltonian $H(t, x, y)$ and a partition $\mathcal{T}$. Let $t_0, t_1 \in \mathcal{T}$ with $t_0 < t_1$. Define the space

$$\mathcal{P}^{\mathcal{T}}(t_0, t_1) = \{c = (x, y) \,:\, x : \mathcal{T} \cap [t_0, t_1] \to \mathbf{R}^n,\ y : \mathcal{T} \cap [t_0, t_1) \to \mathbf{R}^n\}$$

of discrete paths in $\mathbf{R}^{2n}$. These discrete paths are finite sequences of length N and $N-1$ where N is the cardinality of the finite set $\mathcal{T} \cap [t_0, t_1]$. The projection $\pi = \pi_{t_0 t_1} : \mathcal{P}^{\mathcal{T}}(t_0, t_1) \to \mathbf{R}^n \times \mathbf{R}^n$ is given by

$$\pi(c) = (x(t_0), x(t_1)).$$

The **discrete action functional** $\phi^{\mathcal{T}} : \mathcal{P}^{\mathcal{T}}(t_0, t_1) \to \mathbf{R}$ is defined by

$$\phi^{\mathcal{T}}(c) = \sum_{\substack{t \in \mathcal{T} \\ t_0 \leq t < t_1}} \Big(\langle y(t), x(t^+) - x(t) \rangle - H(t, x(t^+), y(t))(t^+ - t) \Big).$$

The vertical critical points of $(\pi^{\mathcal{T}}, \phi^{\mathcal{T}})$ are the solutions of the **discrete Hamiltonian equations**

$$\begin{aligned} x(t^+) - x(t) &= \partial_y H(t, x(t^+), y(t))(t^+ - t) \\ y(t^+) - y(t) &= -\partial_x H(t, x(t^+), y(t))(t^+ - t). \end{aligned} \tag{6}$$

These equations define $(x(t^+), y(t^+))$ implicitly in terms of $(x(t), y(t))$. Let (y_0, y_1) be the Lagrange multipliers in equation (2). Then $y_0 = y(t_0)$ and $y_1 = y(t_1)$ is defined by equation (6).

Remark 3.7 Assume that $c_0^{\mathcal{T}}$ is a nondegenerate fiber critical point in the previous example and define a discrete generating function $S_{t_0 t_1}^{\mathcal{T}}(x_0, x_1)$ as in the continuous case. Now fix a time interval $[t_0, t_1]$, let the mesh $|\mathcal{T}|$ of the partition go to zero, and let $c_0^{\mathcal{T}}$ converge to a nondegenerate fiber critical point of the continuous variational problem. Then we have a limit

$$\lim_{|\mathcal{T}| \to 0} S^{\mathcal{T}} = S.$$

This follows from standard arguments in the discretization of ordinary differential equations.

4 Hessians

We now compare the fiber Hessians in Examples 3.3 and 3.6. We take $X = \mathbf{R}^n$ and $T^*X = \mathbf{R}^n \times \mathbf{R}^n$. To simplify the notation we assume that for each t the Hamiltonian $H(t, x, y)$ is homogeneous quadratic in (x, y):

$$H(t, x, y) = \tfrac{1}{2}\langle H_{xx}(t)x, x \rangle + \langle H_{yx}(t)x, y \rangle + \tfrac{1}{2}\langle H_{yy}(t)y, y \rangle$$

where $H_{xx}(t)$, $H_{yx}(t)$, $H_{yy}(t)$ are $n \times n$ matrices with H_{xx} and H_{yy} symmetric. We abbreviate $H_{xy} = H_{yx}^T$.

Continuous time

In the continuous time case the fiber Hessian $\Phi = \Phi_{t_0t_1}$ is given by

$$\Phi(\gamma) = \langle W\gamma, \gamma\rangle$$

for $\gamma = (\xi, \eta) : [t_0, t_1] \to \mathbf{R}^n \times \mathbf{R}^n$ with $\xi(t_0) = \xi(t_1) = 0$. The inner product on the right is the L^2 inner product and the fiber Hessian $W = W_{t_0t_1}$ is the self-adjoint operator on $L^2([t_0, t_1], \mathbf{R}^n \times \mathbf{R}^n)$ with dense domain

$$\mathcal{W}(t_0, t_1) = H_0^1([t_0, t_1], \mathbf{R}^n) \times H^1([t_0, t_1], \mathbf{R}^n)$$

given by $W(\xi, \eta) = (u, v)$ where

$$u = -\dot{\eta} - H_{xx}\xi - H_{xy}\eta, \qquad v = \dot{\xi} - H_{yx}\xi - H_{yy}\eta,$$

We call $W = W_{t_0t_1}$ the **second variation** from t_0 to t_1. By Proposition 1.3 the Hessian is nondegenerate if and only if the symplectomorphism $\psi_{t_0}^{t_1}$ generated by H admits a generating function.

Discrete time

In discrete time we do the analogous thing. For $t_0, t_1 \in \mathcal{T}$ with $t_0 < t_1$ define

$$\mathcal{W}^{\mathcal{T}}(t_0, t_1) = \left\{\gamma = (\xi, \eta) \in \mathcal{P}^{\mathcal{T}}(t_0, t_1) : \xi(t_0) = \xi(t_1) = 0\right\}.$$

This is a Hilbert space with the approximate L^2-norm

$$\|\gamma\|_{\mathcal{T}}^2 = \sum_{t_0 \le t < t_1} \left(|\xi(t^+)|^2 + |\eta(t)|^2\right)(t^+ - t).$$

In this case the fiber Hessian is the (finite dimensional) symmetric operator $W^{\mathcal{T}} = W_{t_0t_1}^{\mathcal{T}} : \mathcal{W}^{\mathcal{T}}(t_0, t_1) \to \mathcal{W}^{\mathcal{T}}(t_0, t_1)$ given by $W^{\mathcal{T}}(\xi, \eta) = (u, v)$ where

$$u(t) = -\frac{\eta(t) - \eta(t^-)}{t - t^-} - H_{xx}(t^-)\xi(t) - H_{xy}(t^-)\eta(t^-),$$

$$v(t) = \frac{\xi(t^+) - \xi(t)}{t^+ - t} - H_{yx}(t)\xi(t^+) - H_{yy}(t)\eta(t).$$

We call $W^{\mathcal{T}} = W_{t_0t_1}^{\mathcal{T}}$ the **discrete second variation** from t_0 to t_1. By Proposition 1.3 the Hessian is nondegenerate if and only if the affine symplectomorphism $\phi_{t_0}^{t_1}$ generated by the discrete Hamiltonian equations admits a generating function.

Signature

The operator $W^{\mathcal{T}}$ is defined on a finite dimensional space and hence has a well defined index (number of negative eigenvalues), coindex (number of positive eigenvalues), signature (coindex minus index), and nullity. For the operator W the index and coindex are both infinite and hence the signature is undefined. However, the signature of $W^{\mathcal{T}}$ stabilizes when the mesh of the partition gets sufficiently small. It is related to the Maslov index $\mu(t_0, t_1, H)$ of the Hamiltonian flow (defined below) as follows.

Theorem 4.1 *Assume that $W_{t_0t_1}$ is non-degenerate. If the mesh $|\mathcal{T}|$ is sufficiently small then*

$$\operatorname{sign} W^{\mathcal{T}}_{t_0t_1} = 2\mu(t_0, t_1, H).$$

Here is the definition of the Maslov index. Let $\mathrm{Sp}(2n)$ denote the symplectic group and $\widetilde{\mathrm{Sp}}(2n)$ its universal cover. Think of an element of $\widetilde{\mathrm{Sp}}(2n)$ covering Ψ as a homotopy class of paths starting at $\mathbb{1}$ and ending at Ψ. Define the **Maslov cycle**

$$\Sigma = \{\Psi \in \mathrm{Sp}(2n) : \Psi(0 \times \mathbf{R}^n) \cap (0 \times \mathbf{R}^n) \neq \{0\}\},$$

and its complement

$$\mathrm{Sp}_0(2n) = \mathrm{Sp}(2n) \setminus \Sigma$$

and denote by $\widetilde{\Sigma}$ and $\widetilde{\mathrm{Sp}}_0(2n)$ the preimages under the covering map. For $\Psi_{10}, \Psi_{21} \in \mathrm{Sp}_0(2n)$ let $Q(\Psi_{21}, \Psi_{10})$ denote the composition form of the pair $(\Psi_{10}(0 \times \mathbf{R}^n), \Psi_{21}^{-1}(0 \times \mathbf{R}^n))$. If the matrices Ψ_{kj} are written in block matrix notation

$$\Psi_{kj} = \begin{pmatrix} A_{kj} & B_{kj} \\ C_{kj} & D_{kj} \end{pmatrix} \tag{7}$$

then the composition form is given by

$$Q = B_{21}^{-1} B_{20} B_{10}^{-1}.$$

Theorem 4.2 *There is a unique locally constant map $\mu : \widetilde{\mathrm{Sp}}_0(2n) \to n/2 + \mathbf{Z}$ such that*

$$\mu(\widetilde{\Psi}_{20}) = \mu(\widetilde{\Psi}_{21}) + \mu(\widetilde{\Psi}_{10}) + \tfrac{1}{2}\operatorname{sign} Q(\Psi_{21}, \Psi_{10})$$

whenever $\widetilde{\Psi}_{20} = \widetilde{\Psi}_{21}\widetilde{\Psi}_{10}$ and $\widetilde{\Psi}_{kj}$ covers Ψ_{kj}. This is called the **Maslov index.**

The number $\mu(\widetilde{\Psi})$ of Theorem 4.2 is essentially the intersection number of $\widetilde{\Psi}$ with the Maslov cycle. The definition is modified to adjust for the fact that the curve $\widetilde{\Psi}$ begins at the identity (which is an element of the Maslov cycle). For details see [19]. The number $\mu(t_0, t_1, H)$ of Theorem 4.1 is the Maslov index of the curve $[t_0, t_1] \to \mathrm{Sp}(2n) : t \mapsto \Psi^t_{t_0}$ defined by the evolution system generated by H.

Remark 4.3 Suppose that the evolution system generated by the Hamiltonian H is a symplectic shear

$$\Psi_{t_0}^{t_1} = \begin{pmatrix} \mathbb{1} & B(t_1,t_0) \\ 0 & \mathbb{1} \end{pmatrix}.$$

Then $B(t_2,t_0) = B(t_2,t_1) + B(t_1,t_0)$ and $B(t_0,t_0) = \mathbb{1}$. The Maslov index is given by

$$\mu(t_0,t_1,H) = -\tfrac{1}{2}\operatorname{sign} B(t_1,t_0).$$

For any two symmetric matrices A, B such that A, B, $A+B$, and $A^{-1}+B^{-1}$ are nonsingular we have the signature identity

$$\operatorname{sign}(A) + \operatorname{sign}(B) = \operatorname{sign}(A+B) + \operatorname{sign}(A^{-1}+B^{-1}).$$

This proves that the Maslov index as defined by intersection numbers satisfies the composition formula of Theorem 4.2 in the case of symplectic shears. The signature identity is obvious if the matrices are simultaneously diagonalizable. The general case can be proved with a homotopy argument.

Assume that $t_0, t_1, t_2 \in \mathcal{T}$ are such that the Hessians $W_{t_j t_k}$ are nondegenerate and denote by $Q^{\mathcal{T}}$ the corresponding composition form as in equation (4). The composition forms $Q^{\mathcal{T}}$ converge to the composition form Q of the continuous time problem as the mesh $|\mathcal{T}|$ tends to zero. If the mesh is sufficiently small then, by Proposition 2.2,

$$\operatorname{sign} W_{t_0 t_2}^{\mathcal{T}} = \operatorname{sign} W_{t_0 t_1}^{\mathcal{T}} + \operatorname{sign} W_{t_1 t_2}^{\mathcal{T}} + \operatorname{sign} Q^{\mathcal{T}}.$$

Thus the signature of the discrete second variation $W^{\mathcal{T}}$ satisfies the composition formula of Theorem 4.2 and this can be used to prove Theorem 4.1. Alternatively one can prove Theorem 4.1 first in the special case of a symplectic shear and then use a homotopy argument.

Remark 4.4 Theorem 4.2 is essentially due to Leray [16]. Leray's index $m(\widetilde{\Psi})$ is related to ours by the formula

$$m(\widetilde{\Psi}) = \mu(\widetilde{\Psi}) + \frac{n}{2}.$$

5 Feynman path integrals

Heuristically a variational family (π, ϕ) together with some sort of measure on the fibers determines a distribution on the base

$$f(x) = \int_{\substack{c \in P \\ \pi(c)=x}} e^{i\phi(c)/\hbar} \mathcal{D}c. \tag{8}$$

If the base is a product $X = X_0 \times X_1$ the distribution may be interpreted as an integral kernel

$$K(x_1, x_0) = \int_{\substack{c \in P \\ \pi(c)=(x_0,x_1)}} e^{i\phi(c)/\hbar} \mathcal{D}c$$

of an operator from a space of functions on X_0 to a space of functions on X_1:

$$Uf(x_1) = \int_{X_0} K(x_1, x_0) f(x_0) dx_0.$$

Formally the Feynman path integral is an example of this. The composition formula of Proposition 2.1 should correspond to the composition of operators.

Consider a time dependent quadratic Hamiltonian

$$\begin{aligned} H(t,x,y) &= H_0(t) + \langle H_x(t), x\rangle + \langle H_y(t), y\rangle \\ &\quad + \tfrac{1}{2}\langle H_{xx}(t)x, x\rangle + \langle H_{yx}(t)x, y\rangle + \tfrac{1}{2}\langle H_{yy}(t)y, y\rangle \end{aligned}$$

where $H_{xx}(t)$, $H_{yx}(t)$, $H_{yy}(t)$ are as before, $H_x(t), H_y(t) \in \mathbf{R}^n$, and $H_0(t) \in \mathbf{R}$. Let $\phi(c)$ denote the action integral. The Feynman path integral associated to H is the formal expression

$$\mathcal{U}(t_1, t_0, H) f(x_1) = \int_{\substack{c \in \mathcal{P}(t_0,t_1) \\ x(t_1)=x_1}} e^{i\phi(c)/\hbar} f(x(t_0)) \mathcal{D}c.$$

where $c = (x, y)$. Feynman was led to integrals of this type by physical considerations. He assigned a phase $e^{i\phi(c)/\hbar}$ to each classical path c and summed over all paths c. Our goal is to interpret this integral as a limit in the same way Feynman did. The discrete analogue of the path integral is the expression

$$\mathcal{U}^{\mathcal{T}}(t_1, t_0, H) f(x_1) = \int_{\substack{c \in \mathcal{P}^{\mathcal{T}}(t_0,t_1) \\ x(t_1)=x_1}} e^{i\phi^{\mathcal{T}}(c)/\hbar} f(x(t_0)) \mathcal{D}c$$

where

$$\mathcal{D}c = \prod_{t_0 \le t < t_1} (2\pi\hbar)^{-n} \det\left(\mathbb{1} - (t^+ - t)H_{xy}\right)^{1/2} dx(t)dy(t).$$

The order of integration is the time-order, i.e. first $dx(t_0)$, then $dy(t_0)$, then $dx(t_0^+)$ etc. The notation $\mathcal{D}c$ hides the normalization which makes the Feynman product a unitary operator. The integral does not converge absolutely as an integral in all its variables. Interchanging the order of integration requires justification.

Theorem 5.1 *The limit*

$$\mathcal{U}(t_1, t_0, H) = \lim_{|\mathcal{T}| \to 0} \mathcal{U}^{\mathcal{T}}(t_1, t_0, H)$$

exists in the strong operator topology. It is a unitary operator on $L^2(\mathbf{R}^n)$. Here the partitions partition the interval $[t_0, t_1]$.

We now give an explicit formula for the operator $\mathcal{U}(t_1, t_0, H)$. According to the philosophy of Fourier integral operators it should be possible to replace ϕ by any other phase function defining the same symplectic relation provided that $\mathcal{D}c$ is modified appropriately. In the case at hand the symplectic relation is the graph of the evolution system $\psi_{t_0}^{t_1}$ (see Example 3.3) so it is natural to seek a formula in terms of the generating function $S(x_0, x_1)$ from t_0 to t_1. Let $\Psi_{t_0}^{t_1}$ denote the linear part of $\psi_{t_0}^{t_1}$, $\mu = \mu(t_0, t_1, H)$ denote the Maslov index of $[t_0, t_1] \to \mathrm{Sp}(2n) : t \mapsto \Psi_{t_0}^{t}$, and $B = B(t_1, t_0)$ denote the right upper block in the block decomposition (7) of $\Psi_{t_0}^{t_1}$.

Theorem 5.2 *If $\psi_{t_0}^{t_1}$ admits a generating function then $\mathcal{U}(t_1, t_0, H)$ is given by*

$$\mathcal{U}(t_1, t_0, H) f(x_1) = \frac{(2\pi\hbar)^{-n/2}}{|\det B|^{1/2}} e^{i\pi\mu/2} \int_{\mathbf{R}^n} e^{i\,S(x_0,x_1)/\hbar} f(x_0)\, dx_0.$$

The formula is first proved in the case of discrete time and then convergence as well as the continuous time formula are obvious. To prove the analogous formula in discrete time note that Taylor's formula

$$\phi^{\mathcal{T}}(c) = S^{\mathcal{T}}(x_0, x_1) + \tfrac{1}{2}\langle W_{t_0 t_1}^{\mathcal{T}} \gamma, \gamma \rangle$$

is exact (since the action is quadratic). Here $c = c_0 + \gamma$. c_0 is a fiber critical point with $\pi(c_0) = (x_0, x_1)$, so $S^{\mathcal{T}}(x_0, x_1) = \phi^{\mathcal{T}}(c_0)$. Now integrate over γ. Then the Maslov index appears as the signature of $W_{t_0 t_1}^{\mathcal{T}}$ according to Theorem 4.1.

Associated to the Hamiltonian $H(t, x, y)$ is a second order differential operator $H(t, Q, P)$ where Q_j and P_j denote the self-adjoint operators

$$(P_j f)(x) = -i\hbar \partial_j f(x), \qquad (Q_j f)(x) = x_j f(x),$$

and $H(t, Q, P)$ results from $H(t, x, y)$ by making the following substitutions

$$x_j \mapsto Q_j, \qquad y_j \mapsto P_j,$$

$$x_j x_k \mapsto Q_j Q_k, \qquad y_j y_k \mapsto P_j P_k, \qquad x_k y_j \mapsto Q_k P_j - \frac{i\hbar}{2}\delta_{jk}\mathbb{1}.$$

Pay attention to the mixed term: Q_j and P_j do not commute. If the Hamiltonian has the form $H = \frac{1}{2}|y^2| + V(x)$ the equation in the next theorem is the Schrödinger equation.

Theorem 5.3 *The operators $\mathcal{U}(t, t_0, H)$ are the evolution operators of the time-dependent partial differential equation*

$$i\hbar \frac{\partial u}{\partial t} = H(t, Q, P) u.$$

Proof: Assume that $\psi_{t_0}^t$ admits a generating function and let $S(t, x, x_0)$ be given by the action. Let $B(t)$ denote the right upper block in the block decomposition of $\Psi_{t_0}^t = d\psi_{t_0}^t$ and abbreviate $\lambda = e^{i\pi\mu(t,t_0,H)/2}(2\pi\hbar)^{-n/2}$. Then

$$u(t,x) = \mathcal{U}(t,t_0,H)f(x) = \lambda|\det B(t)|^{-1/2} \int_{\mathbf{R}^n} e^{iS(t,x,x_0)/\hbar} f(x_0)\, dx_0.$$

Differentiating with respect to x gives

$$P_j u = \lambda|\det B|^{-1/2} \int_{\mathbf{R}^n} \frac{\partial S}{\partial x_j} e^{iS/\hbar} f$$

and

$$P_j P_k u = -i\hbar \frac{\partial^2 S}{\partial x_j \partial x_k} u + \lambda|\det B|^{-1/2} \int_{\mathbf{R}^n} \frac{\partial S}{\partial x_j}\frac{\partial S}{\partial x_k} e^{iS/\hbar} f$$

Hence the right hand side of the equation is

$$\begin{aligned} H(t,Q,P)u &= -i\hbar\tfrac{1}{2}\mathrm{tr}\,(H_{yx} + H_{yy}DB^{-1})\,u \\ &\quad + \lambda|\det B|^{-1/2} \int_{\mathbf{R}^n} H(t,x,\partial_x S) e^{iS/\hbar} f. \end{aligned}$$

Here we have used the identity $\partial^2 S/\partial x^2 = DB^{-1}$ where $D = D(t)$ is the lower right block in the block decomposition (7) of $\Psi_{t_0}^t$. Now

$$\begin{aligned} \frac{d}{dt}|\det B|^{-1/2} &= -\tfrac{1}{2}\mathrm{tr}\,(\dot{B}B^{-1})|\det B|^{-1/2} \\ &= -\tfrac{1}{2}\mathrm{tr}\,(H_{yx} + H_{yy}DB^{-1})|\det B|^{-1/2} \end{aligned}$$

and hence

$$i\hbar\frac{\partial u}{\partial t} = -i\hbar\tfrac{1}{2}\mathrm{tr}\,(H_{yx} + H_{yy}DB^{-1})\,u - \lambda|\det B|^{-1/2} \int_{\mathbf{R}^n} \frac{\partial S}{\partial t} e^{iS/\hbar} f.$$

Since S satisfies the Hamilton-Jacobi equation $\partial_t S + H(t,x,\partial_x S) = 0$ this proves the statement whenever $\psi_{t_0}^t$ admits a generating function. The general case follows since both sides of the equation depend continuously on H. □

6 Geometric Quantization

A time dependent Hamiltonian H on $\mathbf{R}^{2n}$ determines an evolution system on $W = \mathbf{R}^{2n} \times \mathrm{U}(1)$ via the formula

$$g_{t_0}^{t_1}(z_0, u_0) = \left(\psi_{t_0}^{t_1}(z_0), u_0 e^{i\phi(c)/\hbar}\right)$$

for $(z_0, u_0) \in W = \mathbf{R}^{2n} \times \mathrm{U}(1)$ where $\psi_{t_0}^{t_1}$ is the evolution system generated by H, $\phi(c)$ is the symplectic action integral evaluated at the curve $c(t) = \psi_{t_0}^{t}(z_0)$. If the generating function $S = \phi(c)$ of $\psi_{t_0}^{t_1}$ is defined then

$$g_{t_0}^{t_1}(z_0, u_0) = \left(\psi_{t_0}^{t_1}(z_0), u_0 e^{iS(x_0,x_1)/\hbar}\right) \tag{9}$$

where $z_j = (x_j, y_j)$, $z_1 = \psi_{t_0}^{t_1}(z_0)$. The group $\mathrm{ESp}(W, \hbar)$ of all diffeomorphisms of W of form $g_{t_0}^{t_1}$ where H runs over the time dependent (inhomogeneous) quadratic Hamiltonians $\mathbf{R} \to \mathcal{F}_2$ is called the **extended symplectic group**. The various groups $\mathrm{ESp}(W, \hbar)$ depend set-theoretically on $\hbar$ but are isomorphic as abstract groups. There is a central extension

$$1 \to \mathrm{U}(1) \to \mathrm{ESp}(W, \hbar) \to \mathrm{ASp}(\mathbf{R}^{2n}) \to 1$$

where $\mathrm{ASp}(\mathbf{R}^{2n})$ denotes the **affine symplectic group**; the projection is given by $g_{t_0}^{t_1} \mapsto \psi_{t_0}^{t_1}$ and the U(1) subgroup consists of those $g_{t_0}^{t_1}$ where H is constant.

If the Hamiltonian H is time independent then the corresponding evolution systems $\psi_{t_0}^{t_1}$ and $g_{t_0}^{t_1}$ are flows: denote by X_H and Y_H the vector fields generating these flows. Then X_H is the Hamiltonian vector field of H, and Y_H is a lift of X_H to L. The Lie algebra to $\mathrm{ASp}(\mathbf{R}^{2n})$ is the image of quadratic Hamiltonians under the representation $H \mapsto X_H$ but this representation is not faithful as the constant Hamiltonians map to zero. However the representation $H \mapsto Y_H$ *is* faithful. Differentiating gives the following

Proposition 6.1 *The vector field Y_H on W is given by*

$$Y_H(z, u) = (X_H(z), uis_H/\hbar), \qquad s_H = \langle y, \partial_y H\rangle - H.$$

Souriau [25] and Kostant [15] describe the extended symplectic group as the group of bundle automorphisms of the $U(1)$ bundle $W \to \mathbf{R}^{2n}$ which cover affine symplectic transformations and preserve the connection form

$$\alpha = -\frac{i}{\hbar}\langle y, dx\rangle + u^{-1}du.$$

7 Representations

The group $\mathrm{EMp}(2n)$ of all unitary operators of the form

$$U = \mathcal{U}(t_1, t_0, H) \tag{10}$$

where H runs over the time dependent quadratic Hamiltonians and t_1, t_0 range over the real numbers form a finite dimensional group called the **extended metaplectic goup**. The subgroup $\mathrm{Mp}(2n)$ obtained by taking only homogeneous quadratic Hamiltonians H in (10) is called the **metaplectic group**. The subgroup $\mathrm{HG}(2n)$ obtained by taking only affine Hamiltonians H in (10) is called the **Heisenberg group**. By Theorem 5.2 the map

$$\mathrm{EMp}(2n) \to \mathrm{ESp}(W, \hbar) : \mathcal{U}(t_1, t_0, H) \mapsto \mathfrak{g}_{t_0}^{t_1}(H)$$

is a well-defined double cover (which depends on $\hbar$). This repesentation of the double cover of the symplectic group is called *Siegel-Shale-Weil representation* or the *metaplectic representation.* The restriction of the double cover to the Heisenberg group is injective and the resulting representation is called the **Heisenberg representation.**

Here is a more explicit description of the Heisenberg representation. If H is an affine Hamiltonian with constant coefficients then

$$\mathcal{U}(t, t_0, H) = \mathcal{T}((t - t_0)H)$$

where

$$\mathcal{T}(H) = e^{-iH_0/\hbar - i\langle H_x, x\rangle/\hbar + i\langle H_x, H_y\rangle/2\hbar} f(x - H_y).$$

If Ψ is a symplectic matrix then the map $H \mapsto \mathcal{T}(H \circ \Psi)$ is another such representation corresponding to the same value of Planck's constant $\hbar$. By the Stone-von Neumann theorem these representations are unitarily isomorphic. In other words there exists a unitary operator $U : L^2(\mathbf{R}^n) \to L^2(\mathbf{R}^n)$, unique up to multiplication by a complex number of modulus 1, such that

$$\mathcal{T}(H \circ \Psi) = U^{-1} \circ \mathcal{T}(H) \circ U.$$

Such an intertwining operator U may be taken as a lift of Ψ to the metaplectic group. This is apparently how the metaplectic representation was discovered (see [23]). The elements of the metaplectic representation are thus viewed as intertwining operators of various incarnations of the Heisenberg representation. See [20] for an exposition in terms of co-adjoint orbits and polarizations.

8 Quantum field theory

By generalizing from affine symplectomorphisms to affine symplectic relations it should be possible to generalize the extended metaplectic representation to the *extended metaplectic functor.* An *extended Lagrangian subspace* is a Legendrian submanifold of W which covers an affine Lagrangian subspace of $\mathbf{R}^{2n}$. A quadratic function $S : \mathbf{R}^n \to \mathbf{R}$ determines an extended Lagrangian subspace $L(S)$ via

$$L(S) = \{(x, y, u) \in W : y = \partial_x S(x),\ u = e^{iS(x)/\hbar}\}.$$

An element of the extended symplectic group can be interpreted as an extended Lagrangian subspace of the external tensor product $W^* \otimes W$ over $\bar{\mathbf{R}}^{2n} \times \mathbf{R}^{2n}$. (The bar indicates that the sign of the symplectic form in the first factor is reversed.) More generally given circle bundles $W_0 \to \mathbf{R}^{2n_0}$ and $W_1 \to \mathbf{R}^{2n_1}$ as in Section 6 let $W_{01} \to \bar{\mathbf{R}}^{2n_0} \times \mathbf{R}^{2n_1}$ be endowed with the connection form $\alpha_1 - \alpha_0$. Then an *extended symplectic relation* is an extended Lagrangian subspace of $\bar{\mathbf{R}}^{2n_0} \times \mathbf{R}^{2n_1}$. Extended Lagrangian subspaces appear as the special case $n_0 = 0$. The extended metaplectic functor assigns to each extended symplectic relation a distribution on $\mathbf{R}^{n_0} \times \mathbf{R}^{n_1}$, determined by the relation only up to a sign, and respecting the operation of composition defined in section 2. In the case of an extended symplectomorphism $g_{t_0}^{t_1}(H)$ the distribution is the distribution kernel of $\mathcal{U}(t_0, t_1, H)$. For a quadratic generating function $S(x)$ the distribution is $e^{iS(x)/\hbar}$ multiplied by a normalizing factor. Composition of extended symplectic relations corresponds to composition of distribution kernels; there should be a formula like

$$\mathcal{U}(R_{01} \# R_{12}) = \mathrm{tr}(\mathcal{U}(R_{01}) \otimes \mathcal{U}(R_{12})).$$

The extended metaplectic functor should give a simple model of Segal's axioms for topological quantum field theory. Taking the homology of a Riemann surface as the underlying symplectic vector space should lead to a $(2+1)$-dimensional theory. This is what Atiyah calls the Abelian case (without the lattice).

References

[1] S. Albevario and R. Hoegh-Krohn, Feynman path integrals and the corresponding method of stationary phase, in *Feynman Path Integrals*, ed. S. Albevario et al, Springer Lecture Notes in Physics **106** (1978) 3-57.

[2] M.F. Atiyah, *The Geometry and Physics of Knots*, Cambridge University Press, 1990.

[3] C. Conley and E. Zehnder, The Birkhoff-Lewis fixed point theorem and a conjecture of V.I. Arnold, *Invent. Math.* **73** (1983), 33–49.

[4] C.C. Conley and E. Zehnder, Morse-type index theory for flows and periodic solutions of Hamiltonian equations, *Commun. Pure Appl. Math.* **37** (1984), 207–253.

[5] I. Daubechies and J.R. Klauder, Quantum mechanical path integrals with Wiener measure for all polynomial Hamiltonians, *Phys. Rev Letters* **52** (1984) 1161; *J. Math. Physics* **26** (1985) 2239-2256.

[6] I. Daubechies, J.R. Klauder, and T. Paul, Wiener measures for path integrals with affine kinematic variables, *J. Math. Physics* **28** (1987) 85-102.

[7] P.A.M. Dirac, The Lagrangian in quantum mechanics, Phys. Zeitschr. d. Sovyetunion **3** (1933) 64-72.

[8] J.J. Duistermaat, On the Morse index in variational calculus, *Advances in Mathematics* **21** (1976), 173–195.

[9] R.P. Feynman, Space-time approach to non-relativistic quantum mechanics, *Rev. Mod. Phisics* **20** (1948), 367–387.

[10] R.P. Feynman and A.R. Hibbs, *Quantum Mechanics and Path Integrals*, MacGraw-Hill, 1965.

[11] V. Guillemin and S. Sternberg, *Geometric Asymptotics*, AMS Math Surveys **14**, 1977.

[12] V. Guillemin and S. Sternberg, *Symplectic Techniques in Physics*, Cambridge University Press, 1984.

[13] L. Hörmander, Fourier integral operators I, *Acta Math.* **127** (1971), 79–183.

[14] J.B. Keller, Corrected Bohr-Sommerfeld quantum conditions for nonseparable systems, *Annals of Physics* **4** (1958), 180–188.

[15] B. Kostant, Quantization and unitary representations, in *Modern Analysis and its Applications* Springer Lecture Notes in Math **170** (1970) 87-207.

[16] J. Leray, *Lagrangian Analysis and Quantum Mechanics*, MIT press, 1981.

[17] J.W. Robbin and D.A. Salamon, Path integrals on phase space and the metaplectic representation, Preprint, 1992.

[18] J.W. Robbin, and D.A. Salamon, The spectral flow and the Maslov index, Preprint 1992.

[19] J.W. Robbin, and D.A. Salamon, A Maslov index for paths, Preprint 1992.

[20] P.L. Robinson and J.H. Rawnsley, *The metaplectic representation, Mp^C structures and geometric quantization* , Memoirs of the American Mathematical Society **81**, 1989.

[21] D. Salamon and E. Zehnder, Morse theory for periodic solutions of Hamiltonian systems and the Maslov index, to appear in Comm. Pure Appl. Math.

[22] I.E. Segal, Foundations of the theory of dynamical systems of infinitely many degrees of freedom (I), *Mat. Fys. Medd. Danske Vid. Selsk.* **31** (1959) 1-39.

[23] I.E. Segal, Transforms for operators and symplectic automorphisms over a locally compact abelian group, *Math. Scand.* **13** (1963) 31-43.

[24] D. Shale, Linear symmetries of free boson fields, *Trans. AMS* **103** (1962), 149–167.

[25] J.M. Souriau, *Structures des Systemes Dynamiques*, Dunod, Paris, 1970.

[26] J.M. Souriau, Construction explicite de l'indice de Maslov. *Group Theoretical Methods in Physics* Springer Lecture Notes in Physics **50** (1975), 117–148.

[27] A. Weil, Sur certaine groupes d'operateurs unitaires, *Acta math* **111** (1964), 143–211.

[28] A. Weinstein, *Lectures on Symplectic Manifolds*, AMS Reg. Conf. Ser. Math. **29**, 1977.

Symplectic Mappings which are Stable at Infinity

Eduard Zehnder
Eidg. Technische Hochschule, Zürich

1. A stability problem

If φ is a symplectic diffeomorphism of $\mathbb{R}^{2n}$, i.e.

$$\varphi^*\sigma = \sigma \ , \tag{1}$$

for the standard symplectic form σ on $\mathbb{R}^{2n}$ given by $\sigma(\xi, \eta) = \langle \xi, J\eta \rangle$ for $\xi, \eta \in \mathbb{R}^{2n}$, one would like to study the orbits $\mathcal{O}(z) = \{\varphi^j(z) \mid j \in \mathbb{Z}\}$ of the points $z \in \mathbb{R}^{2n}$ under all the iterates of φ. It is, of course, well known that the structure of all orbits of a symplectic mapping is extremely intricate; for a recent illustration of the complexity of the orbit structure for a symplectic mapping locally near an elliptic fixed point in the plane we point out C. Genecand [17]. It is, therefore, natural to search for special orbits only such as fixed points or periodic points. But instead of looking for special orbits we shall look for mappings φ which have the property that all their orbits are bounded, i.e.

$$\sup_{j \in \mathbb{Z}} \left|\varphi^j(z)\right| < \infty \quad \text{for all} \quad z \in \mathbb{R}^{2n}. \tag{2}$$

The statement (2) can be viewed as a qualitative stability statement namely the stability of a fictitious fixed point at infinity.

The symplectic mappings considered in the following will be rather special. They belong to the flow φ^t of a time dependent Hamiltonian vectorfield on $\mathbb{R}^{2n}$ defined by a time dependent Hamiltonian function $H(t, z)$, where $t \in \mathbb{R}$ is the time and $z \in \mathbb{R}^{2n}$. The flow is the solution of the initial value problem

$$\begin{aligned} \frac{d}{dt}\varphi^t(z) &= J\nabla H\big(t, \varphi^t(z)\big) \\ \varphi^0(z) &= z \ , \end{aligned} \tag{3}$$

so that every map φ^t preserves the symplectic structure σ, i.e. $(\varphi^t)^*\sigma = \sigma$. In case that the timedependence is periodic, for example $H(t+1, z) = H(t, z)$, we take φ to be the time one map φ^1 of the flow, so that

$$\varphi^n = \varphi \circ \varphi \circ \ldots \circ \varphi \tag{4}$$

is the n-times iterated map. If H is independent of the time t then we have $H(\varphi^t(z)) = H(z)$ for all t and all $z \in \mathbb{R}^{2n}$. Consequently the stability (2) follows provided, for example, all the subsets $H(z) = \text{const}$ are bounded. If, however, the function H depends on time, the energy is not conserved anymore and the stability question is rather subtle.

2. Results for $\mathbb{R}^2$

For a symplectic mapping stability under all iterates can, in general, only be expected for maps on $\mathbb{R}^2$ and we therefore consider first the case $z = (x, y) \in \mathbb{R}^2$. Moreover, we look at the restricted class of Hamiltonian functions which describe so called "classical systems", namely

$$H(t, x, y) = \frac{1}{2}y^2 + V(t, x)\,. \tag{5}$$

Setting $y = \dot{x}$, the corresponding Hamiltonian equation (2) is equivalent to the second order equation on the real line:

$$\ddot{x} + V_x(t, x) = 0\,, \qquad x \in \mathbb{R}\,. \tag{6}$$

The quest for potentials $V(t, x)$ having the property that all the solutions $x(t)$ of (6) are bounded for all times was already emphasized by J.E. Littlewood. He constructed counter-examples of the form

$$\ddot{x} + V_x(x) = p(t)\,, \tag{7}$$

which possess unbounded solutions for forcing terms $p(t)$ which are periodic in time but which are not continuous; we refer to M. Levi [3] and Y. Long [4].

If the timedependence is more intimately involved in the nonlinearity it is, of course, not even clear that the flow does exist over an infinite interval of time. Indeed C.V. Coffman and D.F. Ullrich [18] constructed an example, namely

$$\ddot{x} + a(t)x^3 = 0 \tag{8}$$

where $a(t + 1) = a(t)$ is periodic, continuous and even close to a positive constant, such that there are solutions blowing up in a time interval shorter than the period. In their example $a(t)$ is, however, not of bounded variation.

It should be recalled that the first positive result is due to G.R. Morris [7], who proved the stability statement (2) for the equation

$$\ddot{x} + x^3 = p(t) \tag{9}$$

with a continuous and time periodic forcing $p(t)$. His proof, based on J. Moser's twist theorem [14], gave rise to farreaching extensions to more general potentials $V(t,x)$, which are sufficiently smooth and which depend periodically on the time $V(t+1,x) = V(t,x)$, we refer for example to R. Dieckerhoff and E. Zehnder [8], M. Levi [11], S. Laederich and M. Levi [6], J. Norris [9] and Liu Bin [5].

In contrast to the time periodic forcing we are interested in the following in potentials V which depend quasiperiodically on the time. We therefore assume that

$$V(t,x) = F(\omega_1 t, \dots, \omega_N t, x)\,, \tag{10}$$

where $\omega = (\omega_1, \dots, \omega_N) \in \mathbf{R}^N$ is a given constant vector, the so called frequencies, and where $F(\xi_1, \dots, \xi_N, x)$ is periodic with period 1 in every variable ξ_j, so that $F(\cdot, x)$ is a function on the torus T^N. We shall use the abbreviating notation $V(\omega t, x)$.

The frequencies ω are assumed not only to be rationally independent but to satisfy, in addition, the diophantine conditions (D.C.)

$$|\langle \omega, j\rangle| \geq \gamma\, |j|^{-\tau} \quad \text{for all} \quad j \in \mathbf{Z}^N \backslash \{0\}\,, \tag{11}$$

with two constants $\tau > N$ and $\gamma > 0$. Recall that almost every $\omega \in \mathbf{R}^N$ meets these inequalities.

That stability is possible for quasiperiodic potentials will be illustrated for potentials having the following special form

$$V(\omega t, x) = \sum_{j=1}^{2d+2} a_j(\omega t) x^j \,, d \geq 1 \tag{12}$$

where all the coefficients a_j are quasiperiodic in time with frequencies ω, and where, in addition, the leading coefficient $a \equiv a_{2d+2}$ is positive

$$a(\omega t) \geq \min_{\xi \in T^N} a(\xi) > 0\,. \tag{13}$$

For this class of potentials the following rather surprising stability statement holds true:

Theorem 1. [1] Let $\varphi^t(z)$ be the flow of the Hamiltonian system defined by the function $H(t,z) = \frac{1}{2}y^2 + V(\omega t, x)$ where $z = (x,y) \in \mathbf{R}^2$, and where V is quasiperiodic in time t satisfying (11), (12) and (13). Assume that

$$a_j \in C^k(T^N)\,, \quad k > 4\tau + 6$$

for all $1 \leq j \leq 2d+2$, then

$$\sup_{t \in \mathbf{R}} \left|\varphi^t(z)\right| < \infty$$

for every $z \in \mathbf{R}^2$.

The smoothness requirement on V necessary for the stability depends only on the number of underlying frequencies ω but not on the degree of V as a polynomial in x.

3. Sketch of the proof of theorem 1

The proof uses a well known technique: in order to get bounds for all solutions one proves the existence of quasiperiodic solutions using the fact that the phase space is $\mathbf{R}^2$. Geometrically the idea is as follows. One writes the equation (6) as a system on the extended phase space $(x, y, t) \in \mathbf{R}^3$:

$$\begin{aligned} \dot{x} &= y \\ \dot{y} &= -V_x(\omega t, x) \\ \dot{t} &= 1\,. \end{aligned} \tag{14}$$

By X we abbreviate the vectorfield on $\mathbf{R}^3$ defined by the right hand side. The idea now is to construct for every $C > 0$ sufficiently large, an embedded cylinder $w : Z = \mathbf{R} \times S^1 \to \mathbf{R}^3$

$$w : (t, s) \longmapsto \big(u(t, s), t\big) \tag{15}$$

containing the time axis in the interior, contained in $A_C = \{(x, y, t) \in \mathbf{R}^3 \mid x^2 + y^2 \geq C^2\}$ and satisfying

$$C < \inf_Z |u| < \sup_Z |u| < \infty \tag{16}$$

and which, moreover, is tangential to the vectorfield X on $\mathbf{R}^3$. Therefore the surface $w(Z)$ is invariant under the flow $\big(\varphi^t(z), t\big)$ of X. If a solution $\big(z(t), t\big)$ satisfies $|z(t^*)| \leq C$ for some $t^* \in \mathbf{R}$ then it follows by the invariance of the cylinder and by the uniqueness of O.D.E. that the solution $z(t)$ exists for all time $t \in \mathbf{R}$ and satisfies, in view of (16), $|z(t)| \leq \sup u < \infty$ for $t \in \mathbf{R}$, so that $z(t)$ is indeed bounded.

The existence of these invariant surfaces in $\mathbf{R}^3$ is a consequence of the observation that in the region A_C for large C the system (14) can be viewed as a system close to an integrable one. For such systems the KAM theory is applicable. The techniques involved require an excessive amount of smoothness of the equation, as is well known. The near integrability is, on the other hand, not a priori obvious and its proof is one of the main tasks in proving theorem 1.

In order to make the required near integrability visible one uses transformation theory of time dependent Hamiltonian equations: firstly one rescales the time and the x-variable appropriately, then one introduces action and angle variables and, after a further finite sequence of canonical transformations one finds the Hamiltonian in the following form, suitable to apply the analytical techniques. On the annulus $(\varphi, I) \in S^1 \times D$, where $D = \{I \mid a \leq I \leq b\} \subset \mathbf{R}$ is a bounded interval, the Hamiltonian is given by the function

$$H(\varepsilon\omega t, \varphi, I, \varepsilon) = H_0(I, \varepsilon) + \varepsilon^3 H_1(\varepsilon\omega t, \varphi, I, \varepsilon)\,. \tag{17}$$

Here $\varepsilon > 0$ is a small parameter, and the integrable part H_0 satisfies the so called twist condition

$$C < \frac{\partial^2 H_0}{\partial I^2}(I, \varepsilon) < C^{-1}\,, \quad I \in D \tag{18}$$

for a constant $C > 0$ which is independent of ε.

One is confronted with the analytical problem of finding, for every $\varepsilon > 0$ sufficiently small, a quasiperiodic solution contained in $S^1 \times D$ having prescribed $1 + N$ frequencies $(\alpha, \varepsilon\omega) \in \mathbf{R}^{1+N}$ for an appropriate $\alpha = \frac{\partial H_0}{\partial I}(I, \varepsilon)$. To be more precise, recalling that $H = H(\xi, \varphi, I, \varepsilon)$ is periodic in $\xi \in T^N$ and $\varphi \in S^1$ one looks for a differentiable map

$$w : T^{1+N} \longrightarrow S^1 \times D\,, \tag{19}$$

$w(\vartheta, \xi) = \big(u(\vartheta, \xi), v(\vartheta, \xi)\big)$ and $u(\vartheta, \xi) - \vartheta$, $v(\vartheta, \xi)$ periodic in ϑ, ξ, which maps the constant vectorfield V on T^{1+N} defined by

$$\dot{\vartheta} = \alpha\,, \quad \dot{\xi} = \varepsilon\omega\,, \tag{20}$$

into the given Hamiltonian vectorfield; i.e.

$$dw\begin{pmatrix}\alpha \\ \varepsilon\omega\end{pmatrix} = J\nabla H\big(\xi, w(\vartheta, \xi), \varepsilon\big) \tag{21}$$

for all $\vartheta, \xi \in T^{1+N}$. It then follows that w maps the solutions of V into the required quasiperiodic solutions $z(t) = w(\alpha t, \varepsilon\omega t)$ of the Hamiltonian equation defined by the function (16). In particular the cylinder $\hat{w} : S^1 \times \mathbf{R} \to S^1 \times D \times \mathbf{R}$ defined by $\hat{w}(\vartheta, t) = \big(w(\vartheta, \varepsilon\omega t), t\big)$ consists of solutions of the Hamiltonian vectorfield and hence is the required invariant surface leading to the bounds we are looking for.

The existence of solutions of the P.D.E. (21) uses J. Moser's technique in [13] together with the improvements, crucial for our regularity, in D. Salamon [15] and in D. Salamon and E. Zehnder [16]. For the technical details of the proof we refer to [1].

Summarizing, the stability is concluded from the fact that a dominant part of the phase space $\mathbf{R}^2$ for $|z|$ large is covered by quasiperiodic solutions. In particular we can state

Theorem 2. [1] The equation

$$\ddot{x} + V_x(\omega t, x) = 0\,, \quad x \in \mathbf{R}$$

with the potential V satisfying the assumptions of theorem 1 possesses uncountably many quasiperiodic solutions having $1 + N$ frequencies (α, ω) satisfying the D.C.:

$$|\alpha \cdot k + \langle \omega, j \rangle| \geq \gamma(|k| + |j|)^{-\tau}$$

for all $(k, j) \in \mathbf{Z}^{1+N} \backslash \{0\}$ with the constants γ and τ as in (11).

For a related perturbation problem this statement was already conjectured by V. Arnold [21] in the early sixties. Finally it should be said that the ideas of this section lead to the stability for a relatively large class of quasiperiodic potentials, see [1].

4. Dropping the quasiperiodicity

As a sideremark we mention that as soon as the requirement on the recurrent nature of the time forcing is omitted unbounded solutions may show up for forcing terms which are small, smooth and tending to zero as time gives to infinity. The following simple example of instability illustrates this.

Proposition. [1] Given any $\varepsilon > 0$ and $r \in \mathbf{N}$ there exists a function $p \in C^\infty(\mathbf{R})$ satisfying

$$|p|_{C^r} \leq \varepsilon \quad \text{and} \quad \lim_{t \to \infty} D^j p(t) = 0$$

for $0 \leq j \leq r - 1$, such that the equation

$$\ddot{x} + x^3 = p(t)\,, \qquad x \in \mathbf{R}$$

possesses an unbounded solution $x(t)$. Moreover, the growth rate of this solution is given by

$$C \leq t^{-a} E(t) \leq C^{-1}\,, a = \frac{4}{2r+3}\,,$$

$t \geq 1$, for a constant $C > 0$, where $E(t) = \frac{1}{2}\dot{x}(t)^2 + \frac{1}{4}x(t)^4$ is the energy of the solution $x(t)$. On the other hand the decay rate of the forcing is given by

$$\sup_{t>0} t^b \left|D^j p(t)\right| < \infty \, , b = \frac{2(r-j)}{2r+3}$$

for $0 \leq j \leq r-1$.

For the elementary proof we refer to [1].

5. Exponential stability in higher dimensions

Turning to the stability question in higher dimensions we consider on the special phase space $(x, y) \in T^n \times \mathbf{R}^n$ the Hamiltonian system defined by

$$H(\omega t, x, y) = \frac{1}{2}|y|^2 + V(\omega t, x) \, , \tag{22}$$

where $V = V(\xi, x)$ is a function on $T^N \times T^n$. In sharp contrast to the equations in section 2, the system is already given in action and angle variables. Since V is bounded, the system is in the region where $|y|$ is large near an integrable one. This allows again to conclude many quasiperiodic solutions having prescribed frequencies $(\lambda, \omega) \in \mathbf{R}^n \times \mathbf{R}^N$. To this end one solves the nonlinear P.D.E.

$$dw\begin{pmatrix}\lambda\\ \omega\end{pmatrix} = J\nabla H\big(\xi, w(\xi, x)\big) \, , \tag{23}$$

$(\xi, x) \in T^N \times T^n$, for a map $w : T^N \times T^n \to \mathbf{R}^n \times T^n$ which is, for every fixed $\xi \in T^N$, an embedding $w(\xi, \cdot) : T^n \to T^n \times \mathbf{R}^n$. For the existence proof which requires, of course, diophantine conditions on the frequencies and a sufficient amount of smoothness for the potential we refer to L. Chierchia and E. Zehnder [10].

We point out that there is no smallness assumption on the potential V, instead the frequency vector $|\lambda|$ has to be chosen sufficently large.

In the special case $n = 1$ one concludes, as above, immediately the stability of all pendulum like equations with potentials $V(t, x)$ which are periodic in x and periodic or quasiperiodic in the time t, provided only V is sufficiently smooth.

Theorem 3. [10] [11] [12] Let $\varphi^t\big(x(0), y(0)\big) = \big(x(t), y(t)\big)$ be the flow of the Hamiltonian system on $S^1 \times \mathbf{R}$ belonging to the function (22) for $n = 1$. If ω satisfies D.C. and if V is sufficiently smooth, then

$$\sup_{t \in \mathbf{R}} |y(t)| < \infty$$

for every $y(0) \in \mathbf{R}$.

Although there exists an abundance of quasiperiodic and so bounded solutions, they do not lead to bounds for all solutions if $n > 1$, since the invariant surfaces covered by these solutions do not bound open sets. Recall now that N.N. Nekhoroshev [19] discovered estimates for the action variables of all solution not over the whole time but over an exponentially long interval of time for systems near integrable systems, provided the Hamiltonian is not only smooth, but real analytic and provided the corresponding integrable system meets convexity type conditions. One might therefore conjecture that such estimates hold true also for our global problem. Indeed the following statement of exponential stability replaces Theorem 3 in higher dimensions.

In order to formulate the result we assume $V(x,t)$ to be real analytic on $T^n \times \mathbf{R}$ and moreover to have a holomorphic extension to a complex strip $|\mathrm{Im}\, x| < \sigma$ and $|\mathrm{Im}\, t| < \sigma$ for some $\sigma > 0$ such that

$$|V|_\sigma := \sup_{\substack{|\mathrm{Im}\, x| < \sigma \\ |\mathrm{Im}\, \xi| < \sigma}} |V(x,\xi)| < \infty . \tag{24}$$

No periodicity or quasiperiodicity in t is required. The following statement of exponential stability replaces (1.4) in higher dimensions.

Theorem 4. [2] Let $\varphi^t(x(0), y(0)) = (x(t), y(t))$ be the flow on $T^n \times \mathbf{R}^n$ of the Hamiltonian vectorfield corresponding to:

$$H(x,y,t) = \tfrac{1}{2}|y|^2 + V(x,t) .$$

Assume that the potential $V = V(x,\xi)$ is real analytic on $T^n \times \mathbf{R}$ and has, moreover, an analytic extension to a complex strip satisfying (24). Then there are positive constants T^* and R^* depending on $|V|_\sigma$, σ and the dimension n such that for every $\rho \geq (R^*)^{\frac{1}{\alpha}}$

$$|y(t) - y(0)| \leq \rho$$

for all t in

$$|t| \leq T^* e^{\frac{1}{R^*}\cdot\rho^\alpha}, \quad \alpha = \frac{2}{n^2+n} ,$$

where $T^* = \frac{\sigma}{|V|_\sigma} \cdot \left(\frac{1+e^{-\sigma}}{1-e^{-\sigma}}\right)^{-n}$.

The assumptions are met for example if $V(x,\omega t)$ depends quasiperiodically on t provided $V(x,\xi)$ is real analytic on $T^n \times T^N$.

There is no smallness requirement on V, the potential, however, is required to be more than real analytic quite in contrast to theorem 1. The proof uses the underlying ideas of N. Nekhoroshev [19] and of G. Benettin, L. Galgani and A. Giorgilli in [20] and we refer to [2] for the details.

6. References

[1] M. Levi and E. Zehnder: Boundedness of solutions for quasiperiodic potentials, to be published.

[2] A. Giorgilli and E. Zehnder: Exponential stability for quasiperiodic potentials, to be published.

[3] M. Levi: A note on Littlewood's counter example, preprint FIM, ETH Zürich (1989).

[4] Y. Long: Unbounded solution of a superlinear Duffing-equation, preprint Nankai University Tianjin (1989).

[5] Lin Bin: Boundedness for solutions of nonlinear Hill's equations with periodic forcing terms via Moser's Twist Theorem, J. Diff. Eq. 79 (1989), 304–315.

[6] S. Laederich and M. Levi: Invariant curves and time-dependent potentials, preprint F.I.M., ETH Zürich (1989).

[7] G.R. Morris: A case of boundedness in Littlewood's problem on oscillatory differential equations, Bull. Austr. Math. Soc. 14 (1976), 71–93.

[8] R. Dieckerhoff and E. Zehnder: Boundedness of solutions via the twist theorem. Ann. Scuola Norm. Sup. Pisa 14 (1987), 79–95.

[9] J.W. Norris: Boundedness in periodically forced second order conservative systems. Preprint University College of Wales, Aberystwyth (1989).

[10] L. Chierchia and E. Zehnder: Asymptotic expansions of quasiperiodic solutions. Annali della Sc. Norm. Sup. Pisa Series IV, vol XVI (1989), 245–258.

[11] M. Levi: Quasiperiodic motions in superquadratic time periodic potentials. Preprint BU (1990).

[12] J. Moser: Quasiperiodic solutions of nonlinear partial differential equations. Bol. Soc. Bras. Mat. 20 (1989), 29–45.

[13] J. Moser: On the construction of almost periodic solutions for ordinary differential equations. Proc. Int. Conf. Functional Analysis and related topics, Tokyo (1969), 60–67.

[14] J. Moser: On invariant curves of area preserving mappings of an annulus. Nachr. Akad. Wiss. Göttingen, Math.-Phys. Kl. II (1962), 1–20.

[15] D. Salamon: The Kolmogorov–Arnold–Moser theorem. Preprint F.I.M. of the ETH Zürich (1986).

[16] D. Salamon and E. Zehnder: KAM theory in configuration space, Comm. Math. Helv. 64 (1989), 84–132.

[17] C. Genecand: Transversal homoclinic points near elliptic fixed points of area preserving diffeomorphisms of the plane, to appear in Dynamics Reported (1991).

[18] C.V. Coffman and D.F. Ullrich: On the continuation of solutions of a certain non-linear differential equation, Monatshefte für Mathematik 71 (5) 1967, 385–392.

[19] N.N. Nekhoroshev: Exponential estimate of the stability in time of near-integrable Hamiltonian systems, Russ. Math. surveys 32 (6) 1977, 1–65.

[20] G. Benettin, L. Galgani, A. Giorgilli: A proof of Nekhoroshev's theorem for the stability times in nearly integrable Hamiltonian systems, Cel. Mechanics 37 (1985), 1–25.

[21] V.I. Arnold: On the behaviour of an adiabatic invariant under slow periodic variation of the Hamiltonian. Transl. Sov. Math. Dokl. 3 (1961), 136–140.

* * *

Printed in the United States
By Bookmasters